Fuel Cell Technician's Guide

Fuel Cell Technician's Guide

William Gleason

DELMAR
CENGAGE Learning

Australia • Brazil • Japan • Korea • Mexico • Singapore • Spain • United Kingdom • United States

Fuel Cell Technician's Guide, First Edition, International Edition
William Gleason

Vice President, Editorial: Dave Garza
Director of Learning Solutions: Sandy Clark
Acquisitions Editor: Stacy Masucci
Managing Editor: Larry Main
Senior Product Manager: John Fisher
Vice President, Marketing: Jennifer Baker
Marketing Director: Deborah Yarnell
Marketing Manager: Erin Brennan
Marketing Coordinator: Jillian Borden
Production Director: Wendy Troeger
Production Manager: Mark Bernard
Senior Content Project Manager: James Zayicek
Production Technology Assistant: Emily Gross
Art Director: David Arsenault
Technology Project Manager: Joe Pliss
Cover Image: Image Content: water bubbles Copyright: © Racnus/Dreamstime.com

© 2013 Delmar, Cengage Learning

ALL RIGHTS RESERVED. No part of this work covered by the copyright herein may be reproduced, transmitted, stored, or used in any form or by any means graphic, electronic, or mechanical, including but not limited to photocopying, recording, scanning, digitizing, taping, Web distribution, information networks, or information storage and retrieval systems, except as permitted under Section 107 or 108 of the 1976 United States Copyright Act, without the prior written permission of the publisher.

> For permission to use material from this text or product, submit all requests online at **www.cengage.com/permissions**
> Further permissions questions can be emailed to
> **permissionrequest@cengage.com**

Library of Congress Control Number: 2012940581

International Edition:

ISBN-13: 978-1-133-27757-6

ISBN-10: 1-133-27757-8

Cengage Learning International Offices

Asia
www.cengageasia.com
tel: (65) 6410 1200

India
www.cengage.co.in
tel: (91) 11 4364 1111

Australia/New Zealand
www.cengage.com.au
tel: (61) 3 9685 4111

Latin America
www.cengage.com.mx
tel: (52) 55 1500 6000

Brazil
www.cengage.com.br
tel: (55) 11 3665 9900

UK/Europe/Middle East/Africa
www.cengage.co.uk
tel: (44) 0 1264 332 424

Represented in Canada by Nelson Education, Ltd.
www.nelson.com
tel: (416) 752 9100 / (800) 668 0671

Cengage Learning is a leading provider of customized learning solutions with office locations around the globe, including Singapore, the United Kingdom, Australia, Mexico, Brazil, and Japan. Locate your local office at: **www.cengage.com/global**

For product information: **www.cengage.com/international**
Visit your local office: **www.cengage.com/global**
Visit our corporate website: **www.cengage.com**

Printed in the United States of America
1 2 3 4 5 6 7 16 15 14 13 12

TABLE OF CONTENTS

	Preface	vi
CHAPTER 1	A Short History of Fuel Cells	1
CHAPTER 2	What Is a Fuel Cell?	13
CHAPTER 3	Fuel Cells, Codes, and Standards	27
CHAPTER 4	Fuel Cell Classifications and Reactions	51
CHAPTER 5	Fuel Cell Main Components	81
CHAPTER 6	Managing Fuel Cell Components	121
CHAPTER 7	Fuel Cell Systems and Subsystems	147
CHAPTER 8	Fuel Cell Systems	171
CHAPTER 9	Stationary Fuel Cell Applications	187
CHAPTER 10	Mobile Fuel Cell Applications	205
CHAPTER 11	Fuel Cell Systems: Process and Instrumentation	229
CHAPTER 12	Fuel Cell Systems: Power and Control	241
CHAPTER 13	Fuel Cell Systems: Engineering, Operations, and Maintenance	273
CHAPTER 14	Fuel Cell Systems: End of Life	285
	Discussion Questions and Answers	299
	Index	313

PREFACE

The way we make and distribute electricity is changing. Centrally located power plants using large machinery to create electricity and then distribute it over entire continents are becoming less cost effective in the modern world. The vulnerability to disruption of such networks, their inherent instability, and initial capital costs are changing the way people and governments look at electricity. Systems that distribute the generation of electricity rather than distribute the electricity are becoming more attractive, and the options available for distributed generation are increasing. Fuel cells have been around as long as electricity generation itself and are now becoming viable alternatives to large-scale generation and distribution.

Fuel cells are complicated technologies, though, not so much because what they do is complicated but because there are so many options available in the fuel cells themselves and because keeping them supplied and running is no small task. Providing a means to understand and appreciate these and other distributed technologies is the only way to ensure that they successfully take their places in our electric world.

INTRODUCTION

The objective of this text is to provide a base for understanding not only fuel cells but the systems needed to create electricity and supply it in a timely and predictable manner. For students of technology, a basic explanation of the science behind fuel cells and associated balance of plant equipment is meant to lay a strong foundation in operating principles. For more advanced students, the work is presented as a contrast to the compartmentalized teaching generally seen in engineering and technical education. In addition, the goal of this textbook is to not only teach the basic principle involved, but to promote discussion about the means to generate power, the way generation is dealt with on a day-to-day basis, and how both the generation and the availability of power are guaranteed. Distributed generation and use of power is different from grid generation, and the knowledge needed to create and run these systems is equally as distributed as the generation capabilities themselves. It is hoped that this textbook can provide a base broad enough for the student to see the business as a whole rather than as a series of troubleshooting tests or design constraints.

SUPPLEMENTS

An instructor resource CD is available for this text. This is an educational resource that creates a truly electronic classroom. It is a CD-ROM containing tools and instructional material that enrich your classroom and make your preparation time shorter. The elements of the instructor resource link directly to the text and tie together to provide a unified instructional system. With the instructor resource, you can spend your time teaching, not preparing to teach (ISBN 978-1-1113-1822-2).

Features of the instructor resource include:

- **Answers** to end-of-chapter problems.
- **PowerPoint® Presentations.** Slides for each chapter of the text provide the basis for a lecture outline that helps you present concepts and materials as well. Key points and concepts can be graphically highlighted for student retention.
- **ExamView® Testbank.** Chapter testbanks provide instructors with test questions to evaluate students as they work through each chapter. Answers and book page references are also provided. These materials are completely editable, so instructors can delete or add questions to meet the needs of their individual class.
- **Quizzes.** Additional test questions are provided for each chapter.
- **Image Gallery.** This database of key images taken from the text can be used in lecture presentations, as transparencies, for tests and quizzes, and with PowerPoint presentations.

ABOUT THE AUTHOR

Bill Gleason is an engineer. He has been educated by outstanding teachers in a number of schools and has worked for some fine companies providing materials for the modern world. He now teaches metallurgical and materials engineering as the Newmont professor at Montana Tech, where he does research on fuel cells and prepares the next generation of people who will have to keep the lights on and the wheels turning. He is married with two children and five grandchildren, all who mean more to him than anything else that ever was or ever will be.

ACKNOWLEDGMENTS

I want to personally thank two people at Delmar, Cengage Learning who are more than kind and twice as patient:

Stacy Masucci, Acquisitions Editor
John Fisher, Senior Product Manager

The entire book staffs from Delmar, Cengage Learning and PreMedia Global deserve credit for this book, especially those who helped with the art and Divya Tyagi in particular.

I would also like to thank the following individuals who reviewed the manuscript:

J. C. Morrow, Hopkinsville Community College, Hopkinsville, Kentucky
Lakshmi V. Munukutla, Arizona State University at the Polytechnic, Mesa, Arizona

This book is dedicated to Cody, Sam, Sage, Tyler, and Jade, who light up my life more than any fuel cell ever could.

CHAPTER 1

A SHORT HISTORY OF FUEL CELLS

objectives

This chapter will introduce the student to fuel cells. A short history will explain the development of these units over the last couple of centuries and begin the discussion of how and why they work to lay the groundwork for more advanced topics in later chapters. Electricity and the two major methods of producing it will be discussed so the student will better understand the history of fuel cells along with why they are only now gaining widespread acceptance for generating electricity. By the end of the chapter, the student should understand how and why electricity is produced, particularly in fuel cells, how it is supplied to customers, and how scientific and technical advances occur when they do.

INTRODUCTION

Where does **electricity** come from?

Where we live, it comes from the wall, but that was not always the case nor is it the case in much of the world even now. When the early scientists began to use electricity, they had to make their own. They did that using fuel cells. If you look fuel cells up in the Dictionary of Scientific and Technical terms, you will find: "a cell that converts chemical energy directly into electrical energy, with electric power being produced as a part of a chemical reaction between the electrolyte and a fuel. . . ."

Alessandro Volta produced electricity in the early 1800s utilizing **chemical reactions** in his voltaic pile. Not long after, William Henry Grove produced what is considered the original fuel cell and called it the Grove Cell. The early scientists named a considerable number of things after themselves, and so you know most of their names, or at least versions of their names such as **volt**, **amp**, **watt**, and **ohm**. Groves' name is less well known than some, but only because his invention was left behind by your plug in the wall. That was not always the case though.

In the early 1800s, while Volta was creating electricity, William Nicholson and Sir Anthony Carlisle were using his electricity to decompose water into **hydrogen** and **oxygen** in England. Christian Friedrich Schoenbein went the opposite way in Switzerland, producing electricity when hydrogen and oxygen were recombined to make water. He then undertook a systematic investigation of fuel cells not long after, using what is called the **scientific method**. By the 1840s, a basic understanding of what worked in fuel cells, if not exactly why it worked, was widely available to those who cared about such things. And there were quite a few people who did care.

Volta's **voltaic pile** had a blotter soaked with salt water (later sodium nitrate) between a silver **electrode** and a zinc electrode with the two electrodes connected by wire, shown in Figure 1-1. Judge Grove had begun his work in the 1830s by suspending zinc in zinc sulfate and platinum in nitric **acid**, tied several of these cells together and proved he could make electricity, too. Grove produced these industrially, and one of the first demonstrations of the telegraph by Morse and Vail

2 FUEL CELL TECHNICIAN'S GUIDE

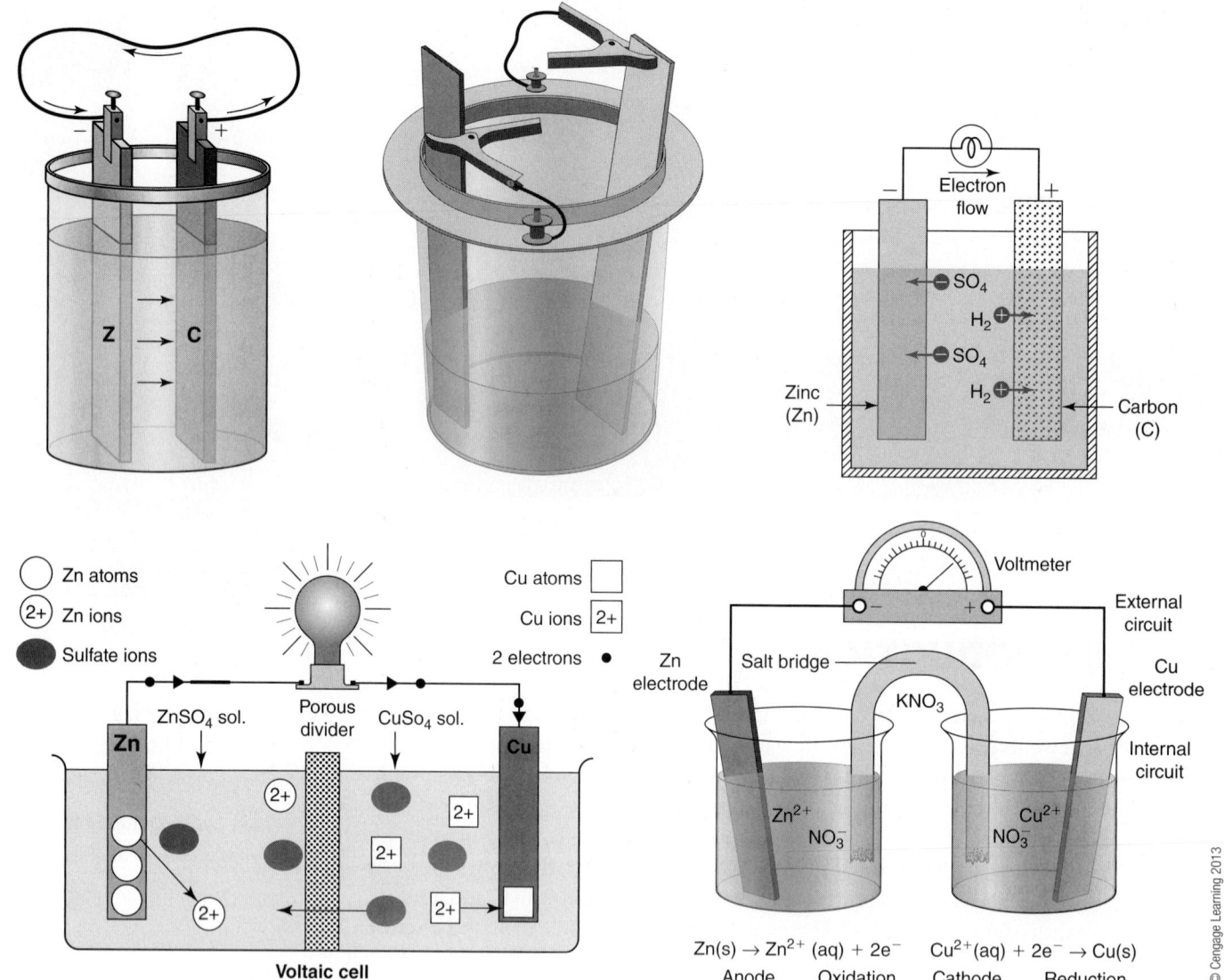

Figure 1-1 The voltaic pile (cell).

used a Grove **battery** to send messages from Washington to Baltimore in 1837. If you watch old western movies, you have seen Grove's work in the telegraph office, since they had to make their own electricity, too, and his industrially produced Grove Cell ran a large part of America's telegraph systems from its beginning until just after the Civil War. It produced 1.9 volts and 12 amps with 1 cell being used to power about 20 miles of line. Unfortunately, the battery also produced a toxic gas, but the risks of gassing your batterymen were far outweighed by the rewards of instant communication across a continent.

That is why so many people cared about batteries and fuel cells during this period. The telegraph had to have electricity or it did not work. There were no plugs in any walls and if you wanted information to pass across continents in minutes, you had to make your own electricity on site. Better cells came along all through the last half of the nineteenth century because other pioneers of electricity had discovered there was money in electricity and so considerable competition erupted. These were for the most part **primary cells** involved in the irreversible creation of electricity. Secondary, or **storage cells**, came along about 1860 when Plante came up with the lead acid system. Even a couple of centuries later, you can still buy versions of both the primary zinc cell to put into your watch and the secondary lead batteries to put into your car, as shown in Figure 1-2.

Grove improved his design over a number of years using the constant stream of discoveries coming out of the emerging laboratories of Europe, and he also moved on from zinc systems to hydrogen/oxygen systems. The basic design of his original hydrogen fuel cell was very simple: two upside-down containers of hydrogen gas and oxygen gas in an **electrolyte** solution. He used acid as the solution, pushed two platinum electrodes that were wired together up inside the gas containers (one in each) and again made electricity. This is considered the first fuel cell, but they were never produced industrially, since Daniel cells, Leclanché cells, Callaud gravity cells, and storage batteries used with dynamos had taken over the telegraph.

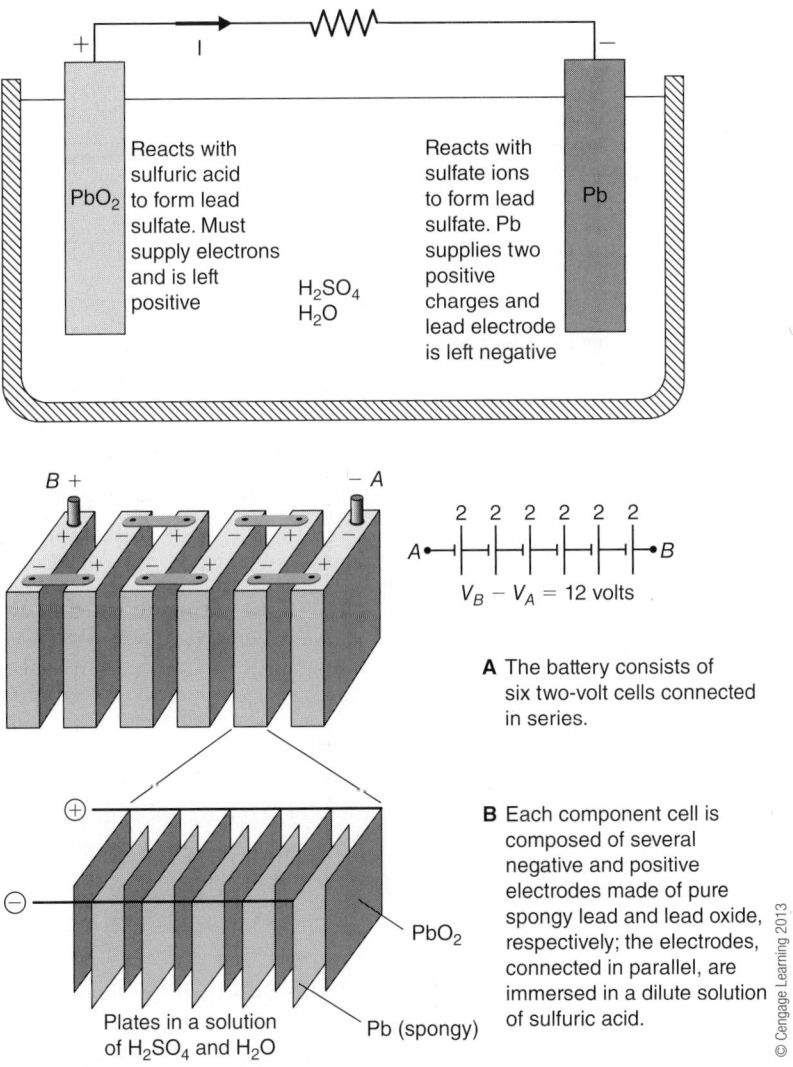

■ **Figure 1-2** The basic lead (secondary) battery.

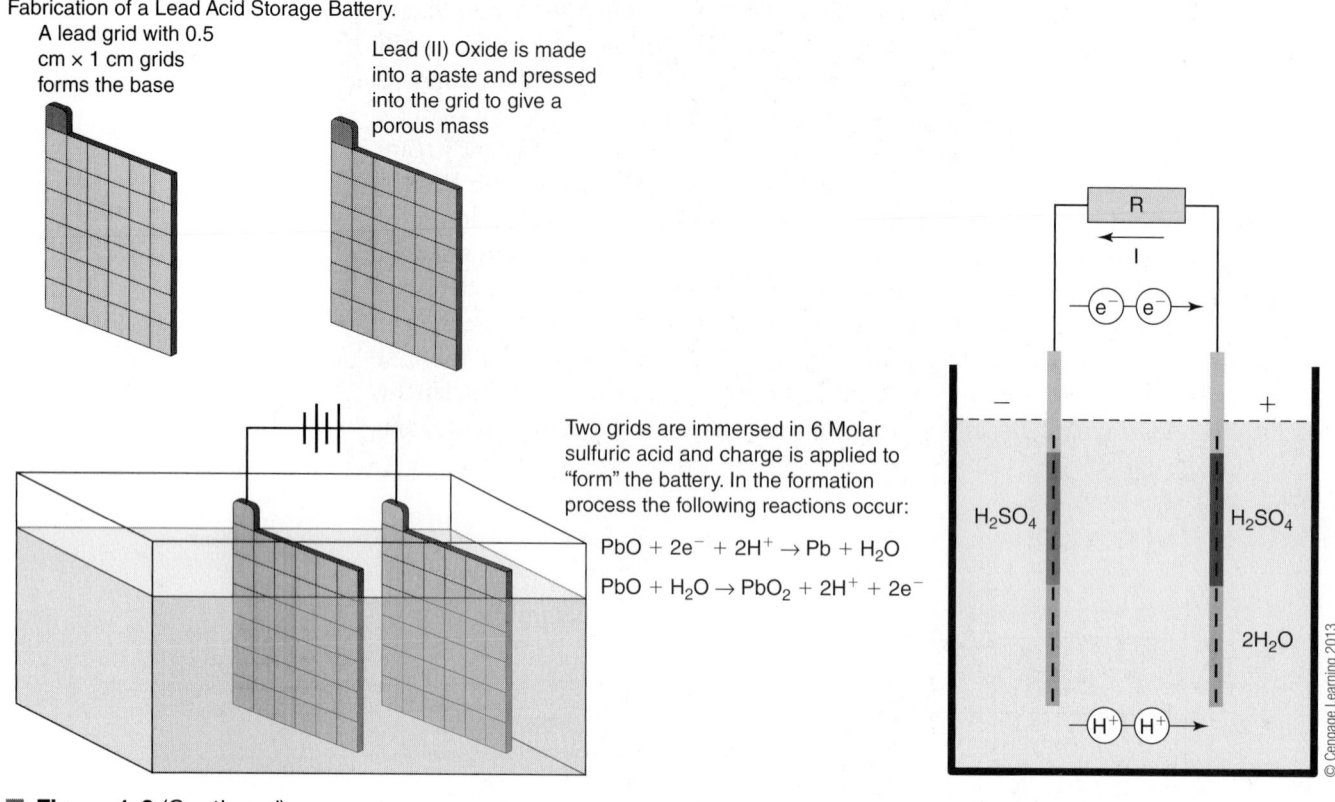

Figure 1-2 (Continued)

MAKING ELECTRICITY

Even though electricity was being used industrially in the early 1800s, there were many arguments over how that electricity was actually being produced. There were two general theories at the time, the **contact theory** and the **chemical theory**. Volta and the original fuel cell industry relied on the chemical theory to produce electricity. In the early 1830s Michael Faraday used the contact theory to produce electricity using a dynamo and no chemicals.

In the chemical theory, electricity is produced when chemical reactions occur, preferably spontaneously. This can be explained using Judge Grove's cell to make water from his two gases, shown in Equation 1 and graphically in Figure 1-3.

$$2H_2 + O_2 \leftrightarrow 2H_2O \quad \text{Eq. 1}$$

Remember that the hydrogen and the oxygen are kept separate and that there is an acid solution between the two gas chambers. Both of those are important to the chemical reactions. At one platinum electrode, the hydrogen breaks down like this:

$$2H_2 \leftrightarrow 4H^+ + 4 \text{ electrons}^- \quad \text{Eq. 2}$$

Two things happen at that platinum electrode: the **diatomic** hydrogen **gas** is broken apart into two positively charged hydrogen **atoms**, and four negatively charged **electrons**. That second part is electricity, which is nothing more than the negatively charged electrons moving from one spot on a wire to another spot on a wire. Remember that electrons are always negatively charged, that being one of the fundamental definitions of electrons.

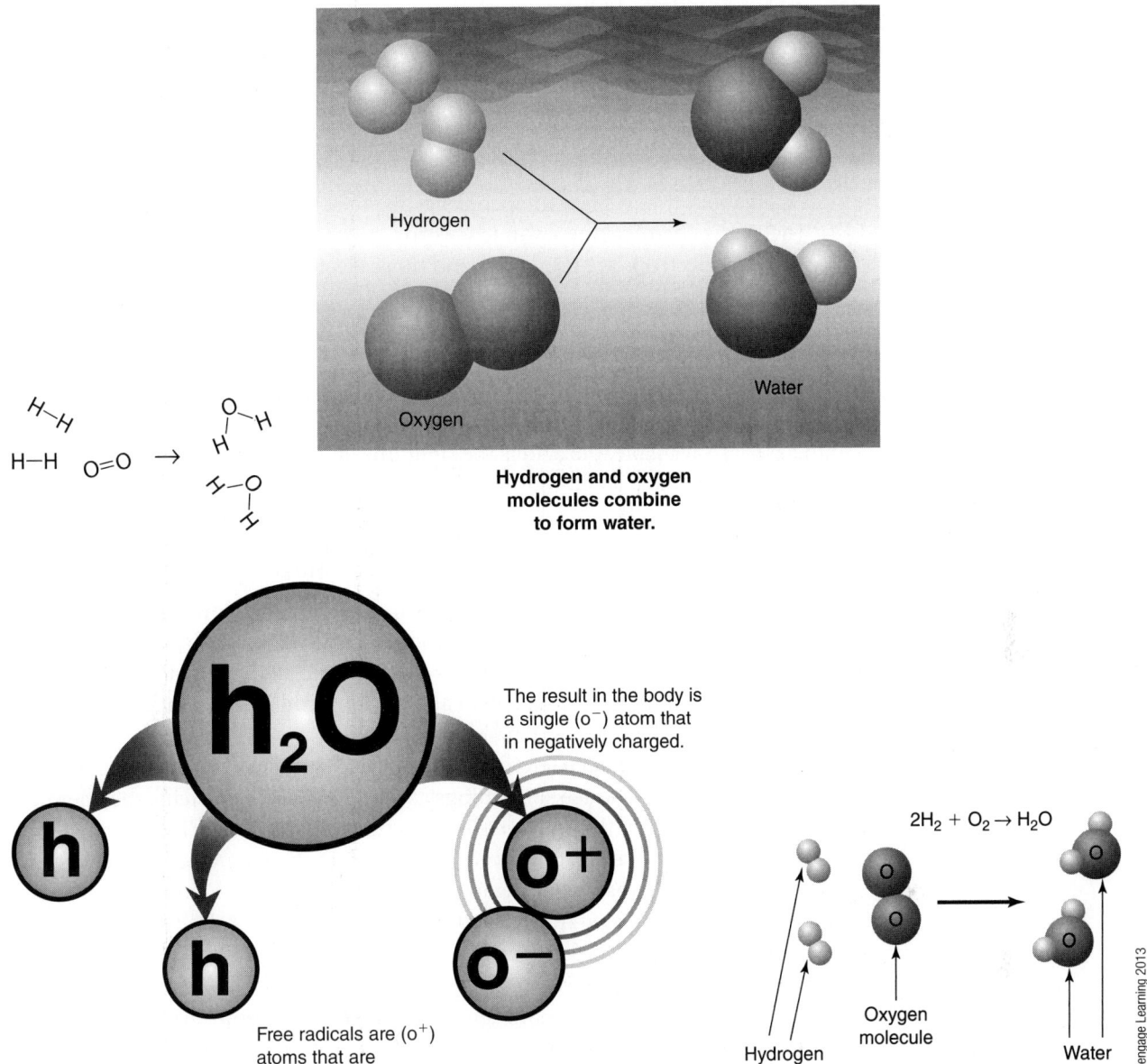

Figure 1-3 Hydrogen and oxygen combine to form water.

At the other platinum electrode, the oxygen combines with the hydrogen and the electrons to make water like this:

$$O_2 + 4\text{ electrons}^- + 4H^+ \leftrightarrow 2H_2O \quad \text{Eq. 3}$$

The important parts are the acid, the two platinum electrodes and the four electrons. An acid is essentially water with positively charged hydrogen ions, called **protons**, in solution. The stronger the acid, the more protons there are. That means the protons being made from the hydrogen gas do not need to go to the oxygen chamber because there are already protons there in the acid. If you take four protons out, they are replaced by four more coming in, and it doesn't matter too much where they happen to be in the solution. The electrons do have to go to the oxygen chamber though. They go there because the two platinum electrodes are wired together, and the electrons then flow from the negatively charged hydrogen cell platinum electrode to the positively charged oxygen cell electrode. Now you have a nice electric circuit, electrons being produced at the hydrogen gas side to flow from the platinum anode through a wire to the platinum cathode and the oxygen gas side where they are being used to make water.

You should understand that platinum not only allows the electrons to flow (electricity) but also acts as a place to break the hydrogen gas up and a place to combine the hydrogen and oxygen. They are called **catalysts**, and there will be a lot more said about that subject in upcoming chapters.

That is the chemical theory in few words. A chemical reaction occurs in such a way that free electrons are produced, and those electrons move along a wire, which results in electricity as we all know it. The problem is not in the way electricity is produced, but in how much is produced and how long it is produced. In our simple design, electricity is only produced where the liquid acid, the gases, and the platinum surface all meet because that is the only place where all the things needed are. This does not produce a lot of electricity, and it only produces electricity when both gases are available, the acid is the right mix, and the surface of the platinum is clean. We will come back to those three issues as well.

Fuel cells were effective in laboratories, where small **currents** and **voltages** could be used to study any number of interesting things ranging from the jumping of dead frog legs to plating gold onto a cheaper metal. Fuel cells were useful in telegraphs where all that was needed was limited current flowing in such a way that it would be altered just enough to produce the dots and dashes of Morse code. Small currents that lasted only as long as the available gas were not of much use in running a cotton mill or a loom day in and day out, and so we come to the contact theory and Michael Faraday's dynamo.

While it might seem farfetched to think that chemically changing one thing into another would produce such things as electrons and electricity, the chemical theory had nothing on the contact theory in that department. When Oersted took the zinc cells and moved a compass close enough to the wire where the electricity was flowing, his compass no longer pointed to true north. When Ampere took two wires from the cells and moved them close together, he found that they repelled each other but only when electricity was flowing. Finally, when Faraday took a metal rod and pushed it back and forth through a coil of copper wire to get electricity in the wire, shown in Figure 1-4, he did away with the chemistry part entirely. That is the contact side of electricity that Faraday investigated in the early 1800s, now more commonly called **electromagnetic** field theory.

Just like the chemical theory, you had to have the correct circumstances. The rod had to be a magnet and the wire in the coil could not touch. The magnet did something in the wire to move electrons. Remember, they did not know there were such things as electrons, only that something was moving through a metal wire that they could measure the effect of. Even on the very simple side, people could understand this. If electrons move through a wire, the wire heats up. Even if someone knew nothing about anything, that person would know something was going on because of one of the following three evident effects: the wire would heat up as Faraday's experiment progressed; the more current (heat) that flowed in Ampere's experiment, the more the two wires would repel each other; or the more current, the farther from true north Oersted's compass went.

Faraday kept his rod stationary and rotated his coil to produce electricity as well. Play with various configurations of that and you get a dynamo, more often referred to now as a **generator**, shown in Figure 1-5. By the 1870s dynamo electricity and arc lights were replacing gas lamps above the streets of cities in Europe and America but only on a small scale. They were **direct current**

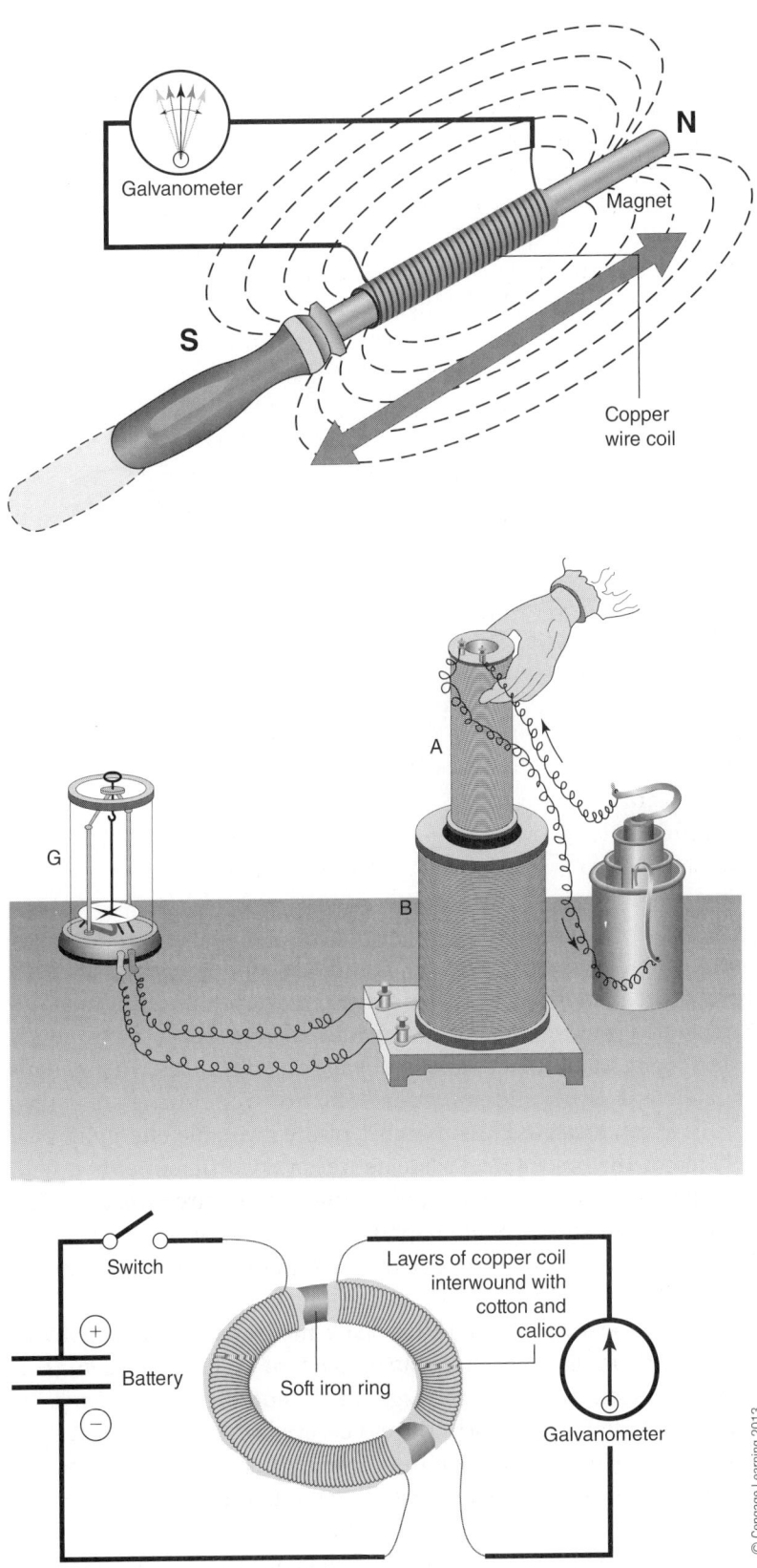

■ **Figure 1-4** Faraday's rod and coil.

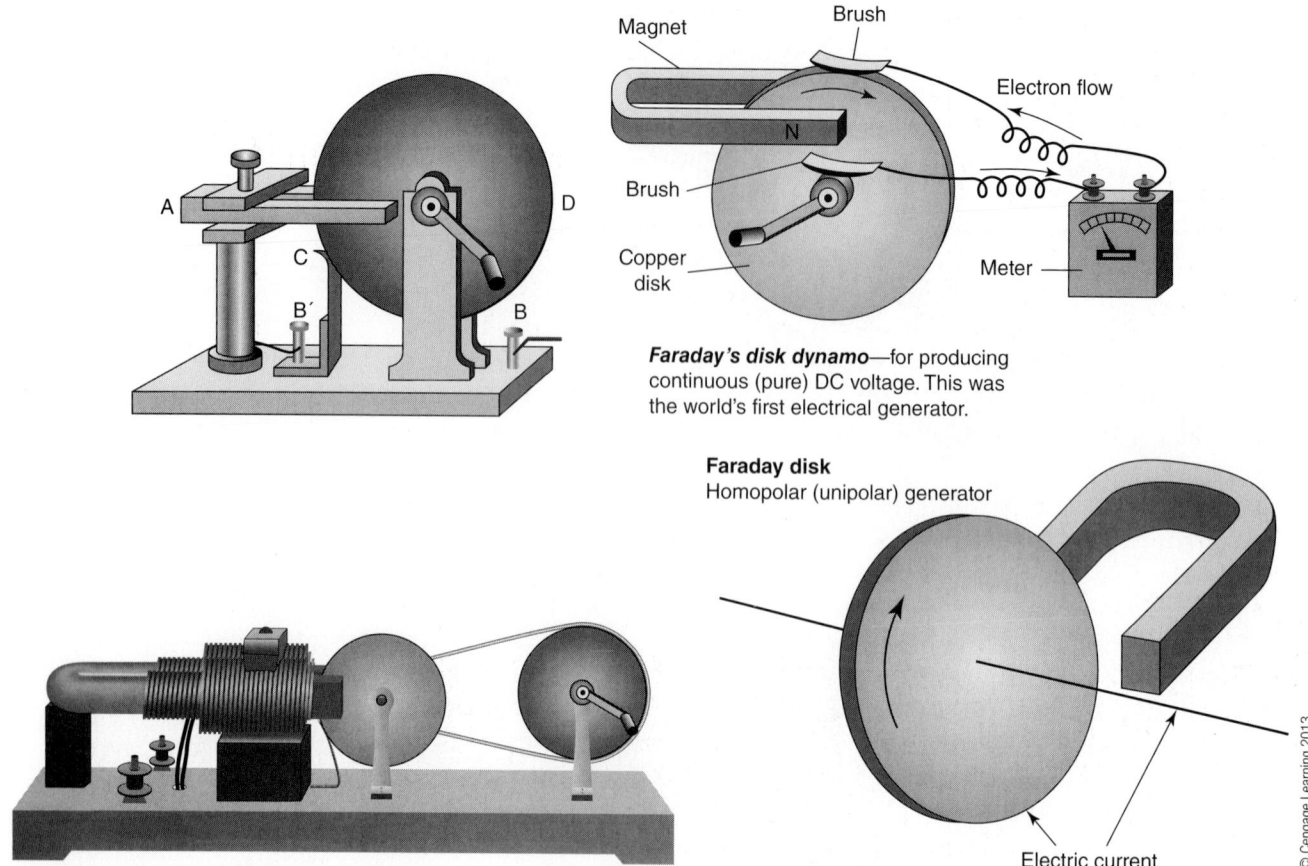

Figure 1-5 Faraday's disc generator.

generation stations that by nature were small and the copper lines to transmit the power were short. The initial DC grid systems were all neighborhood based, with a generating station serving a very limited area, because DC power does not transmit along metal wires very well over long distances. The systems were not very efficient either, Edison's coal-burning generating station to serve lower Manhattan converted less than 3% of the available energy in coal to electricity. None of the mechanical systems were very efficient either, with reciprocating steam engines converting less than 15% and steam turbines less than 20%. We will talk in subsequent chapters about why fuel cells are good, and one of those reasons is how efficient they are in converting the energy available in fuel to electricity.

George Westinghouse and Nikola Tesla solved the transmission problems, or at least enough of them, with the **alternating current** system. By 1896, their high-speed turbines generated power at Niagara Falls and then transmitted AC electricity 26 miles to Buffalo, New York. The era of large-scale power generation and transmission was born, and the plug in the wall was here to stay. Now, AC power transmits over entire continents in systems of astounding complexity.

MODERN FUEL CELLS

We already know who won the commercial contest between the two theories, since there is a plug in the wall and not a fuel cell in the basement. To understand the issues facing fuel cells today, however, you need to understand the reason why fuel cells and the chemical theory fell by the industrial wayside. More than a century has passed since fuel cells disappeared from the market, and the underlying reasons why have not changed. What has changed is the marketplace.

Both Volta and Faraday were right, the chemical theory explains the production of electricity just as the contact theory does. Chemical reactions involve electrons, and electrons moving in a wire cause a magnetic field. Faraday was more right though, not because his science was better but because the economics of the contact theory were superior. Electricity was a gentleman's game for a century while trains were speeding across the world and cotton gins were spinning cotton out because it was a curiosity. It could not power trains or plow fields or run machining tools. In the nineteenth century, when so much of the foundation for today's industrial society was being laid, what was important was the ability to do **work**: to lift something up, to drag something along, or to move something from one place to another.

Electricity can do work, but what was coming from fuel cells was not on the scale that was needed. Electricity found a market in the telegraph because it used so little, but once electricity became commercially available outside of laboratories, people began to find out just how useful it could be. The discovery that knocked the fuel cell down if not out was the one that made electricity capable of doing those critically important things like turning lathes and lifting ore to the surface, the discovery that made electricity capable of doing the heavy work. It was the electric motor, and the electric motor was nothing more than the contact theory in reverse. You could use very large generators to make lots of electricity in a central location, move it hundreds of miles across a wire grid system, and then use hundreds of smaller versions of the generator-in-reverse as motors. The advantage of large scale production is called the **economy of scale**, and corporations began to employ it in earnest when electricity became available in industrial quantities, allowing the mass production of everything from guns to tin cans.

The market for electricity through most of the twentieth century was in two things, heavy work and lighting. Fuel cells could not produce enough electricity to run big motors. Fuel cells could produce enough electricity to run lights, but the issues of poison gas or pure hydrogen and oxygen in the home were a bit problematical when all you needed from the competition was some copper wire and a plug in your wall. Fuel cells lost the first round for a number of reasons: they lacked a fuel supply infrastructure, whereas power plants could be set where the fuel was, at coal mines or gas fields or at the end of reservoirs. They were lacking in safety because if they went bad in a home, they killed people's children, whereas if large-scale generation went bad, it killed workers who got paid to take such risks. They lasted only a short time, whereas a power plant was good for decades. They could not support constant power output. They were expensive. They fouled too quickly. The list went on. And it still does. What has changed is not the fundamental science behind the fuel cell but the marketplace. Heavy lifting and lights are still there, but new markets include laptops and unmanned surveillance airplanes and cell phones and off-grid housing and a list that is longer than the one detailing fuel cells' limitations. The one market that truly resurrected fuel cells was the one market where electricity could not go in a wire.

Keep in mind that fuel cells did not disappear even after the plug showed up in walls courtesy of Westinghouse, Tesla, and millions of miles of copper wire. If they could not compete with industrial generation, they could still compete in the laboratory. Generators and electric motors are relatively simple and work very well, but the plug in the wall requires a massive infrastructure investment that many could not afford and many still cannot afford. In spite of the Western world's adoption of the plug in the wall, small-scale generation continued to be investigated in laboratories.

In the last decades of the nineteenth century, as large-scale generation was ramping up, Ludwig Mond and Charles Langer built a fuel cell that used air and industrial coal gas instead of hydrogen and oxygen. It used thin, perforated platinum to increase the contact area of the electrodes with the other ingredients and could generate 6 amps per square foot of working electrode at 0.73 volts. It worked as long as the platinum was clean, but once it was fouled by such things as sulfur, which you tended to find in coal gas, the electricity stopped. Like Grove's hydrogen cell, it was another interesting fuel cell laboratory experiment that never went anywhere. But science rarely comes out of the laboratory intact. Improvements are made that allow the next generation to go another step; basic knowledge is gained that gives insight into fundamental principles. Mond and Langer made a substantial contribution to science because their system used a porous matrix to hold fluid and powdered platinum-black metal as a catalyst instead of a solid piece of metal. The science of catalysis based on the **surface area** available for reaction took a giant step forward, aiming now at large scale chemical production.

At the same time, Friedrich Wilhelm Ostwald was inventing a whole new branch of chemistry using the same principles as Mond and Langer. Grove had realized that all the action of his cell occurred where the electrode, gas, and liquid acid electrolyte solution all met, but he never went any further than that. Ostwald went further, quite a bit further, and along the way came up with what is now one of the fundamental areas of modern chemistry, called physical chemistry. He explained and began to quantify using mathematics the relationships between the components in the fuel cell—the electrodes, electrolyte, **oxidizing agents, reducing agents, anions,** and **cations**.

William Jacques was also experimenting with the direct coal fuel cell and produced a very large carbon and air system that delivered 1.5 kW in 1896. His system overcame the issue of how much electricity was produced but since he used molten alkali (**base**) electrolytes which interacted with the carbon instead of the acid solution electrolyte, his cells did not last long enough.

At the turn of the century, Walther Nernst went from the liquid acid and molten alkali electrolytes to a solid electrolyte. He had run an electric current through zirconia at high temperatures to obtain light in the lab to prove zirconia would conduct electricity under the right circumstances. He then transferred this concept into a solid electrolyte system.

By this time, large-scale generation had taken over the business, and there was a bit of a lull in fuel cells. It wasn't until the 1920s that Buar and Tobler used molten carbonate as an electrolyte in a hydrogen system to produce electricity at low temperatures, where Nernst had needed high temperatures for his solid electrolyte.

Ten years later, Sir Francis Bacon began his systematic examination of fuel cells in the early 1930s at Cambridge University in England. Benefiting from the substantial increase in basic knowledge of chemical principles that had taken place in the early part of the century, he went back to the earlier designs and began improving them. He returned to hydrogen gas, replaced platinum with nickel gauze and combined that with a potassium hydroxide electrolyte which was less **corrosive** to the nickel. He used a differential gas **pressure** across the cell to improve the electrolyte/electrode contact to achieve a power density of

1.11 A/cm² at 0.6 V. Mond and Langer's best had been 6 amps produced per square foot. His system, known as the **alkali fuel cell (AFC)**, operated at 240°C and at pressures up to 3000 psi. Unfortunately, it did not last very long either, since the nickel corroded too quickly. He worked for a couple of decades on this cell, developing a nickel oxide electrode doped with lithium to both improve conductivity and corrosion resistance along the way, until he was able to produce a 10 inch diameter cell stack capable of producing 5 kW for a British Royal Navy submarine demonstration.

Many of the issues plaguing fuel cells had been addressed, but the main one remained. The cost of electricity produced by the systems was too expensive compared to the plug in the wall. Fortunately, by the late 1950s when Doctor Bacon was producing his fuel cells, there was a market where cost was not an issue and wires could not go. It was called space. NASA would eventually fund over 200 major research projects for fuel cell development, since no other technology was suitable for space. Wires were not long enough, batteries were too heavy to send, and solar power was in its infancy.

In the mid-1950s, General Electric developed a new fuel cell, the **PEM (Proton Exchange Membrane)** cell. Willard Grubb used a sulphonated polystyrene ion-exchange membrane as an electrolyte and then Leonard Niedrach developed a way to deposit platinum onto this polymer. The Grubb-Niedrach fuel cell went with the Gemini flights. General Electric dropped out of the business eventually, but Ballard Systems resurrected PEM cells in the mid-1980s to great success.

Pratt and Whitney went a different route, licensing the patented Bacon Cell instead of developing their own. Their engineering modifications of the cell during the early 1960s were so successful that their AFC system, sold by International Fuel Cells, displaced the GE PEM cell when the Apollo program went to the moon. It provided not only 1.5 kW of electricity but drinking water as well. In 18 Apollo missions, over 10,000 hours of operational time were clocked without a single incident. In the 1970s, International Fuel Cells refined the system for the Space Shuttle. They supplied three cells, each producing 12 kW of continuous power with a 16 kW peak. So reliable were these systems, that battery backups were not included in the Shuttle design.

After that, the history of fuel cells gets a bit too much for an introductory chapter. Thousands of people worldwide now spend hundreds of millions of dollars on fuel cell research, development, and marketing. The fundamental problems of fuel cells have been addressed and to a great degree overcome, but two primary issues remain, cost of electricity produced and infrastructure. The first is a matter of time—as more cells sell, the cost of each goes down, as does the cost of electricity produced. Infrastructure is a political issue driven by societal imperatives and as such is beyond the scope of this topic.

KEY WORDS

Knowing the terminology used is critical when dealing with fuel cells. Following is a list of the important terms in this chapter, which are also in bold typeface within the chapter. It is recommended that students be required to submit definitions of some of these words as homework assignments, in which they are asked to look the terms up in other books, articles, or on the Internet.

acid	currents	PEM (Proton Exchange Membrane)
alkali fuel cell (AFC)	diatomic	
	direct currents	pressure
alternating current	economy of scale	primary cells
amp	electrode	reducing agents
anions	electrolyte	scientific method
atoms	electromagnetic	storage cells
base	electrons	surface area
battery	fuel cells	volt
catalysts	gas	voltages
cations	generator	voltaic pile
chemical reactions	hydrogen	watt
chemical theory	ohm	work
contact theory	oxidizing agents	
corrosive	oxygen	

DISCUSSION QUESTIONS

1. What is the difference between a primary cell and a secondary cell?

2. What is the difference between the contact theory and the chemical theory?

3. Why are electrons negatively charged?

4. What is the difference between a proton and an electron?

5. What is the difference between direct current and alternating current?

6. What is economy of scale?

7. What is the difference between an anion and a cation?

CHAPTER 2

WHAT IS A FUEL CELL?

FUNDAMENTAL CONCEPTS

To understand fuel cells, the student must be at least somewhat familiar with some fundamental chemical principles, rules, and issues.

An **atom** is one element in the periodic table by itself, such as iron or platinum or hydrogen. A **molecule** is when an atom binds to another atom. Molecules can have as few as two atoms bonded together as in hydrogen gas or several thousand atoms bonded together as in plastics.

If something is above a **chemical symbol** to the right, it means there is a charge on the thing. H^+ then means a hydrogen atom that is positively charged because it has lost a negatively charged electron, e^- (electrons are always a single negative charge; that is one of the rules of chemistry). Cu^{2+} is copper that has lost two electrons and thus has twice the positive charge as H^+. I can count these the same as anything else, so $4H^+$ is four positively charge protons (H^+) while $2Cu^{2+}$ is two doubly positive charged copper atoms.

If something is below a chemical symbol and to the right, it means there is that number of atoms or molecules in the thing and they are bonded together. The symbol O_2 then means you have two atoms of oxygen bound together. $2O$ is not the same as O_2. $2O$ means you have two oxygen atoms, but they are not bonded together, while O_2 means you have two oxygen atoms that are bonded together.

Very large numbers or very small are too hard to keep track of. To make it easier, most of the numbers are rounded and then multiples of 10 are used to keep track of how big or small the number is. This use of powers of 10 is known as **scientific notation**. There will be a smaller number times 10 to some power, positive for very large numbers and negative for very small numbers. If the number above and to the right of the 10 is positive, you can multiply the numbers out, so that 9×10^4 is 90,000 ($9 \times 10 \times 10 \times 10 \times 10$). You can just move the decimal point as well, so that you fill in zeroes and move the decimal point that many places to the right, so 9×10^4 is 90,000. You have to deal with the decimal point too, so it can get tricky: 9.1×10^4 is 91,000 and 9.12×10^4 is 91,200. If the number above and to the right of the 10 is negative, you can still multiply, so that 9×10^{-4} is 0.0009 ($9 \times 1/10 \times 1/10 \times 1/10 \times 1/10$). You can also fill in zeroes and move the decimal

objectives

This chapter will introduce the student to chemical and mechanical concepts central to the production and operation of fuel cells. Fundamental concepts required to understand fuel cells will be introduced and discussed as will the basic operations of a particular type of fuel cell, the proton exchange membrane. More advanced concepts will be introduced that will be discussed in greater detail in later chapters. Finally, the student will be introduced to some of the fundamental design concepts involving fuel cells, again with more detailed discussion to follow in later chapters. By the end of this chapter, the student should be able to understand the basic chemical principles behind fuel cells, why some things work while others do not, and be able to discuss the core concepts that define fuel cells.

point that many places to the left, so 9×10^{-4} is 0.0009 but 9.1×10^{-4} is 0.00091 and 9.12×10^{-4} is 0.000912 while 91×10^{-4} is 0.0091 and 912×10^{-4} is 0.0912. We have to use this type of number because there are thousands of trillions of things like protons (H^+) and electrons involved in fuel cells and because electrons are so small.

Atoms consist of **protons** (positively charged), **neutrons** (neutral), and **electrons** (negatively charged). The protons and neutrons are lumped together in the center of an atom and, for the purposes of this book, do not come apart. The electrons are somewhere around the center, but it is hard to tell where for reasons that are well beyond this book. Hydrogen has one proton, and one electron; that is why it is the first element in the periodic table. It is also why we call a hydrogen atom without its electron a proton (H^+). Hydrogen can have one proton, one neutron, and one electron as well, a substance called deuterium. Helium is the second element in the periodic table and has two protons, two neutrons, and two electrons. It goes on from there along the periodic table. The electrical charge on one electron is -1.602×10^{-19} Coulombs; the charge on a proton is exactly the same, but it is a positive charge not a negative charge. If you have the same number of protons and electrons, the atom has no charge (neutral) because the two charges cancel.

There is **mass** and there is **weight**. The two are not the same. Mass is a fundamental property but weight is actually a force (mass times acceleration). The mass of something does not change; the weight does. Your mass on earth is the same as your mass on the moon, but your weight is different because the acceleration due to gravity is different. The mass of an electron is 9.1×10^{-28} grams while the mass of protons and neutrons are 1.67×10^{-24} grams. That is very small, but the important issue for fuel cells is that protons and neutrons have almost two thousand times the mass of electrons. Electrons are easier to move than protons (H^+) for the same reason it is easier to move a bowling ball than a '59 Buick.

There are trillions and trillions of atoms in even a very small mass (12 grams of carbon-12 has 6.02×10^{23} atoms). This is important because electricity needs a lot of electrons. Electrons are less than 1×10^{-13} centimeters in diameter, although their size is still being argued. Electrons are very small but they do not go very fast (in electricity at least). The ones in your wall go about 1/10 millimeter per second.

An **amp** is defined as 6.25×10^{18} electrons moving past a point in one second. One amp is then the charge on one electron multiplied by the number of electrons moving past any given point in one second (1.602×10^{-19} multiplied by 6.25×10^{18} equals 1). One trillion is 1×10^{18} so one amp is six and one-quarter trillion electrons moving past a point in one second. It takes a lot of electrons to get the sort of electricity you are used to.

Voltage is many times explained as pressure or the amount of work needed to move a unit charge between two points. Voltage is the difference in electrical potential between two points in the same way that water pressure in a gravity fed system can be expressed as the difference in height between where the water comes in and where it goes out. Just as water moves downhill, electrons move to the positive electrode in a circuit and away from the negative.

Electrodes are not positively or negatively "charged" so much as they are places where there are either too many electrons or too few electrons. When you walk across a carpet, you are transferring electrons from the carpet to you, and you become an electrode. You are no longer neutral (the same number of negative electrons as positive protons) but instead have an excess of electrons. In fuel cells, the same thing happens, in that electrons are liberated by the chemical reaction and so there are too many of them in one spot. At another spot, there are not enough electrons, since the protons (H^+) have gathered there. The negatively charged electrons go toward wherever there are too many positively charged particles. In chemistry, an electrode is where chemical reactions occur in a voltaic cell. If electrons are going to an electrode, then it must be the positive electrode (**cathode**). If electrons are going away from an electrode, then it must be the negative electrode (**anode**). In fuel cells there are two electrodes. Electrons are made where hydrogen gas is broken apart into protons (H^+) and electrons (e^-). Water is made where protons (H^+) combine with oxygen (O_2) and electrons (e^-).

The **cell potential** or electromotive force (emf) is the voltage between the electrodes of a voltaic cell. The greater the difference between the two electrodes in a fuel cell (the more excess electrons at one electrode and the more excess protons (H^+) on the other electrode), the greater the cell potential and the greater the voltage of the cell. To put it another way, making more electrons at the anode means more protons (H^+) at the cathode. The two combine to make more pressure to drive electrons away or attract electrons forward to equalize the charge. The more electrons and protons being produced, the more pressure behind the movement of electrons and so the more voltage.

A **watt** is the rate of energy conversion. One watt is defined as one ampere flowing through a conductor the ends of which are maintained at a potential difference of one volt. You pay for watts in your power bill instead of amps or volts because you have converted that much electricity to something, be it light or cold air or hot metal. Keep in mind that you do not use electrons up. Some energy difference (voltage) drives electrons down a wire. Electrons are then diverted into say a light bulb, where the energy driving them is converted to light. The electrons then continue to move down the circuit but with less energy driving them. It is like a car. Once the engine is fired and provides energy to the wheels, the car moves. You can remove that energy from the car by hitting a small sign. The energy is converted to work and bends the steel of the sign, but the car is not consumed just as an electron is not consumed.

A **voltaic cell** is a cell which produces free electrons (electricity) through a chemical reaction involving oxidation and reduction in a closed system. A fuel cell is one type of voltaic cell and a battery is another type of voltaic cell.

There are **primary batteries** and **secondary (storage) batteries**. Primary batteries use chemical reactions that produce electrons and so do secondary batteries, but at some point, primary batteries stop creating electrons. They may use up one of the species in the chemical reaction or the **electrolyte** might dry up so there is no longer a circuit. Secondary batteries produce electrons just like primary batteries but can reverse the reaction by flowing electrons back through the circuit. The battery is not "recharged" so much as the chemical reaction is reversed. This is possible because the oxidation and reduction reactions that deal with electrons are paired up and like all chemical reactions can go both ways.

Fuel cells cannot be recharged. Instead a continuous supply of fuel (the chemical that gives up the electrons) such as hydrogen is required.

Chemical **oxidation** occurs when electrons are removed from an element or molecule in a chemical reaction. The hydrogen gas (H_2) in a fuel cell is oxidized to form two protons (H^+) and two electrons (e^-). It is usually not so simple though.

Chemical **reduction** occurs when electrons are added to an element or molecule in a chemical reaction.

Oxidation and reduction in a chemical reaction usually occur together, called a **redox reaction**.

Chemical reactions can occur spontaneously, can create energy when they happen or require energy to happen, and can go forward or backward depending on the circumstances (that is why the arrows in the reactions go both ways).

A **catalyst** lowers the energy needed for a reaction to occur (if energy is needed, that is) but is not used up in the reaction. A catalyzed reaction is one that would not occur spontaneously without a catalyst but that will occur with the assistance of a catalyst. When the reaction is over, the amount of catalyst present remains the same.

Some things let electrons move from one atom to the next atom and then on down the line but some things do not. To put it differently, some things conduct electricity and some things do not. The reasons why are simple but at the same time very complex. In some things like rubber, electrons stay around the proton/neutron center and have to be forced to move away. This is because the electrons are part of the binding holding the atoms together to form a solid or a liquid or a gas and if the electrons move, it destabilizes the material. Materials like that do not conduct electricity unless a lot of energy is used to force electrons into moving, and then they tend to come apart; they are called **insulators**. In metals, the electrons are able to move from one atom to the next as long as there is a steady stream, so any particular atom always has the right number of electrons to maintain a neutral charge. The reasons why some materials can pass electrons and others cannot are complicated but for the most part, the **metallic bonding** system can pass electrons while the **ionic** and **covalent** bonding systems do not.

BASIC PRINCIPLES

Most of the topics listed above have something to do with electrons. That is because electrons are the basis of chemistry, (remember the chemical theory and the basis of electricity, remember the contact theory). Fuel cells liberate electrons from some element or compound and then do something with those electrons. The way they operate is relatively simple. We will review the basics of chemistry and fuel cells in general and use the PEM cell to explain things in particular.

The basic hydrogen fuel cell is shown in Figure 2-1.

Hydrogen gas goes in and contacts a **conducting catalyst** such as platinum. At the surface of the conducting catalyst, two things happen. The first is the catalyzed reaction. In this case, the oxidation reaction (the hydrogen is losing an electron) that occurs at the anode (the negatively charged electrode/catalyst) is shown in Equation 2.

$$2H_2 \leftrightarrow 4H^+ + 4\,electrons^- \qquad Eq.\ 2$$

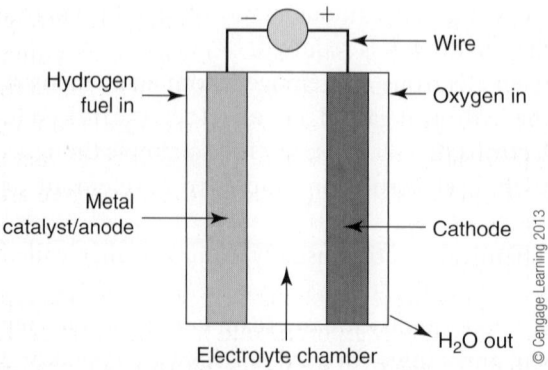

Figure 2-1 The basic hydrogen fuel cell.

In this case, there are two **molecules** of hydrogen gas (H_2) to start with. Each of the molecules has two atoms of hydrogen bonded together. Since each hydrogen atom has one electron, the hydrogen gas molecule (H_2) has two electrons and two proton/neutron centers. The electrons depart the hydrogen gas molecule (H_2) and without those electrons, the molecule breaks up into two positively charged hydrogen atoms. There are 2 hydrogen gas (H_2) molecules in Eq. 2 because Eq. 3 needs four protons (H^+) to combine with the two oxygens that come from the one molecule of O_2 gas. Everything has to balance out in the end. Oxygen gas is two atoms of oxygen bonded together into a molecule just like hydrogen gas is two atoms of hydrogen bonded together into a molecule. Nitrogen is another gas that has two atoms per molecule, and these are called **diatomic gases**. Not all gases are like that. Carbon dioxide gas, for instance, has one atom of carbon and two atoms of oxygen (CO_2).

After the catalyzed reaction, the electrons move into the conducting anode and then are repelled from the negatively charged anode while being attracted toward the positively charged cathode. The electrons flow out of the anode, through the connecting wire and to the cathode. If you put a meter in the conducting wire, you will read current flow that depends on how much hydrogen gas was in contact with how much catalyst/anode which in turn produced so many electrons.

After the catalyzed reaction, the protons (H^+) migrate into the electrolyte. Electrolytes allow ions (atoms or molecules that have a charge, either positive or negative) to pass just like metal allows electrons to pass. The ion, in this case a proton (H^+), can go to the cathode where they can recombine with the oxygen and electrons to make water via Equation 3.

$$O_2 + 4\text{ electrons}^- + 4H^+ \leftrightarrow 2H_2O \qquad \text{Eq. 3}$$

In this case, you start with the oxygen gas that has two atoms bonded together into one molecule. The gas comes in contact with four electrons and four protons (H^+). All these must come together at the same time, and when they do, the oxygen gas breaks up into two oxygen atoms and finds four protons (H^+), while the four electrons also find the protons (H^+). Oxygen and hydrogen and electrons all combine to form water. The four protons (H^+) and four electrons are needed because you must use both oxygen atoms once the oxygen gas breaks apart. You cannot leave one oxygen hanging; that is one of the laws of chemistry. This is important for a fuel cell because it means you must carefully balance the oxygen gas and the hydrogen gas.

You will see many drawings like the one shown above explaining fuel cells, and they would not actually work very well. This drawing is presented here to illustrate a few important points. If the anode/catalyst is a platinum strip, then the protons (H^+) would be produced at the surface on the far left where H_2 gas and metal catalyst meet. That means the protons (H^+) would have to migrate through the platinum to get to the electrolyte on the other side. Now hydrogen is very small and capable of moving through platinum, just not very fast. The same is true of the cathode side, in that the O_2 would have to migrate through the cathode to meet up with the proton (H^+) on the other side. Oxygen will not move through metal, and even if it did, then the water would be produced between the electrolyte and the cathode, which would soon block the proton's access to the cathode and stop the cell from working.

We will go to the PEM cell to see how these problems were solved in one type of fuel cell. PEM stands for Proton Exchange Membrane and it is the heart of the system. The cell uses a solid polymer electrolyte membrane that will pass protons (H^+) but not much else. In particular, it will not pass hydrogen (H_2) or oxygen

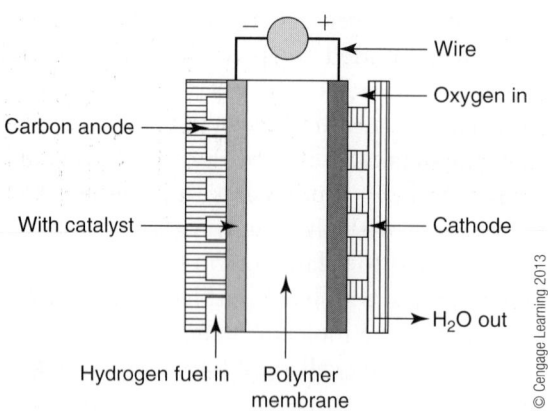

■ **Figure 2-2** The basic hydrogen fuel cell redrawn.

(O_2) gas nor will it pass electrons. Since the membrane will pass only protons, they named it the Proton Exchange Membrane. The actual workings of the membrane are complicated and so well beyond the scope of this book. However, simply put, it passes the hydrogen protons because of a combination of features. First is the size of the proton, then the charge on the proton, and finally that it is hydrogen. The membrane is not a liquid but often a gel type solid. It can come into direct contact with the electrode and not move, so the cell works in all positions. This was pretty important in space.

The gas, though, still needs to get to the membrane so the protons (H^+) can pass through the cell. In the PEM cell, the electrodes are not solid metal; they are porous so gas can flow through. Many are carbon cloth or carbon fiber paper since those conduct electricity well enough for fuel cells and allow gas to pass. We need the catalyst as well though, and modern PEM cells use the same type of methods Niedrach originally did to deposit very fine platinum powder into one side of the porous carbon (where it will contact the polymer membrane and is usually called the active layer). There is an interface where the porous electrode, the catalyst particle, and the polymer membrane all meet. The hydrogen gas (H_2) flows through the porous electrode until it hits this interface to contact the fine catalyst particles so Eq 2 can happen. Since the catalyst particles are in contact with the conducting electrode as well as the electrolyte polymer membrane, the protons (H^+) and the electrons both have paths to follow. The protons (H^+) cannot go through the anode/wire system but can go through the proton exchange membrane to the cathode. The electrons cannot go through the proton exchange membrane but can go through the conducting carbon anode, through the connecting wire and to the cathode. When the oxygen (O_2) gas is put in at the cathode side, it meets with the proton (H^+) as well as the electrons and they combine to make water. Figure 2-2 shows a more realistic view of a basic fuel cell.

ADVANCED PRINCIPLES

That sounds simple enough. It is not really though. A few questions make it complicated quickly enough. Why is there a catalyst? What is a catalyst? Why do positively charged protons go to the positively charged cathode while negatively charged electrons go to the positively charged cathode too? To answer questions like that and to understand the basic operating principles of fuel cells, we have to talk about more complex things such as temperatures, pressures, size, catalysts, heat, and electrodes. Please keep in mind that this discussion is not meant to teach chemistry but to explain why fuel cells have the parts they do, and to do that we have to at least talk about chemistry.

Rules exist that determine whether chemical reactions can happen. For instance, put sodium (Na) and chloride (Cl) together and get sodium chloride, commonly called table salt. Put sodium and hydrogen together, nothing happens, but put hydrogen and chloride together and get hydrochloric acid. Not only that, but the reactions can be stopped or sped up or even reversed.

Here are four rules of chemistry that determine to a large extent why fuel cells are built the way they are:

- if a chemical reactions is possible, it goes from right to left and left to right until it balances out;
- when a chemical reaction happens, it changes in some way the pressure, the volume, or the temperature of its surroundings;
- changing the pressure, the volume or the temperature changes whether a chemical reaction will or won't happen;
- just because a chemical reaction can happen, does not mean it will happen anytime soon.

In the PEM fuel cell, all three equations shown so far have to happen or electricity is not produced. When all the rules that determine whether reactions can occur are taken into consideration, the chemical reactions are possible. Remember that once a chemical reaction is possible, it is either going to go from the left to the right or from the right to the left until a balance is struck (the arrow goes both ways). A fuel cell must continue to free up electrons and move them along the wire. It won't do that if, for instance, Eq. 2 goes from right to left. Instead it will suck up electrons and produce hydrogen gas. That cannot be allowed to happen and is stopped simply enough. First, add more hydrogen gas. Having too much of what is on the left side of the equation makes the equation go from left to right, just as having too much of what is on the right side of the equation makes it go from right to left. Secondly, get rid of the electrons and the protons (H^+) for the same reason. The electrons go into the wire and the protons (H^+) into the polymer membrane. The same applies to the reaction $O_2 + 4\ electrons^- + 4H^+ \leftrightarrow 2H_2O$ (Eq. 3) happening at the cathode. We make sure there are more than enough oxygen gas available as well as more than enough protons (H^+) so that whenever four electrons arrive, we get a reaction. We also get rid of the water.

But, will $2H_2 \leftrightarrow 4H^+ + 4\ electrons^-$ (Eq. 2) happen if we just have hydrogen gas in a container? No, it will not and for a couple of reasons. The first and simplest is that things almost always have a neutral charge or to put it another way, electrons rarely just pack up and hit the road. Secondly, there is a chemistry term and concept called the Gibbs **Free Energy** (named after Gibbs of course) which tells you which way a reaction will go. Issues like this are dealt with in a part of chemistry called **thermodynamics**. In the case of Eq. 2, we want it to go from left to right, but it will not do that because the Gibbs Free Energy is actually zero which is to say the reaction is at equilibrium. That means it goes from right to left in the same amount it goes from left to right, or to put it a different way, the same amount of hydrogen gas (H_2) that breaks apart into protons (H^+) is created from the protons (H^+) that are combining together with electrons to form hydrogen gas. Mind you not many go either way, but in chemistry that is different than saying absolutely none go from right to left or from left to right (the reaction never happens).

So according to the rules of chemistry, Eq. 2 is not really going to happen very often. How is it then that we get protons and electrons moving in fuel cells?

Something has to be done to make it happen. One of the rules is that changing the temperature, pressure, or volume of the system where the reaction is taking place changes whether the reaction happens, so we would change one of those three. Changing one of those changes the Gibbs Free Energy but high temperature or pressure reactions need special precautions, which always increase the price and complicate the equipment. That is not good for fuel cells because they are expensive enough as it is. There is another way, and to put it bluntly that is to cheat the process. We use a catalyst. A catalyst makes a reaction spontaneous that normally would not be. An automobile exhaust catalytic converter does this. The nitrogen and oxygen coming hot out of the engine want to combine to produce pollutants made up of nitrogen and oxygen atoms combined into molecules. The catalytic converter provides a different reaction path for them by using catalysts on a porous substrate so the exhaust will flow much like fuel cells need gas to flow. Instead of combining, the nitrogen reacts with itself to form nitrogen gas (N_2) and the oxygen reacts with itself to form oxygen gas (O_2). That works out fine, since these are the two main constituents in air anyway. In the PEM cell, platinum is used to cheat the process so that when the hydrogen gas contacts the catalytic surface, it breaks up into protons (H^+) and electrons (e^-) via Eq. 2. Then, since the protons (H^+) and electrons (e^-) are taken away on different paths, they cannot recombine to make hydrogen gas. In Grove's cell, the platinum also acted as a conductor for the electrons, but modern cells use some other, cheaper conductor. For $2H_2 \leftrightarrow 4H^+ + 4\,electrons^-$ (Eq. 2) at the anode, the fuel cell will not work without a catalyst.

What about $O_2 + 4\,electrons^- + 4H^+ \leftrightarrow 2H_2O$ (Eq. 3) at the cathode? Think about that reaction. What happens if you boil water at 100°C (212°F)? You get steam and not hydrogen and oxygen gas. In general if the reaction is possible and you know it will not be going from right to left (water to gas), then it probably goes from left to right (gas to water). The Gibbs Free Energy tells us the same thing as common sense does in this case, the reaction goes from left to right even at room temperature. If that is the case though, then why did Grove use a platinum catalyst as the cathode as well as the anode?

One part of chemistry is to make a reaction happen but there is also the issue of time. The reactions have to happen in the correct amount of time, a part of chemistry called **kinetics**. For instance in an internal combustion engine running at 1000 rpm, the combustion of gasoline has to happen in one-thousandth of a minute; if the engine is running at 2000 rpm, then 2000 reactions have to happen in one minute. If the combustion reaction of gasoline with oxygen takes 30 seconds to go to completion, then the reaction would not be of much use in an internal combustion engine. On the other hand, if the reaction to take place combines flour and water to make a cake at an elevated temperature, then half an hour is fine. The reaction has to fit the circumstances. It is hard to cheat the process with kinetics; you have to change the temperature, pressure, or volume of the system where the reaction is taking place. A catalyst is used at the cathode because Eq. 3 is slower than Eq. 2. A catalyst lowers the energy needed for Eq 3 to happen but it does not really increase how much is reacting in any given time. In this case, though, it makes the reaction more likely to happen and so if you do have to raise the temperature to make it happen fast enough, it can be raised by a smaller amount. Many fuel cells have to run at elevated temperatures so that the protons (H^+) and electrons are then used up at the cathode in the same amount they are produced at the anode.

Keep in mind that hydrogen gas must be broken apart, electrons and protons transferred and brought back together, oxygen added, water made and then removed. Gas pressure has to remain steady, water cannot build up, and the number of electrons flowing has to be steady so the current remains steady. There are many processes occurring, and they all have to balance out, just like the chemical equations do. Both catalysts and elevated temperatures are used to exactly balance the two reactions in time as well as in the amount being consumed and produced.

Heat is an important aspect of fuel cells as it is in most chemical equations, and no matter if it is added or taken away, it must be carefully controlled. One of the most important parts about balancing the system in fuel cells as in most electronic devices is in thermal management. This is particularly true where chemical reactions are occurring. When chemical reactions occur, heat is either created in what are called **exothermic reactions**, or heat is consumed in what are called **endothermic reactions**. If heat is created, it has to be removed; if heat is consumed, more must be added. Heat builds up in fuel cells because when you add everything up, the reactions are exothermic. That is fine if you need heat to balance the reactions, but unfortunately, the longer you run the system, the more reactions happen and the more heat builds up. Most fuel cells have either a passive heat management system such as fins to remove heat or an active system such as fans controlled by thermostats. The size of the systems many times determines whether the thermal management systems are **active** or **passive**.

Finally, we get to an issue that is pure chemistry. Electrons go to the cathode because electrons always go to a cathode, that is part of the definition of a cathode. Why do the protons (H^+) go to a cathode when they are positively charged? They do so because of what is called a concentration gradient. Remember electrons move because you have more of them in one spot (the anode) than in another (the cathode); that is an electron gradient that by definition is called voltage. There is also a buildup of protons (H^+) at the anode that cannot go back to being hydrogen gas because the electrons are already gone. The longer the reaction goes on, the more protons (H^+) build up at the anode. The proton (H^+) gradient along with more protons (H^+) coming in every second, makes the protons (H^+) move toward the area where there are less protons. The area around the cathode has the least number of protons (H^+), since that is where they are being used up to make water. The protons will go from the area of high concentration to the area of low **concentration**.

This is a common chemistry principle used every day. Several of the principles being discussed can be seen in action if you slowly pour a colored juice powder into a glass of water and do not stir. The color will move out into the water, finally turning the whole glass of water a pale shade of whatever color it is, with the darker color at the bottom where the powder has fallen. The powder falls through the water because its mass is greater than that of water and gravity is accelerating it toward the center of the earth. It stops moving and stays because the force of gravity is not strong enough to break the glass. The color moves because the powder reacts with water to form molecules of flavoring and water, sugar and water, or coloring and water. The very high concentration of flavoring, sugar, or coloring must move into that area of the water where there is no concentration of flavoring, sugar, or coloring. To stir a drink like that, be it iced tea or powdered lemonade or juice is to change the kinetics by adding energy into the system. The energy of concentration gradients is small, so it takes time to move the molecules (**diffusion**) through the water, but the energy you add by stirring is large, so things move very quickly. The powder can also be distributed back up into the glass because the energy you add is stronger than the force of gravity.

THE DESIGN OF FUEL CELLS

Fuel cells are simple arrangements that promote chemical reactions in the right proportions, the right time and the right places so that a steady flow of electricity is produced. Their simplicity compared to the electromagnet methods of making electricity such as generators is one of their strengths. Having discussed many of the issues that force fuel cells to have the parts they have, it is time to discuss the basic mechanical systems needed in fuel cells.

The list of required design elements includes such things as fuel gas supply, fuel gas distribution at the anode, gas treatment at the catalyst, electron transport, proton transport, proton and electron distribution at the cathode, water production, water distribution at the cathode, and thermal management. The list of preferred design elements include such things as fuel gas purity, anode service life, catalyst cost, electrolyte efficiency, pressure and temperature constraints, size, and service life. Keep in mind that these are by no means the entire list of elements used to design fuel cells, but they are enough to gain a basic understanding of why a cell is built the way it is.

It starts with the fuel gas. PEM cells can use hydrogen gas but also liquids like methanol. Quality of fuel must be guaranteed. The fuel must be either stored or delivered via in-place infrastructure. If the fuel is delivered as part of a supply system, then the cell must be stationary. If the cell needs to be portable, then the fuel must go with the cell. Fuel flow must be controllable. Fuel gas temperature must be steady. Once the fuel is moving into the cell, it must distribute equally onto the catalyst. A system to spread liquid fuel such as methanol will not work for a gas fuel, meaning that fuels will not be interchangeable. Fuel cannot bleed through the cell nor can it escape from around the edges. Since individual fuel cells do not supply enough electricity, they are stacked until enough are available to generate the electricity required. Fuel must be supplied to each cell in the stack. Fuel that is not used must be either recycled in the case of hydrogen or discarded in the case of secondary fuels.

The electrical circuit must be carefully designed. The electrons must be moved from the catalyst to the anode, requiring some type of uniform collection system for each cell in a stack to maintain constant current. Each cell in the stack must be wired to a central bus so electrons flow uniformly to the cathode. The electrons must have access to the catalysts on the cathode. The electrolyte must be maintained to ensure the overall electrical circuit remains intact. Control systems must be integrated into the cell and the stack. Power conditioning must be done, for instance, to change the DC power of the fuel cell to AC power if that is needed.

The anode electrode assembly must stand up to the gas flow and pressures. It must have sufficient catalyst, spaced at appropriate intervals so that the fuel gas is used but not have so much that the cost becomes excessive. It must not expand or contract during operational cycles or the gas seals and electrical connections may be compromised. It must not degrade from contact with hydrogen gas, the electrolyte, current flow, proton (H^+) flow, or electrical connections.

The catalysts, both anode and cathode, must be affordable. Platinum, a common catalyst, is expensive, over $1000 per ounce, so it does not take much to add substantial cost to a fuel cell. This is particularly true for personal consumer units where costs must be kept very low. The original GE cells used over 5 mg/cm^2 of platinum, with later typical designs going down below 4 mg/cm^2. Many of the design goals for platinum loading on the PEM electrodes are 0.4 mg/cm^2 (0.00082 $pounds/feet^2$) which is now rather commonly achieved, with some experimental

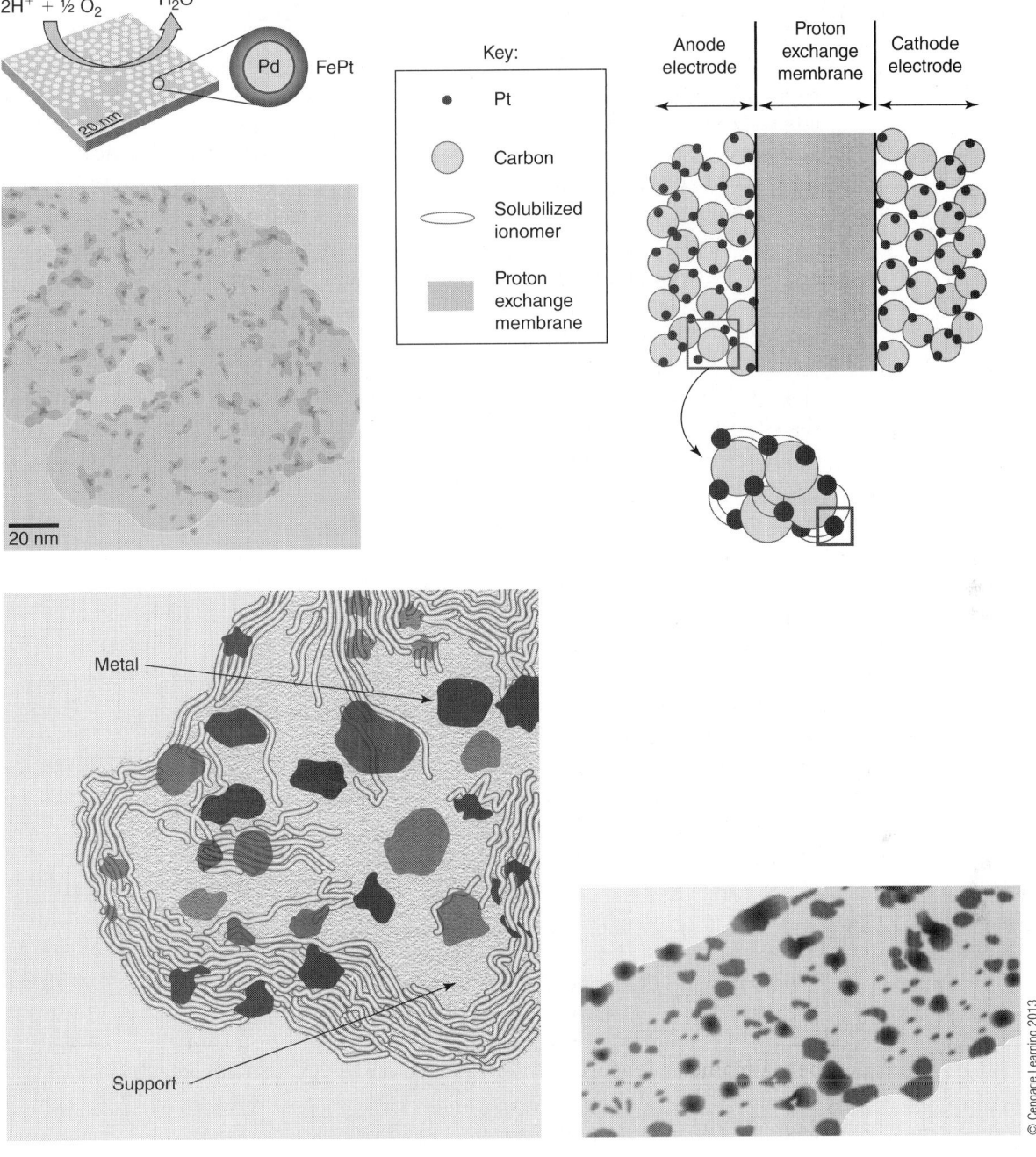

Figure 2-3 Platinum dots on an electrode.

loading going down as far as 0.02 mg/cm² loading. Figure 2-3 shows how much platinum might be loaded onto an electrode.

The energy must be produced at a reasonable cost as well. The efficiency of the system combined with the cost of production, distribution, required infrastructure and a number of other issues determine the cost of energy produced. At 25°C (77°F or "room temperature") and atmospheric pressure (called standard temperature and pressure or STP), the hydrogen/oxygen fuel cell will produce 1.23 volts. We do not need to do the calculations that prove this since we have to draw the line at how complex we are going to go, but this can be calculated if you know the Gibbs Free Energy (a measure of whether a chemical reaction is spontaneous), how many electrons are involved in the reaction (there are two electrons in Eq. 2) and how many coulombs are in a mole of electrons (just let that one go by for now). The 1.23 volts you are supposed to get never shows up, though, because nothing is 100% efficient. Fuel cells lose efficiency because there is resistance within the wire and electrodes to the passage of electrons; there is resistance within the electrolyte to the passage of protons (H^+); there are stray currents in every electrical circuit; the reactants must get to the anode, cathode, and catalysts which takes time and is never uniform across the board; not to mention several other issues having to do with chemistry and electrical engineering that are beyond the scope of this book. Calculated using these chemical methods, the theoretical efficiency of fuel cells is 83%, which is quite a bit better that the 3% Edison first managed when he produced electricity from coal.

Fuel cells are complicated, not only in the way they are made, but in the chemical reactions being used, the raw materials needed to supply them, and the cost of the energy they produce. To understand fuel cells, the student must understand what is important in a number of different systems associated with the cell itself.

KEY WORDS

Knowing the terminology used is critical when dealing with fuel cells. Following is a list of the important terms in this chapter, which are also in bold typeface within the chapter. It is recommended that students be required to submit definitions of some of these words as homework assignments in which they look the terms up in other books, articles, or on the Internet.

- active
- amp
- anode
- atom
- catalyst
- cathode
- cell potential
- chemical symbols
- concentration
- conducting catalyst
- coulombs
- covalent bonding
- diatomic gases
- diffusion
- electrolytes
- electrons
- endothermic reactions
- exothermic reactions
- free energy
- insulators
- ionic bonding
- kinetics
- mass
- metallic bonding
- molecule
- molecules
- neutrons
- oxidation
- passive
- primary batteries
- protons
- redox reaction
- reduction
- scientific notation
- secondary (storage) batteries
- thermodynamics
- voltage
- voltaic cell
- weight

DISCUSSION QUESTIONS

1. How many atoms of oxygen are in each of the following molecules?

2. In H_2SO_4 (sulfuric acid), are the oxygen atoms bonded to something or are they individual atoms in a group that are not bonded to any other atom?

3. Express the following numbers in scientific notation.

4. The electrical charge on one electron is -1.602×10^{-19} coulombs and one amp requires 6.25×10^{18} electrons per second moving past your plug. Since your house has a 150 amp service, how many electrons would it take flowing through your main breaker in one second to trigger an overload and shut your power off?

5. Why would the catalyst in the fuel cell shown in in this chapter have to be conducting?

6. Which of the following is an atom and which a molecule?

CHAPTER 3

FUEL CELLS, CODES, AND STANDARDS

objectives

This chapter will introduce the student to some of the safety issues inherent in fuel cells as well as some of the requirements of safety programs governing the workplace. Definitions of common hazards and control schemes will be discussed. Codes and standards that govern fuel cell design, fabrication, construction, and maintenance will be presented with relevant sections shown. Methods and manners of inspection will also be discussed. By the end of the chapter, the student will be familiar with the complex and formal manner many such installations must be dealt with.

INTRODUCTION

Fuel cells will burst into flames, explode, and electrocute you. The fuels used in fuel cells can be **toxic**, **flammable**, and **explosive** as well. Of course, the same can be said of your car, your hot water tank, and many of the semitrucks going 70 miles an hour on the interstate next to you. What makes fuel cell **codes** and **standards** different than many current codes is that there is limited experience in these systems. Most codes are a result of years of effort resulting from design, use, and maintenance of existing infrastructures; however, the infrastructure and experience required for comprehensive code development for fuel cells is not yet in place. For that reason, many of the codes and standards in place or being considered are incomplete while others are intentionally restrictive to account for this inexperience. A considerable number of the codes and standards are also being based on like systems, such as natural gas or large scale emergency generators, so that at least some regulation is in place, even though it may not be totally appropriate. Those are issues being addressed on a number of different fronts all over the world as fuel cells move toward mass commercialization. This chapter is meant to familiarize the student with chemical and electrical safety issues, review some international, national, state, and local codes and discuss some practical safety concerns that would be encountered in dealing with fuel cells. Students should be aware that information concerning codes and standards may be out of date for a number of years to come due to ongoing code development. A program of reviewing current information codes and standards on a regular basis is strongly recommended for any person or company involved in fuel cells.

GENERAL OSHA INFORMATION

In the United States, safety and industrial hygiene are governed nationally by the Occupational Safety and Health Administration (**OSHA**) with corresponding units in state and sometimes local governments as well. OSHA was created in the early 1970s by the federal government to provide uniform legal requirements across the country for industry. Many of the general requirements for OSHA will directly impact the student's future career in fuel cells. A general review of safety is meant to provide the student a solid foundation for subsequent training on the job.

The base used in OSHA is the **General Duty Clause**, which explains the foundation of the requirements for both employer and employee. In part, it reads:

- Each employer: (1) shall furnish to each of his employees employment and a place of employment which are free from recognized **hazards** that are causing or likely to cause death or serious physical harm to its employees (2) shall comply with occupational safety and health standards promulgated under this Act.
- Each employee shall comply with occupational safety and health standards and all rules, **regulations** and orders pursuant to this Act which are applicable to their own actions and conduct.

Safety in the workplace is a legal requirement not only of the employer but also of the employee. Failure of an employer to follow safe practices can result in fines, sanctions, and even jail time, while failure of an employee to follow safe practices can result in fines and dismissal.

The Hazard Communication Standard details how workplace hazards are to be communicated to the employee by the employer but also how the employee is to communicate with the employer. The base of the **HazCom standard** is that the employee has the right to know about the hazards in the workplace and how to work safely in spite of those hazards.

Employees have the right to be informed about the requirements of the OSHA Standard itself, any hazardous materials present in the work area, and the location and availability of required documents such as **Material Safety Data Sheets**. Employees are responsible for reading and obeying all posted warnings, cautions, or other information, complying with applicable OSHA standards, following all employer safety and health rules and regulations, using required protective equipment while working, promptly reporting any hazardous conditions or accidents to a supervisor, cooperating with any OSHA compliance officer during an inspection, and knowing all applicable emergency procedures.

Workplace hazards are also defined in the HazCom standards. For fuel cells, chemical hazards in particular are important. There are seven classes of **chemical hazards**. **Flammable** hazards are liquids that have a flash point (will ignite) at or above 100°F, solids that will burn faster than one-tenth of an inch per second and aerosols that project a flame that exceeds 18" at full valve opening. **Combustible** hazards are all liquids with a flash point at or above 100°F but below 200°F. **Explosive** hazards are anything where there is an instantaneous release of pressure, gas, and heat when it is subjected to sudden shock, pressure, or high temperature. **Pyrophoric** hazards are any chemical that will ignite spontaneously in air at a temperature below 130°F. **Compressed gas** hazards are any gas that is in a container that has an absolute pressure greater than 40 psi at 70°F. **Water reactive** hazards are any chemical that reacts with water to release a flammable gas or health hazard. **Unstable** hazards are any chemicals that in the pure state will vigorously change under adverse conditions.

Chemical hazards will produce **health hazards** that employers must **mitigate** and employees have the right to be aware of. Health hazards are any chemical that may produce acute (short duration, usually to relatively high concentrations or amounts of material) or chronic (extending over a long period, usually applies to relatively low concentrations or amounts of materials) health effects in exposed employees. There are five categories of health hazards. **Carcinogens** are any materials that are known or suspected to cause cancer. **Toxic** agents are any substances that can cause adverse health effects. **Reproductive toxins** are any substances with adverse effects on reproductive systems. **Irritants** are any substances that are capable of inflaming living tissue. **Corrosives** are any chemical that causes visible destruction of or irreversible alterations in living tissue.

There are four groups recognized to produce health hazards. The **Chemical group** consists of gases, vapors, dusts, fumes, mists, and smoke. The **Physical group** consists of radiation, noise, vibration, as well as extreme temperatures and pressures. The **Ergonomic group** consists of workstation design, repetitive motion, improper lifting/reaching, and poor visual conditions. The **Biological group** consists of insects, mold, yeast, fungi, bacteria, and viruses. Figure 3-1 shows some of the warning symbols associated with the Chemical Group.

These groups can enter the body through four recognized Routes of Entry. The **Inhalation** route is followed by any airborne contaminants. The **Absorption** route is through the skin but does not require a break in the skin for entry to occur. The **Penetration** route is through the skin as well but requires a break in the skin for entry to occur. The **Ingestion** route is from eating, drinking, or swallowing. Figure 3-2 shows the routes of entry as they are typically presented in safety training.

Recognized hazards and routes must be controlled and there are four schemes allowed. First, **engineering controls** must be set in place if at all possible where equipment and instrumentation control or eliminate the hazard. There are several acceptable routes, such as substituting a less hazardous material or process or changing a process to isolate or eliminate a hazard. If engineering does not provide a means of control or elimination, then **work practices** must be installed so that standard operating procedures limit access to a hazard or controls are in place that shut systems down when hazards exceed limits. If neither engineering controls nor work practice controls are feasible, then **administrative controls** can be set up so workers exposure is limited by rotating shifts around the hazards. Finally, **personal protective equipment** can be used to protect workers from the hazard.

■ **Figure 3-1** Warning symbols.

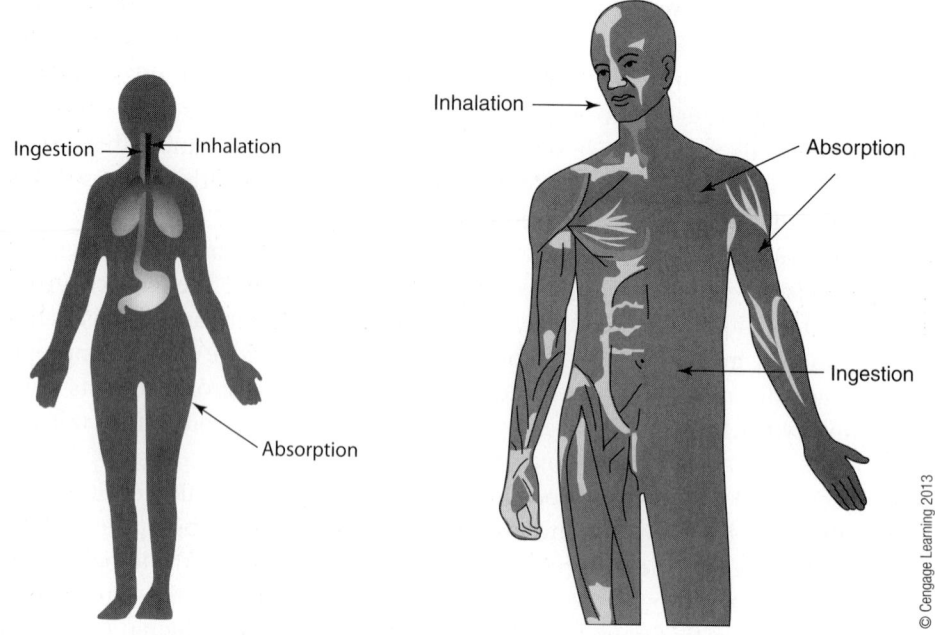

■ **Figure 3-2** Routes of entry into the body.

FUEL CELLS CODES AND STANDARDS

Numerous codes and standards govern fuel cells. These range from internationally recognized codes to local regulations. A partial list is presented. **General installation** siting (indoor and outdoor) codes are covered by National Fire Protection Association (NFPA 853 and 70), the International Fire Code (304, 305, 3112, 2704), the International Building Code (1609, 1612, 1621, 1805) and the International Mechanical Code (102, 303, 306, 401, 501) as well as state and local codes which vary by region. Fuels associated with fuel cell installation are generally covered by the National Fire Protection Association code (NFPA 853) as well as the International Fire Code (27) for supply and storage. These codes covers such things as the use of control areas, signs and labels, requirements for emergency response, piping, tanks spill control and monitoring equipment. If **pressure piping** or vessels are part of the system, the American Society of Mechanical Engineering (Boiler and Pressure Vessel Code, section VIII) in conjunction with The International Fire Code (32), the International Mechanical Code (1001, 1003), the American Welding Society codes (D1.1), and American Society of Non-Destructive Testing codes detail design, installation, and testing requirements. For **process piping** (lower pressure) within an installation, American National Standard Institute (Z 21.83) and ASME (B31.3) codes detail requirements for gas, steam, air water, cryogenic gases as well as other associated systems. Electrical connections and wiring are governed by the National Fire Protection Association code (70, 490, 692), United Laboratories (UL 1741), the national Electrical Code with grid connections by International Electrical and Electrical Engineering Code (IEEE P1547).

Associated codes within those listed as well as others govern fuel cell design, installation, and operations, including such things as **markings**, **fire suppression**, **fuel shut-off systems**, ventilation, exhaust, foundations, support structures, live and dead loading, seismic, wind, floods, lightning, **access**, entrances and exits,

service access, air flow, roofing materials, **fire walls**, automated shut-down sequences, **gas detection**, alarms, **interlock systems**, testing and power-up, **emergency services management**, electrical tie-in systems, **circuit overcurrent protection**, service disconnects, de-energized system monitoring and detection, **system isolation**, wiring methods and materials, **grounding methods**, **output compatibility**, switch gear, high voltage isolation, and inspection access among other things. Figure 3-3 shows commonly used markings that might be encountered in the workplace.

It is not possible to present a discussion of even a small fraction of the codes now in place or being considered. Instead, the NFPC National Electrical Code will be looked at to give an idea of what is involved when dealing with codes and standards.

Electrical

Fuel cells contain a number of different electrical systems, depending on the applications being served. Both high and low voltages are present in fuel cells, ranging from personnel systems of less than half a volt DC to transit systems of over a kilovolt and even to stationary industrial systems serving power grids. The risk of electrocution when working with fuel cells is substantial.

In general, there are seven parts to a fuel cell electrical system, as with most such systems: disconnects, wiring methods and materials, grounding schemes, markings, connections to other circuits, outputs (usually differentiated between less than 600 volts and more than 600 volts), and ancillary equipment.

Three general areas of concern are present in fuel cell systems: the DC electrical output of the fuel cell stack itself, any **grid mains** connected to the system, and the AC output of a DC/AC inverter connected to the cell stack. Figure 3-4 shows what a typical grid main connection area might look like.

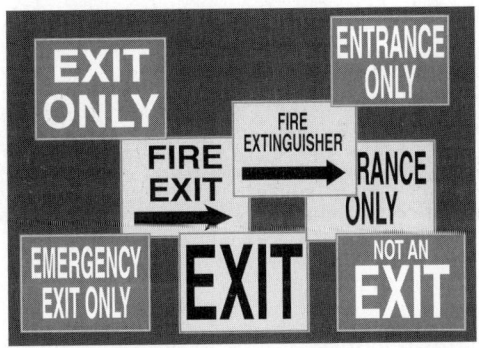

■ **Figure 3-3** Safety and fire markings.

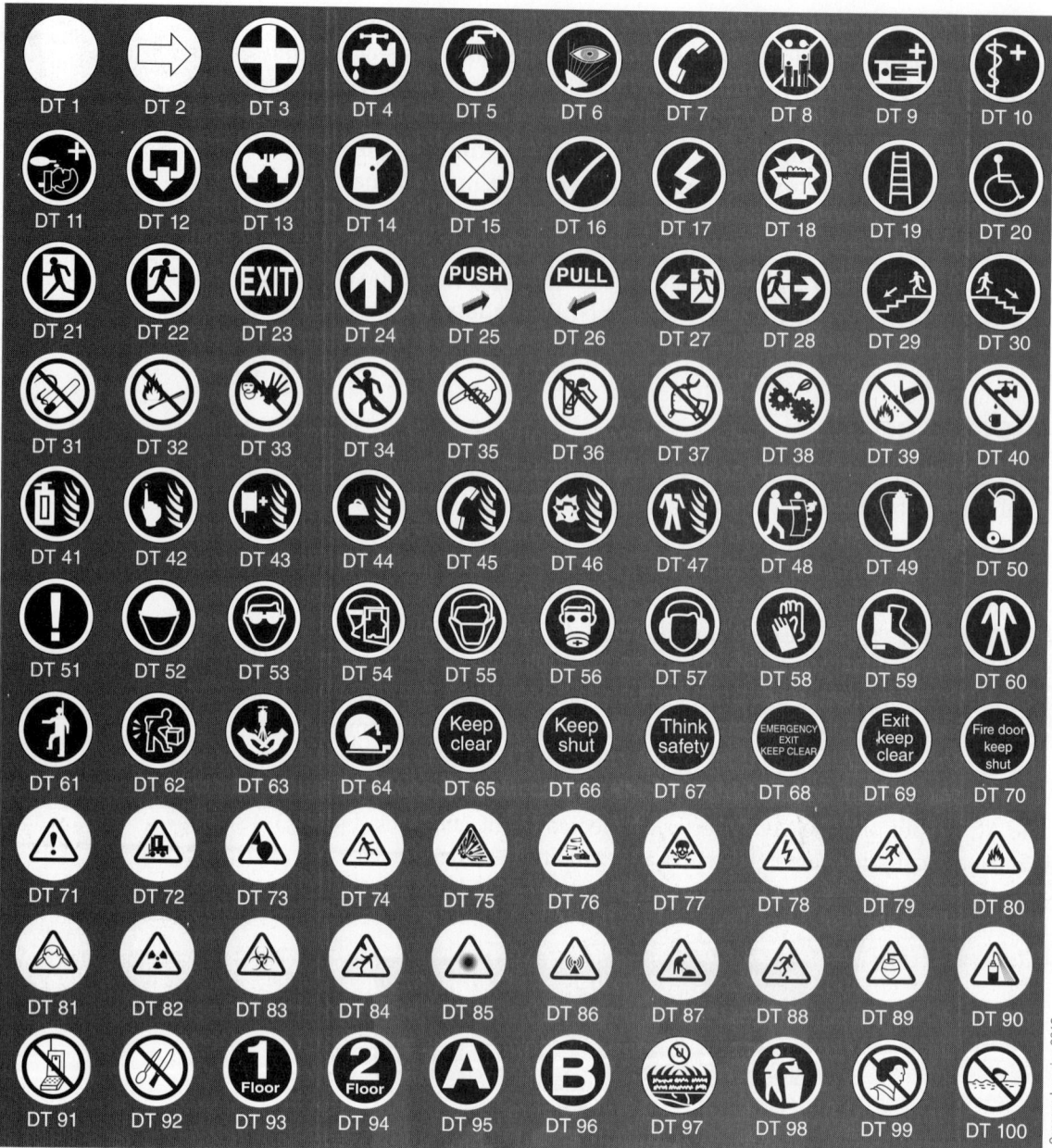

■ **Figure 3-3** (Continued)

Secondary systems of concern depend on the applications being served. If a DC supply is required, there are generally current and voltage control systems in place, whereas AC systems tend to be more complex, with DC/AC conversions, current voltage and frequency controls systems, transformers to step-up or step-down voltages, and harmonic control systems. If fuel cell systems are tied directly into the power grid a number of other units will be in place, such as synchronization, watt ramping, and volt-amp reactive systems. All of these pose particular threats and care must be taken when connecting to or isolating fuel cells from these systems. Many systems also contain what are termed "**dump choppers**," where occasionally excess capacity or energy from residual fuel during shutdown sequences is bled off, so that even if a fuel cell is shut down (fuel supply turned off), it may still be energized and great care must be taken to ensure proper **de-energizing** protocols are followed.

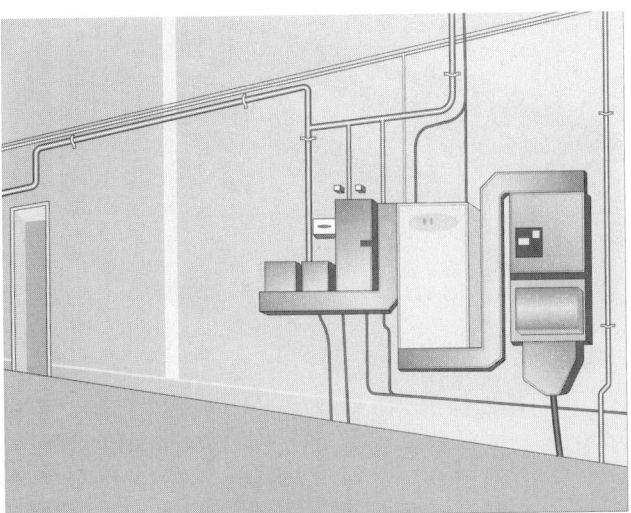
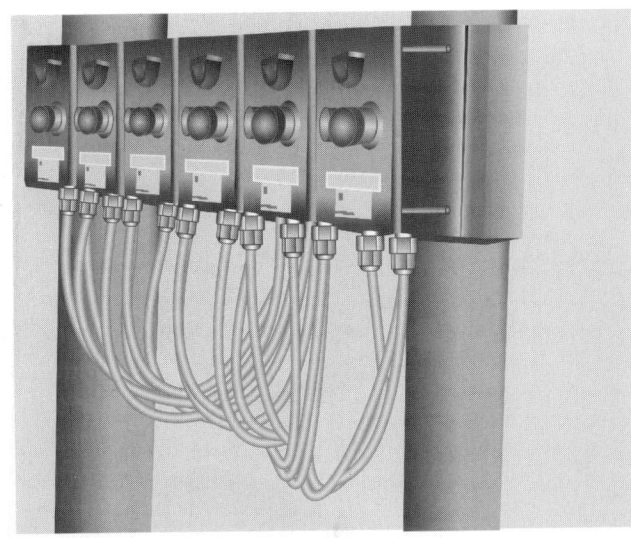
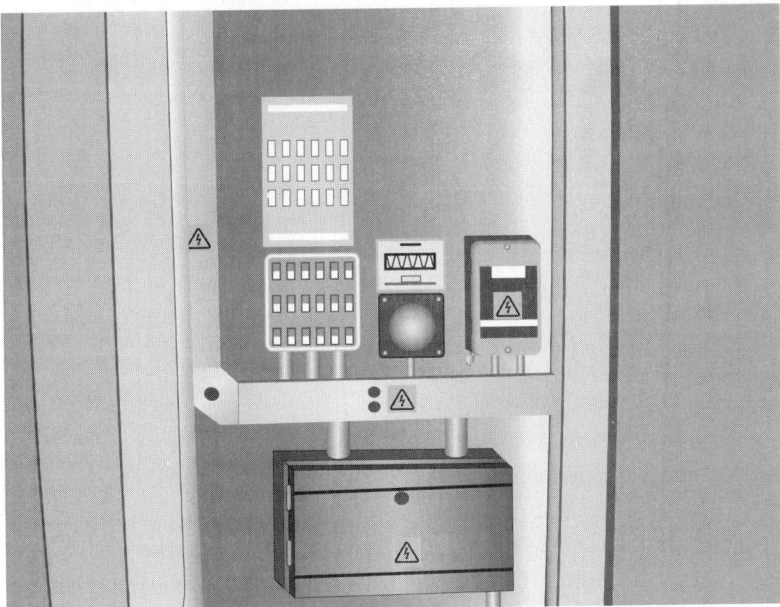
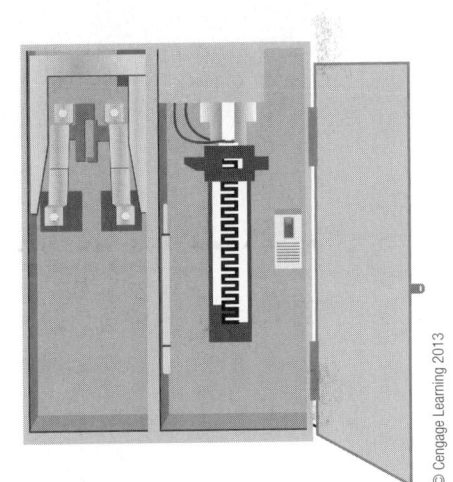

Figure 3-4 Typical main connects.

Most fuel cells contain stacks of cells, and individual fuel cells within the stacks may still be energized even though the stack is not. Each cell will by itself produce electricity but many of the failsafe systems act on the fuel cell as a whole, so that disrupting the fuel cell system circuits may disable failsafes such as dump choppers while at the same time enabling individual cell within the stack to continue producing electricity, recharging part of the system.

Many fuel cell systems work in tandem with batteries and DC/AC inverters. These secondary systems may be of lower voltage but they can remain charged even if fuel cells are physically removed for service. Figure 3-5 shows what an inverter typically might look like.

Systems tied to structures, such as mobile or building systems, can be isolated from the structural components, but they can also become grounded through the structure if problems arise. Cooling systems in particular represent potential areas of concern, and even though ground fault systems are used in many applications, care must be taken to ensure proper **system grounding** is present.

Figure 3-5 Inverter.

The National Fire Protection Association's Nation Fire Code deals specifically with the electrical systems for fuel cells in NFPA 70, Article 692. Subsequent references are from NFPA 70® 2010. A brief summary of the section is presented. Any direct quotations from the code are contained within quotation marks.

As with most codes, the Article 692 begins with definitions of various components it deals with. The NFPA definition of a fuel cell: "An electrochemical system that consumes fuel to produce an electrical current. The main chemical reaction used in a fuel cell for producing electrical power is not combustion. However there may be sources of combustion within the overall fuel cell system such as reformer/fuel processors."

Fuel cell system installation is allowed to "supply a building or other structure" but must be identified with a "permanent plaque or directory denoting all electrical power sources on or in the premises."

Circuits are sized according to the required nameplate where "the **ampacity** of the feeder circuit conductors from the fuel cell systems(s) to the premises wiring circuit shall not be less that the greater of (1) nameplate(s) rated circuit current or (2) the rating of the fuel cell system(s) overcurrent protection device." Many codes and standards provide leeway along these lines, with choices being provided so that design criteria can be met. Provision are also made here for different ground connections, with the ampacity of the grounded or neutral conductor not being allowed to exceed that of the sum of "the maximum unbalanced neutral load current plus the fuel cell system(s) output rating.

Overcurrent protection is dealt with in Section II (Circuit Requirements) Article 692.9. Most codes are arranged in like manner, with subsections called out to deal with particular issues. In this case, "should a fuel cell system be provided with overcurrent protection sufficient to protect the circuit conductors that supply the load, additional circuit overcurrent devices shall not be required."

Most codes allow for exceptions. 629.10, for instance, allows for modifications that do not meet code requirements in stand-alone systems under three circumstances: "a stand alone system shall be permitted to supply AC power to the building or structure disconnecting means at current levels below the rating of that disconnection means"; circuit conductors' connection of the fuel cell to the building or disconnects must be based on the fuel cell output rating, but "these conductors must be protected from overcurrents in accordance with 240.4." Many codes refer to themselves or to other codes as well. Since codes have the force of legal contracts in many cases, care must be taken to understand the entire body of requirements when dealing with codes and standards, acquiring all the relevant sections.

Disconnect means and requirements are addressed in Section III of Article 629, requiring means to disconnect all current carrying conductors in the form of "readily accessible manually operable switch(es) or circuit breakers." Figure 3-6 shows what a disconnect would typically look like.

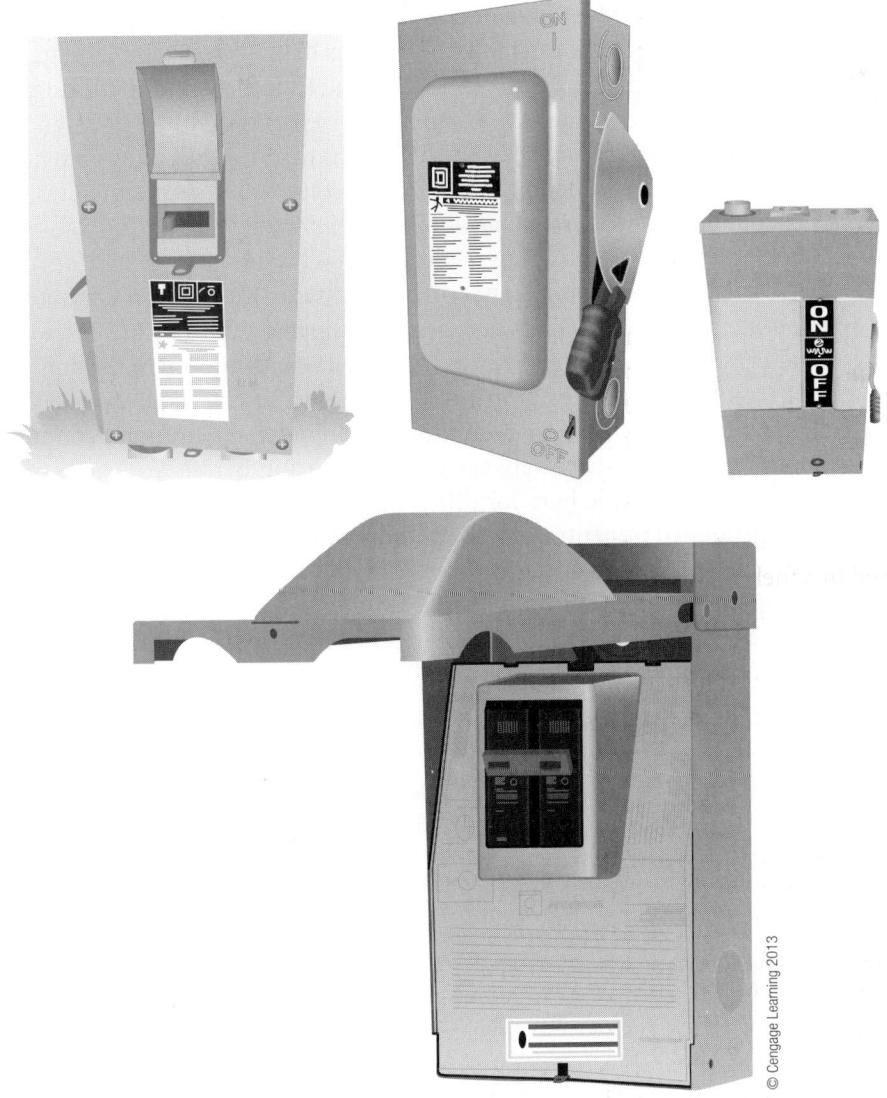

■ **Figure 3-6** Electrical disconnects.

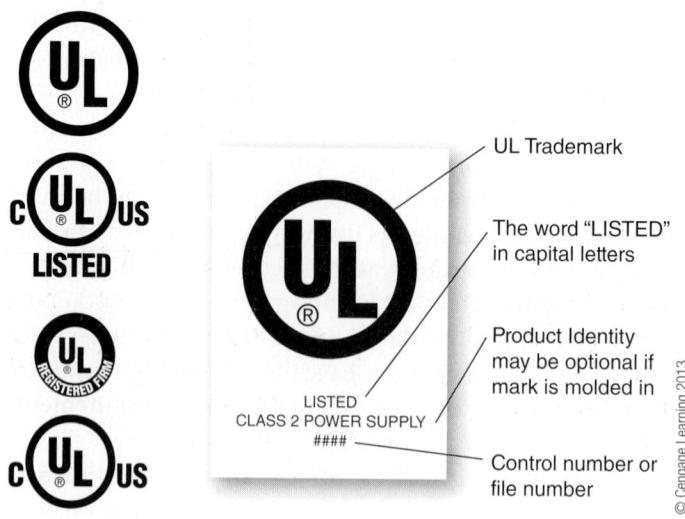

Figure 3-7 United Laboratories (UL) markings.

Section IV deals with Wiring Methods, allowing "wiring systems and fitting specifically intended and identified for use with fuel cell systems." While not specifically calling out another code or standard, sections like this require a fundamental understanding of material and systems manufactured for specific applications. In electrical systems, Underwriters Laboratories (UL) listings and ASTM standards are but two of a substantial listing of specifications dealing with allowable manufacturing, packaging, and labeling requirements. Figure 3-7 shows the markings typically encountered on UL listed products.

Section V deals with grounding for AC and DC systems as well as for mixed systems. In a **mixed system** for instance, the DC systems are required to be bonded to the AC systems with a single grounding bar allowed. The size of the grounding conductor is dealt with in 250.122 which a table of Minimum Size Equipment Grounding Conductors, where, for instance, a 200 amp service requires a #8 (AWQ or kcmil) copper conductor or #6 aluminum or copper-clad aluminum conductor. Figure 3-8 shows the difference between copper wire numbers.

Provision are also made here for different ground connections, with the ampacity of the grounded or neutral conductor not being allowed to exceed that of the sum of "the maximum unbalanced neutral load current plus the fuel cell system(s) output rating. Manuel fuel shut-off valves must be marked where the primary disconnect is and any electrical energy storage device must be clearly marked.

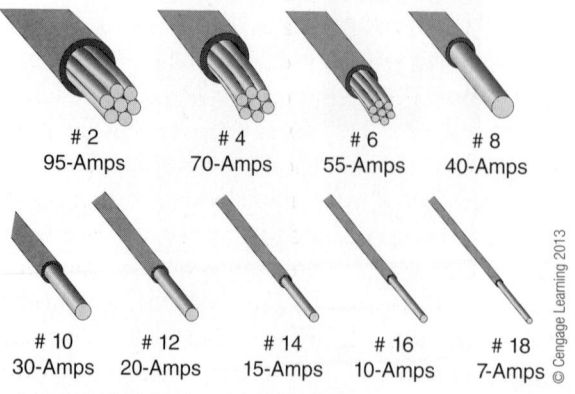

Figure 3-8 Copper wire sizes.

Section VII deals with connections to other circuits. It is one of the larger sections of 692 and details requirements for transfer switches, identifying interactive equipment, output characteristics, loss of power, and unbalanced interconnections.

Installation

The discussion of electrical issues above is an example of a code dealing with a specific part of some greater whole. In the case of the electrical subsystems, connections and safety systems, codes are many times extremely detailed. This is particularly true of codes dealing with construction or infrastructures such as bridges or buildings. Other codes are much more general, meant to provide a framework for general issues which are then dealt with in subsequent codes, standards, or regulations. An example of this is NFPA 853, Standard for Installation of Stationary Fuel Cell Power Systems. All mentions of this code in the following section are taken from the 2010 edition.

NFPA 853 applies to the design, construction, and installation of **stationary fuel cells** as regards fire protection. Keep in mind that this is one part of a collection of codes that would be employed in designing and building a fuel cell installation.

As in most codes, the initial sections explain the scope of the code. These sections are very important, in that they inform the reader as to what can be expected of the code. For NFPA 853, the code covers "**singular prepackaged self-contained power**" systems or any combination of such units, "power system units comprising two or more factory-matched modular components intended to be assembled in the field," or "engineering and field-constructed power systems that employ fuel cells." Thus, these codes would not cover mobile units in automobiles, charging units meant for personal use, or custom constructed units such as those meant for major power generation. Again, this is an important issue with codes, in that they deal with specific situations, equipment, or machinery and using them outside their scope is unwise.

Using Multiple Codes and Standards

The purpose of NFPA is to "provide fire prevention and fire protection requirements for safeguarding life and physical property associated with buildings or facilities that employ stationary fuel cell power systems." The code then is a guidebook for ensuring facilities and personnel are protected from known hazards, in this case fire. In that respect, it and all other codes should be viewed as a source of current best known practices for achieving a goal, in this case, fire protection.

Keep in mind as well that codes and standards have the **force of law** only when they are granted that right by regulatory agencies or contract. NFPA specifically addresses this issue for instance in its Chapter 1.4 Retroactivity sections, where it exempts facilities, structures, and equipment from its requirements if they were built or approved before the effective date of the standard; however, jurisdictional authorities such as local boards are given the right to "apply retroactively any portions of this standard deemed appropriate." All codes deal in some way with this issue, in that codes are almost always created and maintained by professional societies, individuals well known in the industries served and laboratories dealing with the code issues on a daily basis. The members of the technical committees that oversee the codes are given within the codes themselves. The codes then are meant to be advisory but can be granted the force of law by governmental agencies such as zoning boards or by specific contracts between individuals, such as a private company when contracting out construction.

Another issue along those lines is in what is called **equivalency** in NFPA 853, where section 1.5 states "nothing in this standard is intended to prevent the use of systems, methods or devices of equivalent or superior quality, strength, fire protection resistance, effectiveness, durability and safety over those prescribed by this standard." The issue here is in proving equivalency, something that is usually done through engineering testing and scientific evaluation of competing systems. Using nonstandard parts, procedures, or equipment is allowable, provided proof of their effectiveness is available and would be generally recognized by qualified personnel. Chapter 1.5.1 states that "technical documentation shall be submitted to the authority having jurisdiction of demonstrate equivalency." The final decision as to the effectiveness of any system is thus laid on the jurisdictional authority as opposed to those developing and maintaining the codes. As descriptions of best practices, codes are meant to protect individuals working around these systems from known hazards as well as to protect the owners of the systems from claims of substandard preparation or operation, not judge the design and implementation of every possible system.

All codes will at some point include a listing of complimentary codes that deal with issues that are to be expected in the scope of work. NFPA 853, Chapter 2, Referenced Publications, lists these as "portions thereof listed in this chapter are referenced within this standard and shall be considered as part of the requirements of this document." In almost all cases, no single code or standard completely covers the requirements needed to construct and operate buildings or facilities such as fuel cells systems due to their complexity. The complimentary references for NFPA 853 lists 16 NFPA publications ranging from NFPA 30 (Flammable and Combustible Liquids Code 2008) to NFPA 101 (Life Safety Code). It lists two American National Standards Institute (ANSI) publications, ANSI A13.1 (Scheme for Identification of Piping systems 1996) and ANSI CSA FC.1 (American National Standard for Fuel Cell Power Systems 2004). It lists the American Society of Mechanical Engineers (ASME) ASME/ANSI B31.3 Process Piping 1996. It lists the American Society for Testing and Materials (ASTM) ASTM E108 (Standard Test Methods for Fire Tests of Roof Coverings 2007). Finally, it lists Underwriters Laboratories (UL) UL 790 (Standard for Test Methods for Fire Tests of Roof Coverings 2004). Other publications used in NFPA 853 include the Merriam Webster Dictionary and five other NFPA standards where extracts from these are used in mandatory sections.

Two important issues relate to the above lists. The first has to do with the difference between a code and a standard. The American National Standards Institute (ANSI) has federal force under law, in that it governs what can be claimed in the United States generally for manufactured or fabricated items, in the case of ANSI A13.1, when identifying piping systems. That is to say that the in-place identification systems for piping systems in the United States must conform to these standards or they are not allowed. An example of this is the color of oxygen lines, required to be green. Codes are references meant to provide a view of best practices currently in use, but standards many times have the force of law behind them and must be followed to the letter. The other issue of importance is that codes overlap. ASTM E108 is much like UL 790, detailing tests used in evaluating roof coverings, but they may well differ because their missions differ. Underwriters Laboratories (UL) deals with insurance coverage while ASTM deals with materials testing and may not test for the things insurance companies have an interest in.

Definitions

Next in NFPA 853 as in many codes are definitions. Given the complexity of constructing and operating major systems such as fuel cells, it is important that terms be used so that the possibility of misunderstanding is minimized. For instance

NFPA 853 defines "**Shall**" as indicating "a mandatory requirement" while "**Should**" indicates "a recommendation or that which is advised but not required." Definitions are critical because many codes are international in nature and the exact meaning of words used in contractual obligations may determine liability. It is important for those working with codes to understand the language used, and to understand the differences in what might at first glance be a common word. For instance, according to OSHA, Flammable hazards are liquids that have a **flash point** at or above 100°F, solids that will burn faster than one-tenth of an inch per second, and aerosols that project a flame that exceeds 18" at full valve opening. The NFPA 853 definition of flammable is more technical, "a liquid that has a **closed-cup flash point** that is below 37.8°C (100°F) and a maximum absolute pressure of 2068 mm Hg (40 psi) at 37.8°C (100°F)." The NFPA definition includes a second component that the OSHA definition does not, that of pressure in a closed cup. It is critical that the student understand that definitions vary across codes and assuming that a word used in one code means the same when you come across it in another code is unwise.

Siting, Construction, and Installation Requirements

NFPA 853 Chapter 4 begins the more technical aspect of this code, detailing **siting** and construction requirements in some detail. However, as in most codes, exceptions to the code are given up front in the detailed discussions. Since NFPA 853 deals with prepackaged, self-contained units, the exception is for those units which do not meet the definition of **prepackaged** and **self-contained**. This definition is not given in NFPA 853 but rather in the American National Standard ANSI CSA FC.1, which details the design, testing, and listing requirements of a unit sold in America as prepackaged, self contained. Again, the codes and standards are not single units but rather a vast collection that require substantial effort to properly understand. If the units are outside the scope of ANSI CSA FC.1, then NFPA 853 allows for their use by specifying they must meet the intent of the ANSI standard or be "evaluated based on data from operational experience in the same or comparable service or test records covering the performance of equipment or materials." That is to say that the systems must have been in use in a similar situation with detailed operational records available, which is the case for units developed outside the United States, or for units not already in place, have testing data meeting the ANSI requirements. All codes contain nonstandard clauses that allow for exceptions.

While it is not possible to go over NFPA 853 in detail, excerpts from the code can provide a good introduction to the requirements of design, installation, operation, and repair. Keep in mind that repairs must maintain all code and standard integrity.

NFPA 853 Chapter 5 details siting and interconnection requirements. 5.1 General Siting lists 12 criteria for siting and installation, including firm foundations capable of supporting all equipment, protection from the elements, seismic events, floods, unauthorized **access**, or hazardous atmospheres. Systems cannot restrict entrance or exit, exhaust into working areas, be accessible for service or emergencies, or be located in such a way that hazardous materials are stored nearby. If the units are to be located outdoors, then **air intakes** must be properly located, exhaust must be "4.6 meters (15 feet)" from HVAC intakes or any openings into adjoining buildings and cannot be directed onto pedestrian walkways. If units are located indoors, they must "be separated from the remainder of the building by floor, wall and ceiling construction that has at least a 1 hour **fire resistance rating**," as must any electrical or piping penetrations. Any "openings between the room and other occupied spaces shall be protected by **fire doors** and dampers," with both doors and dampers defined by NFPA in different codes. For rooftop installations, the requirement for outdoor installations must first be met, then any roofing material "under and within 30.5 cm (12 in) horizontally from a fuel cell power system or component shall be noncombustible or shall have a Class A rating when tested in accordance with ASTM E 108." In the case of roofing then, either a commercial product tested to that standard and sold as such must be used or the material has to be tested individually by a **qualified** laboratory with records maintained for the life of the installation. Figure 3-9 shows the components that might be required in a fire door.

As can be seen even in the limited discussion of codes above, they are very detailed, contain substantial information and are part of a larger network of requirements ranging from zoning regulations to international maritime law. This textbook is not intended as a course in codes and standards, but rather as an introduction to them and so rather than repeat what is in the NFPA, which amounts to several thousand pages by itself, the discussion will be taken up again in the next section.

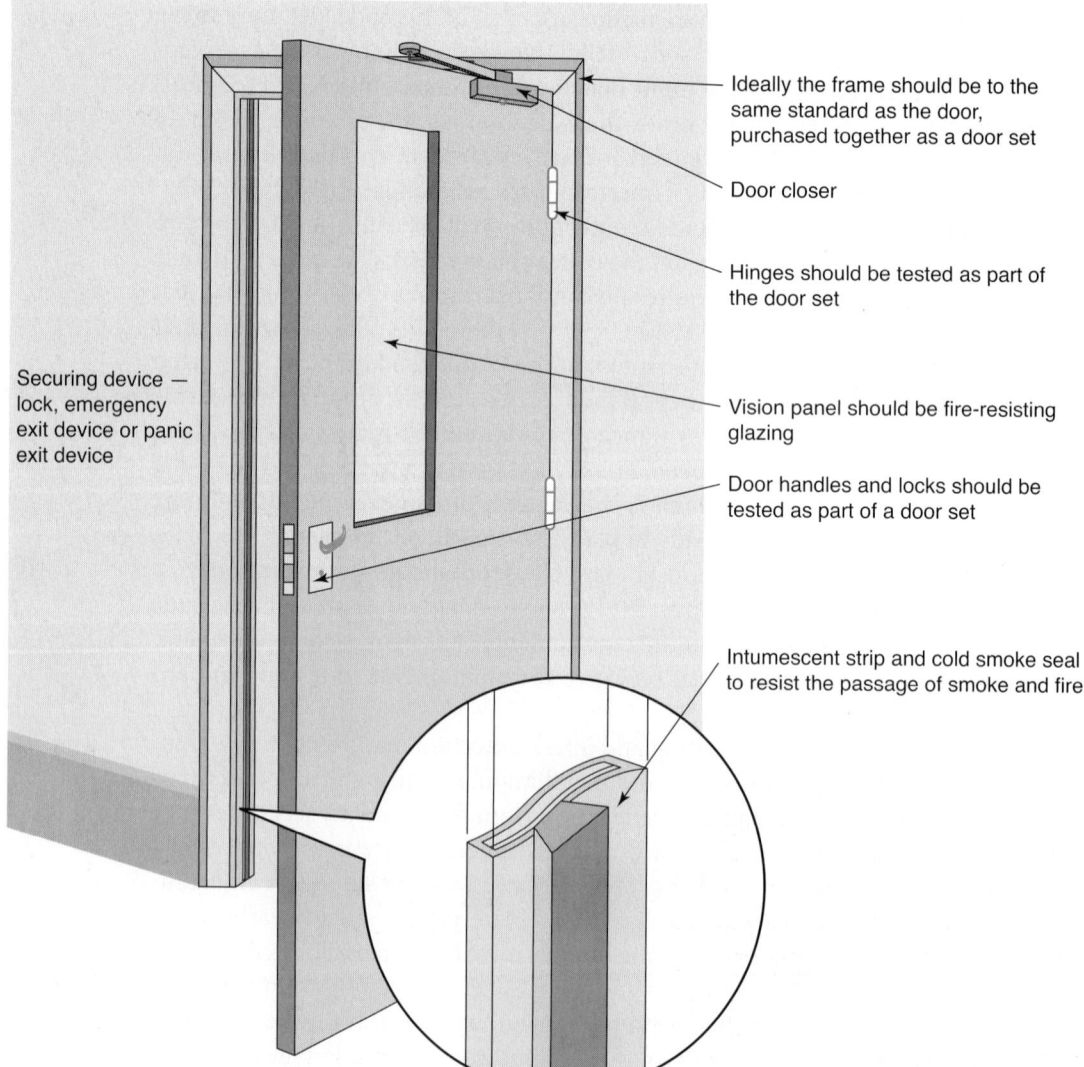

Figure 3-9 Fire door.

CHAPTER 3 • FUEL CELLS, CODES, AND STANDARDS 41

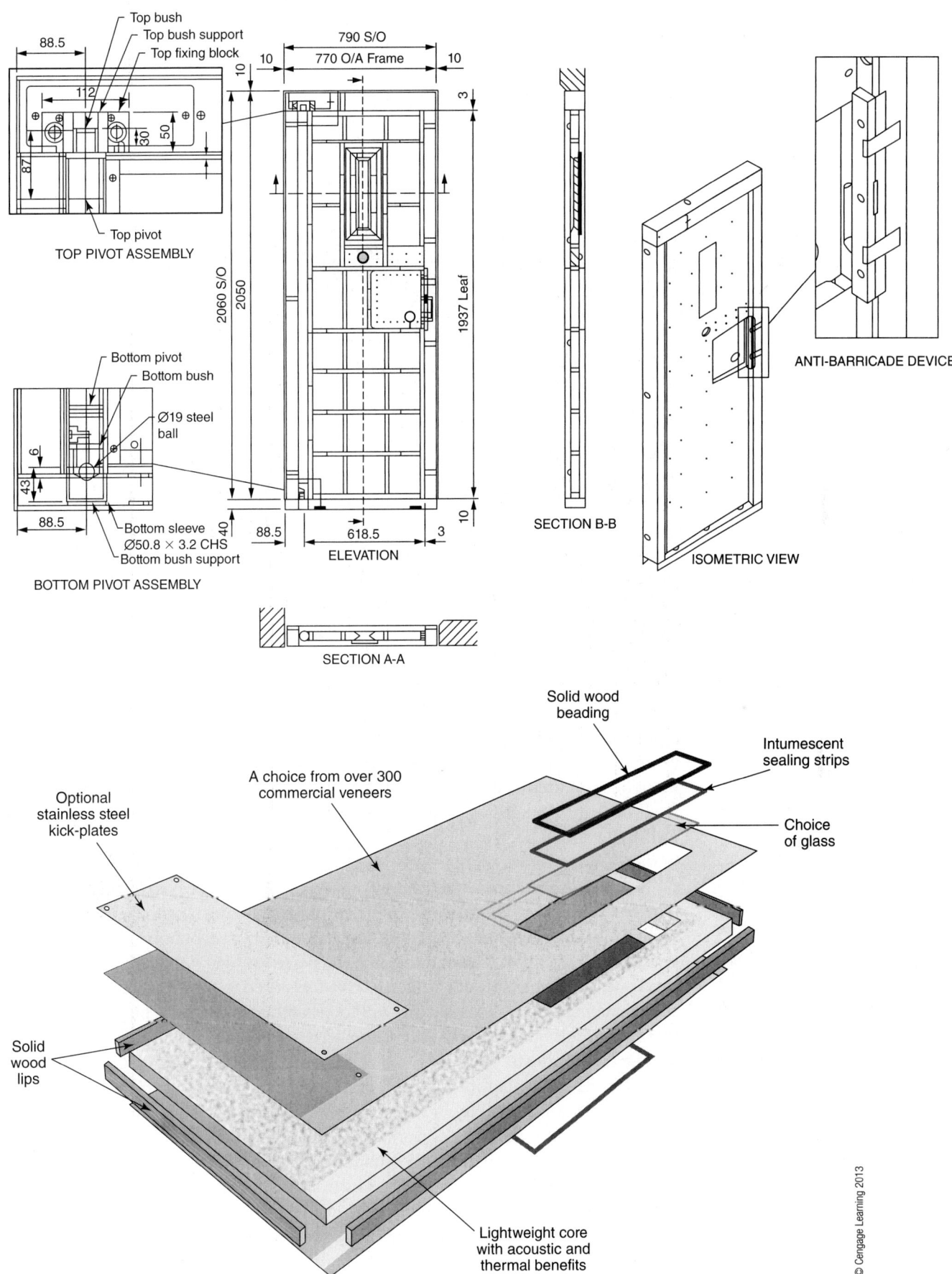

Figure 3-9 (Continued)

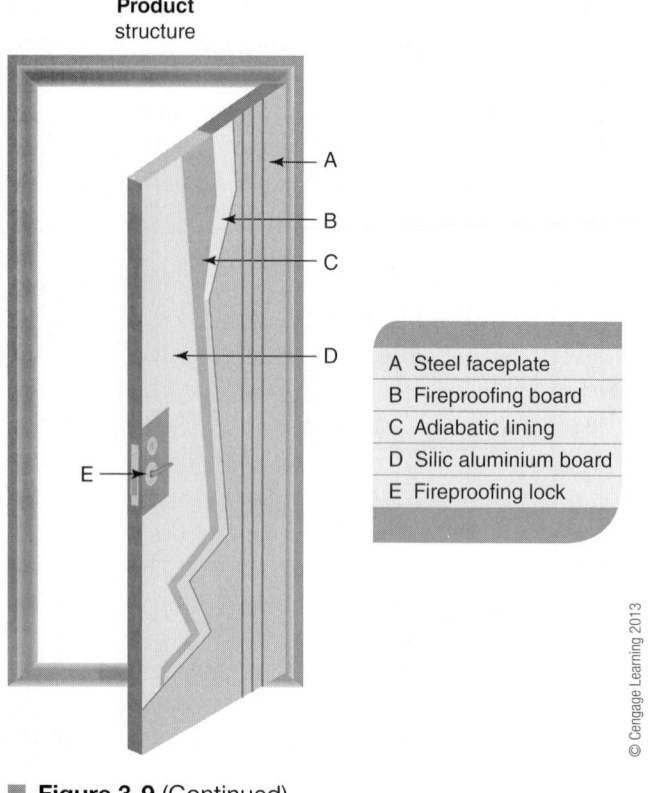

Figure 3-9 (Continued)

Fuel

Fuel cells have particular hazards associated with them, just as any process or equipment does. In discussing specific hazards, each part of the fuel cell must be examined, beginning with **input streams** and ending with **output streams**. The basic fuel requirement of fuel cells is hydrogen. That can be supplied as a raw fuel itself or from virtually anything that contains hydrogen, most commonly from well-known fuels such as methanol, liquefied petroleum gas, natural gas, gasoline, or diesel.

Infrastructure such as buildings, piping, and electrical systems are generally **static systems** that change little once in place. The design phase is of critical importance to infrastructure as is the construction phase, but once in place, these systems are generally only maintained. Other parts of fuel systems are much more **dynamic**, requiring constant attention. The fuels used in these installations are one such dynamic system. This section will focus on hydrogen fuels, but keep in mind that other fuels are routinely used in fuel cell systems, and each will have its own unique requirements.

All fuels are dangerous in that they are chemically reactive. Hydrogen is considered both a flammable and an explosive hazard, and these are its principal safety hazards. It is flammable in small amounts, down to 4% in an air mixture and in large amounts as well, up to 75% in an air mixture. That is to say that hydrogen will burn with an open flame when as little as 4% of a gas mixture (hydrogen and atmospheric air) is hydrogen, and it will continue to burn even when three-quarters of the mixture is hydrogen and the other quarter is air. By comparison, natural gas has a flammability range of 5.3 to 15% in air. Of great importance is the fact that hydrogen burns so cleanly that a flame is virtually invisible so hydrogen fires are not easily spotted until damage is done. Hydrogen's **explosive limits** in air range from 18.3% to 59% (5.7% to 14% for natural gas), but is generally considered to be explosive only when it is enclosed.

Hydrogen will disperse quickly so that if a fuel cell is in an open space, the danger of open-air flammability/explosivity is lessened. By comparison, it will diffuse 3.8 times faster than natural gas and rise in air 6 times faster than natural gas. Many fuel cell designs make use of these features to deal with hydrogen gas releases.

Care must be taken when testing systems as well, since the very low **viscosity** of hydrogen promotes leaking. Systems successfully pressure tested with different gases, such as air or nitrogen, routinely leak when hydrogen is used.

Hydrogen is not considered a hazardous chemical in that it is neither toxic nor poisonous in either **acute** or **chronic** dose levels. It poses few health risks in that it is not a carcinogen, is not an irritant, and is not corrosive.

Hydrogen's other principal safety concern is that it is a gas and as such can cause asphyxiation. Since it is odorless, colorless, and tasteless; travels quickly in air; and can act to displace oxygen, particularly in confined spaces, it will lower the oxygen content of the air, killing those in the immediate vicinity. Unlike natural gas, where odors are introduced into the gas so that people can recognize a leak, hydrogen fuels need to be very pure, so **detection equipment** is required.

Fuel cells do not use hydrogen gas exclusively. Some fuels, such as methanol in a direct methanol fuel cell or borohydrides used to store fuel present unique hazards based on their properties.

The fuel delivery system is also important. For direct cells such as methanol (CH_3OH), where the hydrogen is supplied as a fuel from within the fuel cell, the delivery system is generally a closed system where fuel is delivered from a **plug-in**, disposable (but perhaps recyclable) container. Plug-in systems have limited risk, most having to do with being very close to the end user. In particular, canisters are all explosive risks, whether they contain gas or volatile liquids such as methanol. Chemical hazards are also present should a cylinder be breached and skin contact occur. Methanol in particular is considered toxic with **acute exposure** through inhalation, dermal contact, or ingestion producing central nervous system depression followed by visual problems up to and including blindness. The **chronic effects** of methanol are similar to the acute effects and both can result in death even in relatively low doses.

Fuel cells using hydrogen as a direct source require either storage containers or a direct feed from a hydrogen fuel line. Hydrogen fueling systems can leak from flanges, gaskets, seals, and welds as well as from a number of other places. These risks are the same as mentioned above, having to do with hydrogen as a gas. One particular risk involved in hydrogen delivery systems lies in hydrogen's ability to embrittle metals, steel in particular. **Hydrogen embrittlement** is a well-known and destructive process that can cause catastrophic failures in delivery systems. Routine monitoring and replacement schedules are required to guarantee hydrogen fuel delivery systems. Figure 3-10 shows the types of damage that can result from hydrogen.

Larger systems requiring substantial hydrogen reservoirs generally involve liquid hydrogen. Considered a cryogenic gas since it will boil at −253°C (−423°F), liquid hydrogen has a number of safety and **industrial hygiene** requirements that must be in place to guarantee a safe workplace. Since it will boil (turn into a gas) at such a low temperature, proper venting of the system must be in place to ensure gas pressure does not build up. Since hydrogen is vented, systems must be in place to guarantee the gas does not build up in surrounding areas. Since hydrogen gas

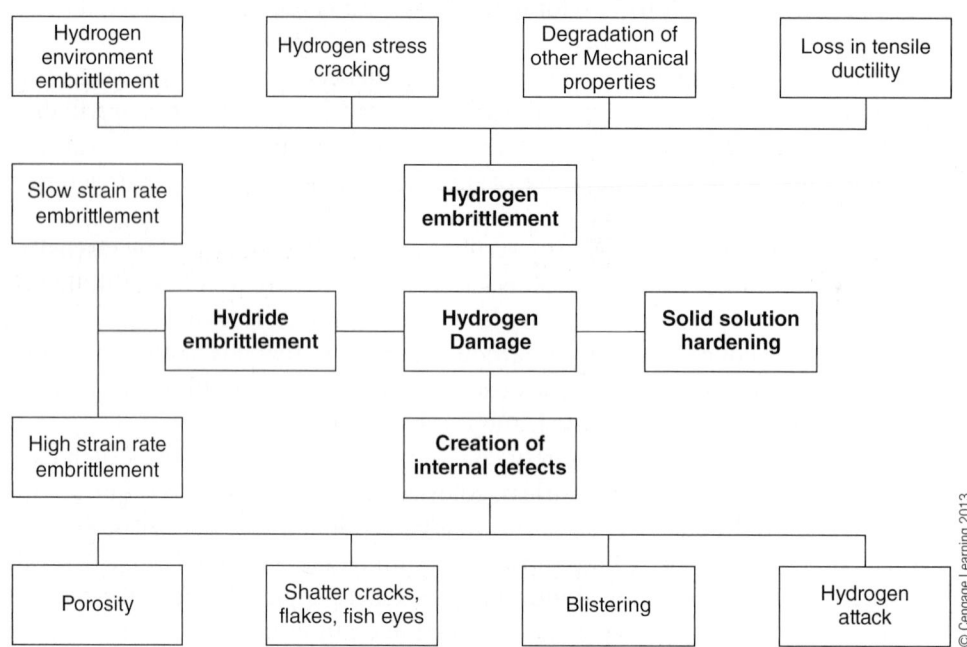

Figure 3-10 Hydrogen.

is lighter than air, most of these systems vent to atmosphere away from enclosed structures where gas may build up. Ice will build up on lines and tanks when moisture in the air contacts the cold pipes, valves, or tanks containing liquid hydrogen. Most of these problems are taken care of with proper insulation. Even though hydrogen is lighter than air and will rise, a hydrogen mist (hydrogen gas saturated with liquid droplets) is slightly denser than air and will initially flow at relatively high rates of speed both vertically and horizontally, and then rise as the gas warms, causing acute **asphyxiation hazards** in numerous locations. The most common personnel hazard in liquid hydrogen use is the possibility of **cryogenic burns** should the liquid contact skin. Larger cryogenic systems of hydrogen, like other industrial gases such as compressed natural gas and liquefied petroleum gas, must be located and controlled to mitigate conventional pressure vessel failure hazards as well as Boiling Liquid Expanding Vapor Explosions (BLEV). Figure 3-11 details what a BLEV explosion is.

Smaller systems may use high pressure tanks, in some cases up to 5000 psi of pressure, with depleted tanks possibly containing over 100 psi of pressure. In that case, a number of different rules and regulations govern the location, storage and use of such high pressure tanks, with stationary systems having different requirements than mobile systems. All **high pressure** tanks are regulated and certified as pressure vessels, governed under several codes, in particular the Boiler and Pressure Vessel Code. Maintenance and inspection of such tanks and any piping directly attached to the tanks must be done on a regular basis, and in fuel cells would be the responsibility of maintenance personnel. Proper pressure **shutdown procedures** must be in place to vent hydrogen gas and depressurize the system during any maintenance or emergency operations as well. The internal workings of fuel cells typically run from 10–350 psi, with the higher end pressures qualifying the fuel cell itself as a **pressure vessel**. The safety, repair, and inspection requirements of pressure vessels are substantial and well beyond the scope of this

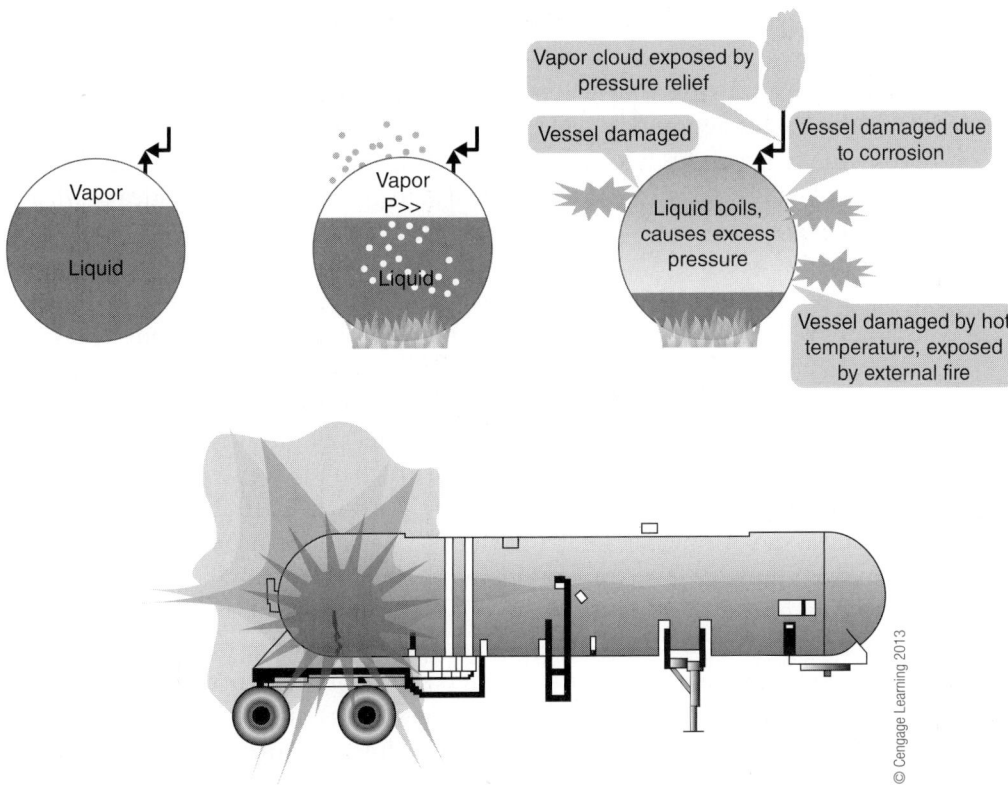

Figure 3-11 BLEV explosion.

book, requiring certified personnel to design, perform, and evaluate. Any pressure system over approximately 35 psi requires certified personnel and should not be touched unless personnel are working under the direction of certified personnel and familiar with high pressure systems.

Gas is dispersed within the fuel cell to guarantee uniform access to the catalysts and the anode structure. Gas within the fuel cell can leak for a number of reasons, ranging from seal failure to stack damage. These leaks are smaller than supply system leaks but potentially just as dangerous, in that build up within fuel cell enclosures is possible, as are pressurization and ignition.

Inspection

Hydrogen piping systems for fuel cell installations are governed by a number of different codes. NFPA 853 6.4 (Hydrogen Fuel Systems and Storage) is one of the principal code sections used in the Untied States. It differentiates between gaseous and liquid hydrogen systems where NFPA 55 deals with the design, location, and installation of storage systems, but the piping valves and fittings from the storage system to the fuel cell power system is governed by ASME/ANSI B31.3, Process Piping and NFPA 853 Sections 6.4.3.1 through 6.4.3.7. These two codes will be looked at to highlight the reasons why there are different codes. The seven NFPA sections deal with the number of **manual shut-off valves** (2) and their locations (within 1.8 m of the storage container and the fuel cell power system) unless there is not enough room or the fuel supply is indoors. In addition, piping and **valves** must be safely located and proper ventilation ensured. Grounding must also be to NFPA 70 (National Electrical Code). There is very little to section 6.4 of the NFPA code, only seven paragraphs. By comparison, ASME/ANSI B31.3 is hundreds of pages because it is a design code, used by engineers to assist in specifying components in such a way that they will meet the operational requirements of a fuel cell installation. Since designing a piping system using sound engineering principles is beyond the scope of this book, Chapter VI (Inspection, Examination and Testing) of ASME/ANSI B31.3 will be briefly looked at to give the student an idea of what is involved in guaranteeing an installation is ready for service.

To begin, Section 340.2 says "it is the **owner's** responsibility, exercised through the owners **Inspector**, to verify that all required **examinations** and **testing** have been completed and to inspect the piping to the extent necessary to be satisfied that it conforms to all applicable examinations of the code and of the engineering design." Once a contract has been written and is in force, construction or repair of a piping system in a fuel cell installation is done; however, it is the owner's responsibility to make sure that the construction or repair is done according to code and to design, using materials up to standard. This is an important issue. An owner, be it a power company or a hospital using fuel cells for auxiliary power or a company using fuel cells as primary power, has the ultimate responsibility for the fuel cell. If a construction or repair company fails to properly perform work, the owner is still liable. For that reason, owners employ engineers and inspectors to make sure contractors or maintenance personnel do the work correctly. In the United States, such oversight is required by code and in many cases by regulations. Those repairing or installing fuel cell systems must be willing and capable of interacting with inspectors and engineers to ensure quality work is done. A number of codes detail inspection requirements, among them ASME, AWS, and ASNDT codes.

When **inspection** is done can be critical, for instance 342.3 (b): "for a welded branch connection the examination of and any necessary repairs to the pressure containing weld shall be completed before any reinforcing pad or saddle is added." Inspection and repair, whether in construction or in maintenance, is done as part of the overall work plan more often than as a final step so that critical parts are available for inspection.

How inspection is done is also critical, for instance, 341.4.2 (b)(I): "Not less than 5% of circumferential butt and miter groove welds shall be examined fully by **random radiography** in accordance with paragraph 344.5 or by random **ultrasonic examination** in accordance with paragraph 344.6." Inspection by qualified personnel is dictated by codes and must be done, since codes are usually either required by law or by contract. Personnel who work on fuel cell installations must recognize the importance of not only the codes, but also in the type of qualifications required of those codes. For instance, welders must be certified as well as inspectors and examination personnel so that they work together to ensure quality construction or repair. Figure 3-12 shows how an ultrasonic examination is done.

Follow-up testing is also required once inspection and examination is done. 345.1 (Required Leak Test) states "Prior to initial operation, each piping system shall be tested to ensure tightness. The test shall be a **hydrostatic leak test** in accordance with paragraph 345.4 except as provided herein." There are a number of issues in final testing since that is the point where contractors or maintenance personnel formally transfer the responsibility and authority for such systems to the owner's operations personnel. Anyone who works on such systems needs to be aware of the importance of such a transfer and the requirements for it to be successful.

CHAPTER 3 • FUEL CELLS, CODES, AND STANDARDS   47

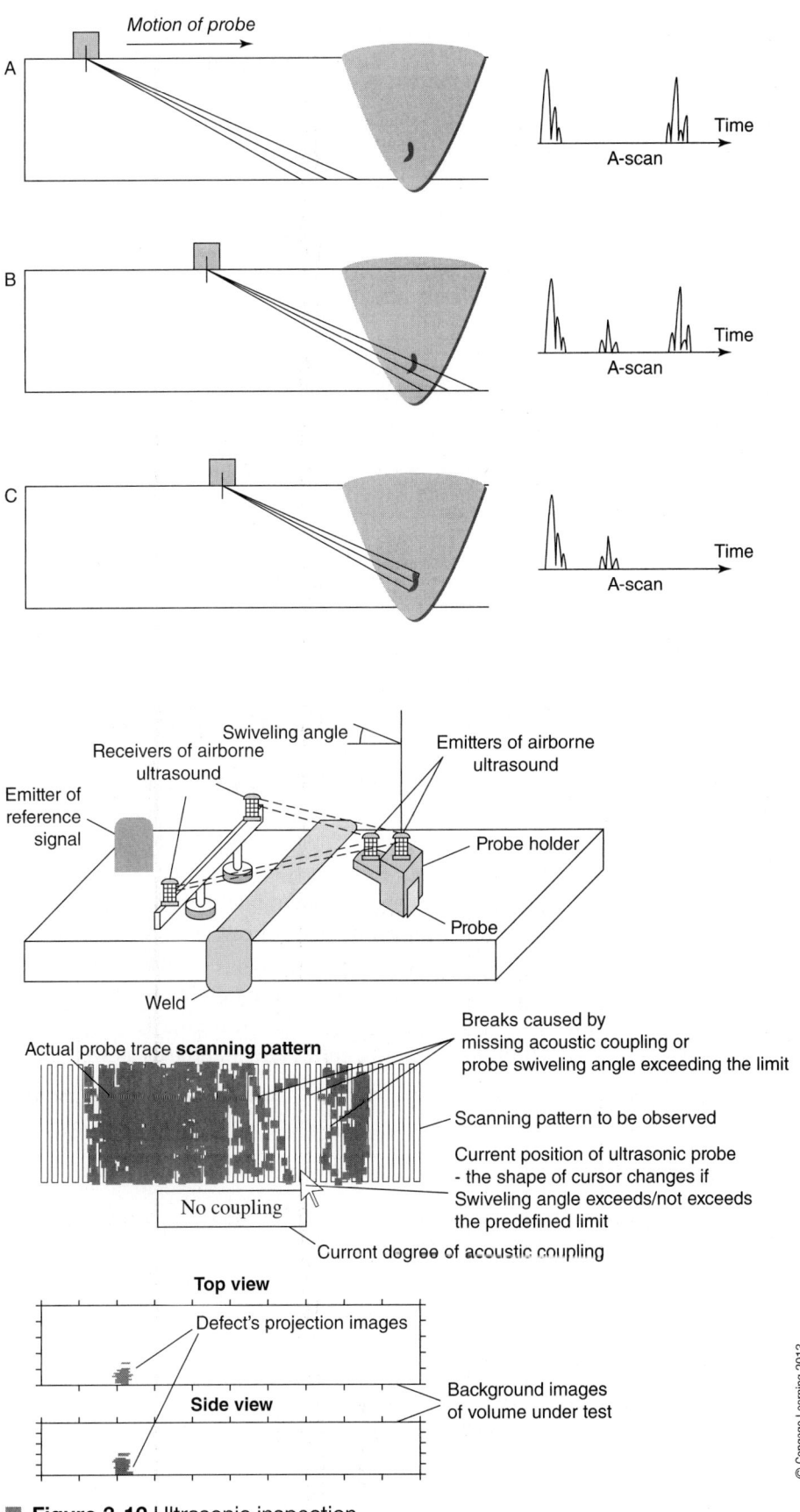

Figure 3-12 Ultrasonic inspection.

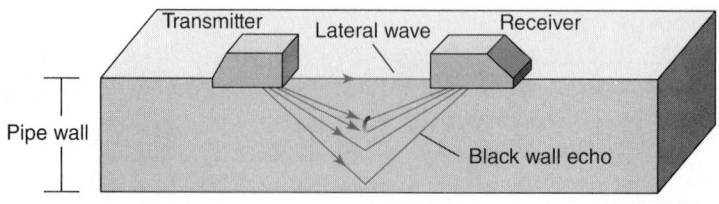

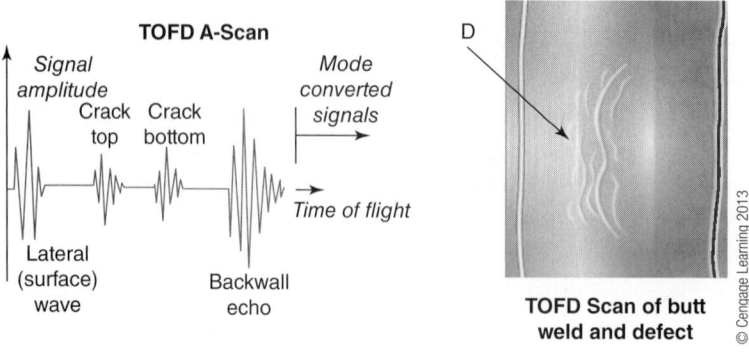

Figure 3-12 (Continued)

OTHER SYSTEMS (BALANCE OF PLANT)

Complex systems such as fuel cells have numerous components that must withstand not only continual use but also rigorous inspections and repair schedules. These systems will be examined in more detail in subsequent chapters, but the student should understand that all of these systems involve and many times require formal systems of design, construction, inspection, and maintenance. A successful career in fuel cells will require at the least familiarity with these codes, standards, and regulations if not a mastery of their contents. It should be noted that many of those who inspect or examine under various codes began their careers as hourly employees working on the equipment and progressed to higher level positions simply because they were willing to learn the codes and standards as well as the methods used in them.

KEY WORDS

Knowing the terminology used is critical when dealing with fuel cells. Following is a list of the important terms in this chapter, which are also in bold typeface within the chapter. It is recommended that students be required to submit definitions of some of these words as homework assignments in which they look the terms up in other books, articles, or on the Internet.

absorption
access
acute
acute exposure
administrative controls
air intakes
ampacity
asphyxiation hazards
biological group
chemical group
chemical hazards
chronic
chronic effects
circuit overcurrent
   protection
closed-cup flash point
codes
combustible
compressed gas
continuous output
   current
corrosives
cryogenic burns
de-energizing
detection equipment
disconnect
dump choppers
dynamic
emergency services
   management
engineering controls
equivalency
ergonomic group
examinations
explosive
explosive
explosive limits
fire doors
fire resistance rating

fire suppression
fire walls
flammable
flash point
force of law
fuel shut-off systems
gas detection
General Duty Clause
general installation
grid mains
grounding methods
hazards
HazCom standard
health hazards
high pressure
hydrostatic lead test
industrial hygiene
ingestion
inhalation
input streams
inspector
inspection regulators
interlock systems
irritants
manual shut-off
   valves
markings
Material Safety Data
   Sheets
mitigate
mixed system
hydrogen
   embrittlement
OSHA
output compatibility
output streams
overcurrent protection
owners
penetration

personal protective
   equipment
physical group
plug-in
prepackaged
pressure piping
pressure vessel
process piping
pyrophoric
qualified
radiography
random
regulations
reproductive toxins
carcinogens
self-contained
service access
shall
should
shutdown
   procedures
singular prepackaged
   self-contained
   power
siting
standards
static systems
stationary fuel cells
system grounding
system isolation
testing
toxic
ultrasonic
   examination
unstable
valves
viscosity
water reactive
work practices

DISCUSSION QUESTIONS

1. Under the HazCom standard, does the employer have the right to withhold information about hazards in the workplace until a person has accepted a job and discovers the hazard?

2. Where would an employee look to find information about the hazards of a specific chemical used in the workplace?

3. Combustible liquids can be either a solid or a liquid. Are explosive hazards a solid, liquid, or gas?

4. Are irritants toxic?

5. Does ingestion require penetration?

6. If signs are required for a fuel cell installation detailing hazards, what codes would give guidance as to the types of signs?

7. Name the parts to a fuel cell electrical system?

8. In a mixed system (AC and DC combined), are grounding systems required for both forms of current?

9. What is the purpose of the NFPA?

10. Are codes the same as laws?

11. Why are definitions critical in codes and standards?

12. Are exceptions allowed under most codes?

13. How far away from an air intake can exhaust fans be located?

14. Look up several of the chemicals mentioned in the preceding chapters and find their MSDS sheets. Do this as new chemicals are introduced in future chapters as well.

CHAPTER 4

FUEL CELL CLASSIFICATIONS AND REACTIONS

INTRODUCTION

Fuel cells must produce free electrons in order to provide electricity. Previous chapters have dealt with one particular type of fuel cell, the PEM cell, which gets electrons from hydrogen, but that is only one way of creating electricity. Other methods exist which use not only other fuels but other ways of moving charged ions from one side of the cell to the other. This chapter will discuss the PEM cell in more detail, then introduce the four other main types of cells and discuss their characteristics, similarities, and differences as well as introduce the types of applications each is used for.

Fuel cells have much in common with conventional **power generating** equipment, whether **batteries** or **generator**s, but they also have some important differences. Two basic differences need to be appreciated by the student in order to understand the appeal of fuel cells as well as their limitations. The first is that fuel cells are pressure and **temperature sensitive**. Unlike batteries or generators which generally operate at **atmospheric pressures** and can tolerate a wide range of **operating temperatures**, fuel cells are more like **internal combustion engines**, where in many cases a narrow range of internal pressure and temperature needs to be maintained for the system to even work. The second is that unlike many, if not most, power systems, fuel cells can be more **efficient** at lower temperatures than at higher temperatures. Gas **turbines** or internal combustion engines tend to increase efficiency (producing more power per unit of fuel consumed) as their operating temperatures increase until some upper limit is reached. Fuel cells do the opposite, losing efficiency as the **reaction chamber** temperature increases. It should be understood that we are not talking about the heat of an internal combustion engine itself, but rather the heat in the piston chamber where the chemical reaction occurs. Internal combustion engine reactions are combustion reactions at elevated temperatures, combining gasoline or diesel with oxygen in the air to produce carbon monoxide and water. It should also be noted that fuel cells tend to operate at high **efficiencies** even at low loads, unlike turbines or internal combustion engines, which see maximum efficiency at a set design point.

objectives

This chapter is will introduce the student to the various types of fuel cells that are currently in use: Alkaline (AFC), Solid Polymer, (SPFC), Phosphoric Acid (PAFC), Molten Carbonate (MCFC), and Solid Oxide (SOFC). It will discuss the core concepts used in fuel cells and how they are dealt with in the five major type. A general description of each type will be given as well as a discussion of the chemical reactions in use to create electricity. The differences and similarities between the major types will begin to be explored.

Unlike many types of power generation schemes though, current fuel cells are many times application driven. Batteries are standardized, with the equipment that uses them making use of standard sizes and power outputs. Backup generators are similar in design, whether they are powered by diesel or propane. Coal-fired power plants are of relatively uniform design, whether they are in Wyoming or Pennsylvania. Fuel cells do not belong to a mature industry so that outside of output channels for electricity and input channels for fuel, the type of fuel cell used can be substantially different from one application to another, even if the applications are similar.

There are two main classifications for fuel cells, **low temperature** and **high temperature** operations. The operating temperature of fuel cells can cover a wide range, and in fact some designs such as the Solid Polymer (PEM) Fuel Cells can be used in either range. Low temperature operations have certain advantages as well as distinct disadvantages over high temperature operations, just as high temperature cells do over low temperature operations.

Advantages for low temperature systems include:

- Quick start-up and shutdown times.
- The ability to respond to operational changes quickly and efficiently due to the dynamic nature of the systems.
- Commercial availability of both systems and the parts used in systems.
- **Scalability**.
- The ability to be used in mass market personal devices.
- Reasonable production and operations costs.
- The ability to use more common materials in fabrication.

Disadvantages for low temperature systems include:

- Precious metal **catalysts** are generally required for their reactions to proceed.
- Relatively pure hydrogen is required as a fuel gas.
- The heat that is generated can be high enough to require cooling schemes but not high enough to be useful.
- Maintenance is difficult.

Advantages for high temperature systems include:

- The ability to use a wider range of fuel gases, including most **hydrocarbons** like gasoline or diesel.
- Precious metal catalysts are not generally required at the higher operating temperatures.
- Operations temperatures are high enough that excess heat can be used in cogeneration processes.
- They can be used with existing systems such as gas turbines to improve efficiencies.

Disadvantages for high temperature systems include:

- Production, operations, and maintenance costs are high.
- Start-up and shutdowns sequences are complicated, requiring extended times.
- High temperature operations require high cost raw materials not only for the fuel cells themselves, but also for the secondary systems such as piping.
- Equipment used in high temperature operations generally have less of an operations lifetime than that used in low temperature operations.

The five major design types of fuel cells are: **Alkaline** (AFC), **Solid Polymer** (SPFC), **Phosphoric Acid** (PAFC), **Molten Carbonate** (MCFC), and **Solid Oxide** (SOFC). Within each of these designs are variations that depend on the application or markets served and cost targets. Fuel cells in fact form a rather large class of products that seems to be growing with every advance made in the technology and every claim made by manufactures as well as proponents of the technology. Each of the main designs will be introduced in this chapter, with subsequent chapters adding more information and more detail. Many of the discussions will be similar, since most fuel cells operate in much the same way and the concepts needed to understand their operations are essentially the same. Some of the core concepts are:

- the fuel used
- the pressure within the system while under operation
- the scheme used to distribute fuel evenly across the fuel cell
- the **anode** electrode where the reaction to release electrons occur
- the **electrolyte** or **ion exchange membrane** that transfers the ion left when the electron is stripped off
- the **cathode** electrode where the electrons and ions are recombined
- the various catalysts used at the anode and cathode
- the operating temperature
- **water management** within the system
- cooling methods
- wiring systems
- the way the individual cells are connected within a fuel cell system.

Table 4-1 shows some of the differences in the five designs.

Alkaline Fuel Cells (AFC)

Alkaline Fuel Cell (AFC) technology was mentioned briefly in the first chapter as having been around since the early 1900's and as having taken the place of the original General Electric PEM fuel cells used in the Gemini space flights. They are essentially the first example of a commercially available fuel cell, although due to the costs involved, few could afford them except for the space program. One of the main advantages of the AFC is that it produces water at the anode using hydrogen and oxygen. If you look at the chemical equation in the next section, you will see that for every two molecules of water it uses, it produces four. In space, that is an important issue, since it produces the water needed by astronauts as a **by-product**. Like SPFC (PEM) systems, the long history of producing AFC systems means the

Design Type	Acronym	Operating Temperatures	Possible Fuels
Alkaline	AFC	60°C–250°C	Hydrogen (H_2)
Molten Carbonate	MCFC	500°C–700°C	Hydrocarbons, Carbon Monoxide (CO)
Solid Oxide	SOFC	700°C–1000°C	Hydrocarbons, Carbon Monoxide (CO)
Solid Polymer (low temp)	SPFC (PEM)	70°C–110°C	Hydrogen (H_2), Methanol (CH_3OH)
Phosphoric Acid	PAFC	150°C–250°C	Hydrogen (H_2)

■ **Table 4-1** Fuel Cell Design Differences

industry, production methods, and product costs are relatively stable, although in its space program form, it is rather expensive. Figure 4-1 shows what a fuel cell system in space looks like.

The **mature** nature of these cells can be judged by the fact that they are so well developed that no back-up units are used in the space program. In fact, the units are stable enough that the principal issue is not that the fuel cells degrade but that the electrolyte eventually corrodes the support structure of the internal cells.

Even though the AFC systems have been replaced in space by newer SPFC (PEM) systems, they continue to be produced by a number of manufacturers and occupy a notable place in the current industry. In great part, this is due to

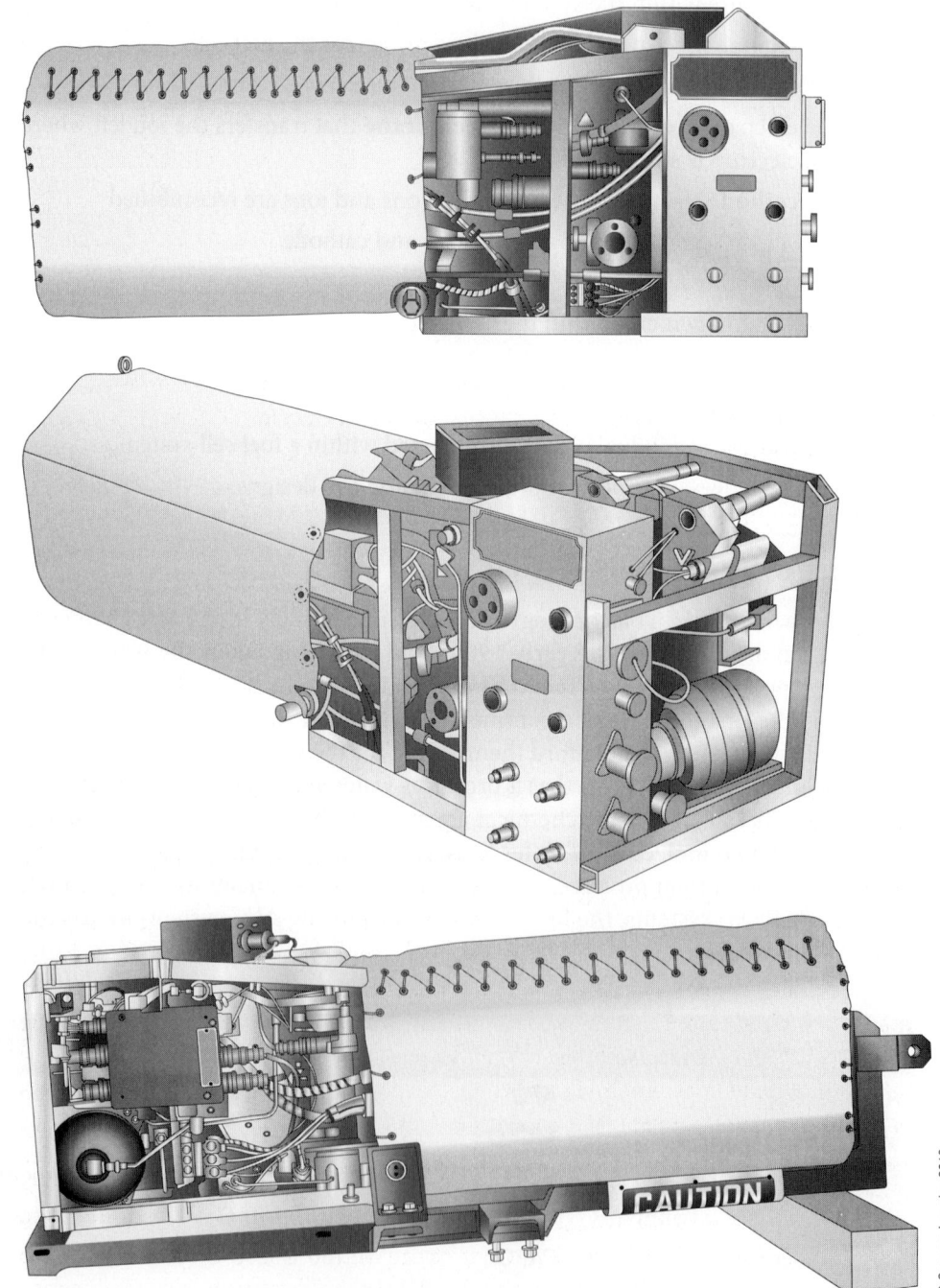

■ **Figure 4-1** Fuel cells in space vehicles.

raw material costs. AFC systems do not require precious metal catalysts at the cathode, making them attractive from a cost and manufacturing point of view. In addition, where SPFC (PEM) systems use a very expensive **polymer** electrolyte membrane to conduct ions from anode to cathode, AFC systems primarily use **potassium hydroxide** (KOH), a considerably less expensive material than polymer membranes.

There are several aspects of the AFC systems that make them fundamentally different from SPFC (PEM) systems as well. The electrolyte can be in a liquid state and thus can be **recirculated** or it can be in a solid form capable of being used in any orientation. The hydrogen fed to the anode produces more water at the cathode, which is useful in space applications where water is needed by astronauts or in fuel cell water heating/cooling loops. AFC systems can operate at lower temperatures to take advantage of commercial opportunities at those temperatures or at higher temperatures, in order to take advantage of efficiency improvements offered by higher pressures and temperatures.

Alkaline fuel cells are highly efficient, handle **peak loading** and unloading **cycles** very well, and can be easily maintained. They are relatively simple systems that can operate under both low and high temperatures, producing under normal circumstances at room temperature 1.2 volts and operating at very respectable efficiency of around 83%.

Alkaline Fuel Cell (AFC) Reactions

The way electricity is produced in fuel cells is always the same, more or less. A fuel comes in as a gas. It undergoes a chemical reaction at an electrode, usually with the help of a catalyst. The reaction produces an **electron**. The electron transfers into a wire and travels to another electrode to produce electricity. At the other electrode, another chemical reaction takes place to recombine the electron with an ion to balance the system out. While the general scheme of things is the same, the nuts-and-bolts of the processes differ, sometime by quite a bit. The Alkaline Fuel Cell (AFC) differs by quite a bit from the PEM cells discussed in the previous chapters.

The **reactions** occurring in AFC systems are not the same as in SPFC (PEM) systems. At the anode, the reaction in AFC systems occurs as follows:

$$2H_2 + 4OH^- \leftrightarrow 4H_2O + 4e^- \qquad \text{Eq. 4}$$

Notice the difference between Eq. 4 and Eq. 2 for PEM cells. In the PEM cells, the hydrogen loses an electron and then is conducted through the membrane to the cathode to find the electron once again and combine with oxygen to make water. In the AFC, the hydrogen does not travel but immediately combines with the **hydroxide ion** (OH^-) to make water at the anode instead of at the cathode. The anode electrode then acts as the site for two reactions, one to break the electron off to produce the proton (H^+) and another to combine the proton with the hydroxide ion (OH^-) to make water. Remember that chemical equations must **balance** out. The progression has two hydrogen gas **molecules** ($2H_2$) breaking up into four hydrogen **atoms** (4H) and then this combines with four hydroxide ions ($4OH^-$). This makes four water molecules ($4H_2O$), which balances the atoms out with four oxygen atoms and eight hydrogen molecules on both sides. This leaves the four minuses on the OH^-. In chemistry, **electrically neutral species** like water do not contain excess electrons; thus, the extra electrons that are attached to the two hydroxide ions ($2OH^-$) have nowhere to go in the chemical reaction. Like the electrons in Equation 2, they go looking for the path of least resistance, in this case the wire connecting the anode to the cathode.

Where do the hydroxide ions (OH⁻) come from? To understand what is going on, the student needs to be aware that an Alkaline Fuel Cell uses an **alkaline-based electrolyte** where PEM cells use an acid based electrolyte. The difference between **acid** and **alkaline** (also called a **base**) is what happens in water. An acid has an extra proton (H⁺) in water where an alkaline (base) has an extra hydroxide ion (OH⁻). Potassium hydroxide (KOH) is a strong base, which means that all the potassium (K) and the hydroxide (OH) come apart in water. This results in another equation involving the electrolyte.

$$KOH \leftrightarrow K^+ + OH^- \qquad \text{Eq. 5}$$

There is usually 30%–35% potassium hydroxide (KOH) in the electrolyte solution of water with KOH. There are then two ions in a water **solution**, one that has lost an electron (K⁺) and one that has gained an electron (OH⁻). The hydroxide ions (OH⁻) combine with the hydrogen fuel gas (H₂) to create more water molecules. Eq. 5 is one of the reasons why AFCs are so versatile. The two ions in solution (K⁺ and OH⁻) allow for good **conduction** in the cell. Keep in mind that for **current** to flow, the negatively charged electrons must move in the wiring system from the anode to the cathode, but that by itself is not enough for a **circuit** to be established. The positively charged ions must also flow from the cathode to the anode, and the potassium and hydroxide combination is very effective.

The hydroxide ions (OH⁻) are being consumed in this process, and if the original ions in the electrolyte are not replaced, the electricity will soon stop. Luckily, chemistry provides the answer to this problem at the cathode. When the extra electrons move across to the cathode, they find oxygen gas and water which results in Equation 6.

$$O_2 + 4e^- + 2H_2O \leftrightarrow 4OH^- \qquad \text{Eq. 6}$$

The reaction combines oxygen gas and water to make more hydroxide ions (OH⁻). Those move back into the electrolyte because there are not enough of the negatively charged hydroxide ions (OH⁻) to balance the positively charged **potassium ion** (K⁺) left over from Equation 5. Remember that one of the most important concepts in the chemistry of fuel cells is that if there is an area where there is more of one thing, like the hydroxide ions (OH⁻) at the cathode, near an area where there is less of the thing, such as the anode where the hydroxide ions (OH⁻) has been used to make water, then more moves toward less. The potassium ion (K⁺) by the way will move away from the anode as well, in essence meeting the hydroxide ions (OH⁻) halfway to balance the electrolyte electrically. Eq. 6 is one of the reasons why AFCs have been in business for so long. This **oxidation/reduction reaction** occurs extremely fast (rapid **kinetics**) so the hydroxide ions (OH₂) can be replenished as fast as the fuel can be supplied, making for good power production characteristics. It also means that a catalyst does not have to be used at the cathode for this reaction to happen in sufficient quantity and time to run the system.

At this point, we will add a bit more information about chemical reactions. In general, when a chemical reaction happens, it either produces heat (**exothermic**) or consumes heat (**endothermic**). This has nothing to do with whether a reaction happens or how long it takes to happen, but involves only what happens while the reaction is occurring. The set of reactions in an AFC results in heat being produced, as do most of the reactions in fuel cells in general. Because heat is being produced, most fuel cells operate at internal temperatures that are higher than the ambient temperature around them. In the case of low temperature AFCs, the heat generated must be dealt with to maintain the low temperature of the system. In the next chapter, ways to deal with this heat will be examined.

There are a number of variations used in AFC systems, such as that shown in Figure 4-2 and Figure 4-3. The simplest consists of a liquid electrolyte (KOH) mixed with **fuel gas**. The electrolyte (KOH)/fuel mix is in contact with both the anode and cathode. The anode has a precious metal catalyst to improve the reactions that both make hydrogen from the fuel gas and then break the hydrogen up into protons (H^+) and electrons. There is no catalyst at the cathode, in part to stop the same type of reactions that are occurring at the anode where there is catalyst. The cathode also has access to air in order to supply oxygen. This type of cell has been the object of considerable interest since a form of it is used in the **sodium borohydride** ($NaBH_4$) cells being developed. These will be discussed in upcoming chapters when the AFC cells are described in greater detail.

The two most common designs share many features but have one distinct difference: one uses a liquid electrolyte and the other uses a solid electrolyte. As in PEM cells, both bring the fuel gas (H_2) in and distribute it as evenly as possible over the anode surface. Unlike PEM cells, AFCs must remove excess water from the anode side, since that is where it is produced. Many do this by using more fuel gas than is required so the excess gas is humidified to remove water and then the excess gas and water stream is removed from the cell. The gas and water stream is then dehumidified with the fuel gas returned to the fuel cell, leaving the clean

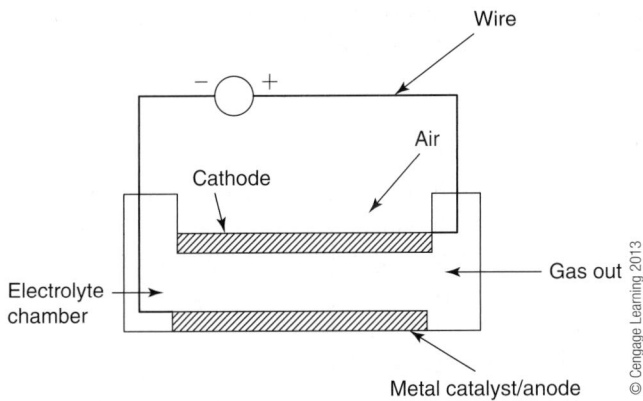

■ **Figure 4-2** Simple horizontal alkaline fuel cell.

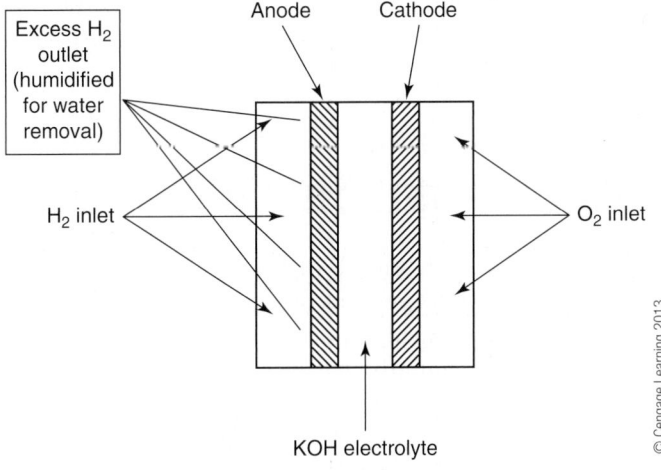

■ **Figure 4-3** Vertical alkaline fuel cell.

water available for other uses. The cathode side requires oxygen, either as a pure gas or as air. The cathode then produces the hydroxide ions (OH^-) of Eq. 6 and those ions move into the electrolyte, drawn in part by the excess K^+ ions there. The electrolyte provides the path for the hydroxide ions (OH^-) between the cathode and anode.

The two different forms of the electrolyte have come about because the potassium (K^+) and the hydroxide ions (OH^-) will interact with any carbon dioxide (CO_2) in the system to form **carbonates**. These carbonates then block the **pores** on the **porous electrodes** (usually carbon) to shut the system down, since any gas transfer is blocked. In fuel gases where hydrogen is produced from a different source such as methanol (**reformate** systems), CO_2 is always present unless it is removed in a separate process. Removing CO_2 is expensive. In systems using air at the cathode, CO_2 is always present as well, since atmospheric air contains around 350 **parts per million** (0.035%). The potassium hydroxide reacts with the carbon dioxide according to Eq. 7 to produce potassium carbonate (K_2CO_3) and water (H_2O). This carbonate then builds up in the pores, finally blocking them.

$$2KOH + CO_2 \leftrightarrow K_2CO_3 + H_2O \qquad \text{Eq. 7}$$

The potassium carbonate (K_2CO_3) causes several other problems in the electrolyte as well. It consumes the hydroxide ions (OH^-), interfering with the system as a whole. The carbonate particles make the electrolyte solution more **viscous** (causing a thick fluid like honey), interfering with ionic diffusion which in turn lowers current flow. The carbonate fills the pores in the porous electrodes, physically blocking access and chemically interfering with oxygen in particular. The carbonate crystals can also destroy the electrode pores in extreme cases.

The easiest way to deal with these issues is to use a fully liquid electrolyte and pump it out so that new electrolyte is sent into the cell on a regular basis. This removes the carbonate and solves the problem. Potassium hydroxide (KOH) is cheap enough that supplying a **loop** of new solution does not increase the cost all that much. The loop has other advantages as well, such as cooling the fuel cell and regulating the water/electrolyte balance.

There are disadvantages to the liquid electrolyte systems though, in particular the complications involved in making the cell function in positions other than horizontal. Fully sealed liquid systems are used with hydrogen (H_2) fuel gas and oxygen (O_2) that eliminate CO_2 in the gas input streams. Efforts are also underway to use solid electrolytes along the lines of PEM cells that retain the alkaline (basic) electrolyte or use fluidized bed systems that mix electrode particles in with the electrolyte to eliminate the porous electrodes altogether.

SOLID POLYMER FUEL CELLS (SPFCs)

Solid Polymer Fuel Cells (SPFCs) are the PEM fuel cells discussed in the first chapters. They were initially called SPFC before the mechanisms of the cell were well known. The technology was developed out of the original designs of fuel cells from the nineteenth through the twentieth centuries. It is one of the most mature fuel cell designs because it has been in production, either in the lab or in industry, for well over half a century. The student should be familiar with many of the concepts involved with solid polymer (PEM) fuel cells from the previous chapters.

SPFC (PEM) Reactions

As mentioned in previous chapters, SPFC (PEM) have two separate reaction occurring. At the anode is:

$$2H_2 \leftrightarrow 4H^+ + 4 \text{ electrons}^- \qquad \text{Eq. 2}$$

At the cathode is:

$$O_2 + 4 \text{ electrons}^- + 4H^+ \leftrightarrow 2H_2O \qquad \text{Eq. 3}$$

You can add chemical equations just like numbers. Using the double arrow as the dividing line between the left and right sides, we add what is on the same side of the two equations. That gives us:

$$2H_2 + O_2 + 4 \text{ electrons}^- + 4H^+ \leftrightarrow 4H^+ + 4 \text{ electrons}^- + 2H_2O \qquad \text{Eq. 4}$$

You can then remove anything that is on both sides of the equations. In this case, that is four electrons$^-$ and $4H^+$. The final addition of the two chemical equations is then:

$$2H_2 + O_2 \leftrightarrow 2H_2O \qquad \text{Eq. 1}$$

You can make sure it all balances by adding up all the atoms. In this case on the left side, the H_2 is two hydrogen atoms and there are two of those, making four atoms altogether. There are two oxygen atoms in the O_2 and only one of those, so that makes two. On the right side, there are still two hydrogen atoms from the H_2O and two of those which still makes four, so that balances. There is only one oxygen in the H_2O but the two water molecules in the equation make two oxygen atoms, which balances the two oxygen atoms in the O_2 on the left side. The SPFC (PEM) system then uses four of everything—hydrogen, oxygen, and electrons—to create electricity. You may in fact find these referred to as the **4-electron reactions**, because of that.

All of these reactions can occur at low temperatures, near room temperature for instance, but only with the help of a catalyst, in most cases that being platinum. That is one of the reasons why SPFC (PEM) cells are so attractive for personal and transportation uses. The low temperature operation guarantees a quick start-up cycle since the system does not have to be brought up to temperature in order to operate. This is not true of many fuel cell systems. In industrial or commercial uses, start-up time is many times much less important that in consumer uses, where instant availability is critical. Some commercial uses require instant start-ups as well, such as emergency generators where interruptions in power can be catastrophic in places like hospitals and computer data centers.

MOLTEN CARBONATE FUEL CELLS (MCFC)

Molten Carbonate Fuel Cells (MCFC) operate at high temperatures with 1202°F (650°C) usually mentioned as the optimal temperature. This operating temperature results in significant differences from the lower temperature operations discussed previously. Some of these differences are basic issues dealing with high temperature operations, such as the use of materials in fabrication that can withstand higher temperatures, heating input streams and cooling output streams, longer start-up and shut-down sequences, and safety concerns. Other issues are specific to molten carbonate fuel cells (MCFC), such as the ability to use lower cost catalysts like nickel and the ability to internally **reform** gases (break the hydrogen out from mixed gases) so that other fuels like carbon monoxide (CO), natural gas (a mix of gases, primarily methane [CH_4], ethane [C_2H_6], propane [C_3H_8], and butane [C_4H_{10}]), biogases (primarily methane produced from landfills or other organic material when no oxygen is present), marine diesel, and coal gas can be used. The temperature is chosen to balance all these issues but also so that the carbonate will be molten so the ions are conducted in sufficient quantity yet at the same time not be so hot that special (expensive) alloys have to be used in construction. In general, MCFCs produce in excess of 25 kW of power (currently about 2500 kW maximum) and thus are meant for larger installations such as industrial plants, cogeneration facilities, landfills, hotels, office buildings, and hospitals. It is not expected that MCFCs will be developed for household or personal use.

The higher operating temperature also means higher efficiencies, so that not only can cheaper fuels be used, but more electricity is produced per unit of fuel consumed than with lower temperature fuel cells. If just methane (CH_4) is used as a fuel, for instance, all of it can be converted to hydrogen gas with nearly two-thirds of this utilized at the anode to produce free electrons and thus electricity. This is offset by a higher initial cost, since the materials used in fabrication must be able to withstand the higher temperatures. The high temperatures combined with the molten carbonate used for the electrolyte subject the internal parts of these fuel cells to an extreme environment that can rapidly destroy the cell.

MCFCs also make use of different chemical reactions to produce the electrons needed for electricity. While still using hydrogen (H_2) at the anode, the cells do not simply free electrons from the H_2 as in a PEM cell, but instead combines the hydrogen H_2 with a **carbonate ion** (CO_3^{-2}) and thus the name. As a review, recall what (CO_3^{-2}) means: one carbon (C) atom bonded to three oxygen (O_3) atoms, with this combination having two extra electrons (CO_3^{-2}). This reaction at the catalyst sites on the anode then produces water (H_2O), carbon dioxide (CO_2), and two free electrons ($2e^-$). Thus, unlike the SPFC (PEM) cells but like the AFCs, water is produced at the anode and not the cathode. The carbonate ion (CO_3^{2-}) must then travel from the cathode through the molten carbonate electrolyte to the anode while the carbon dioxide (CO_2) must travel from the anode to the cathode where it meets oxygen to create the carbonate ion (CO_3^{2-}). This is an important issue with these cells, since CO_2 is not an ion (it is not charged with extra electrons), so it will not travel through the electrolyte and must have a separate (recirculation) route of its own.

The high temperature reactions still need catalysts to occur on schedule, but they do not need very expensive precious metals, instead making do with nickel as a catalyst. In most MCFC systems, nickel (Ni) catalysts are used at the anode and nickel oxide (NiO) catalysts are used at the cathode. These **non-noble metals** are the reason why fuel gases other than pure hydrogen can be used, since these catalysts are not **poisoned** by contaminants in the gas stream to the extent the noble metals like platinum are. In general, the important variables for these cells are the operating temperature, the gas pressures within the cell, the fuel gas used, and the percent of feed gas used at the anode (utilization). If the carbon dioxide (CO_2) pressures (partial pressure) are the same at the anode and cathode with the electrolyte kept constant between them, then the **cell voltages** will depend on the other gases in the cell, hydrogen (H_2), oxygen (O_2), and water (H_2O). Remember that since these are **high temperature cells**, water will not be a liquid but a **vapor**. The cell will usually produce from 100 to 200 **milliamps** per square centimeter of area at 750 to 900 millivolts. Where any specific cell actually operates in these ranges is basically dependant on the type of gas used as the fuel, and whether the fuel is reformed as part of the cell (**Direct Internal Reforming**) or as a separate module attached to the cell (Indirect Internal Reforming). Hydrogen can also be used as a fuel without reforming cheaper gases, but this is rarely done, since it eliminates one of the main advantages of the MCFC.

The electrolyte is very important in the MCFC, probably more so than in any other fuel cell. Usually, the carbonate is a mix of the electrolyte in a porous, ceramic matrix, usually made of fibers. The ceramic most commonly used is $LiOAlO_2$, a mix of lithium, oxygen, and aluminium. The electrolyte is a mixture of alkaline carbonates (CO_3^{2-}), most commonly lithium (Li), sodium (Na), and potassium (K). Recall that alkaline is another word for basic solutions as opposed to acidic solutions. Typically there is 60% by weight of carbonates contained in 40% by weight of ceramic matrix. The chemical formulas for the alkaline (basic) electrolytes are Li_2CO_3, $NaCO_3$, and K_2CO_3. These are usually mixed together depending on whether the cell is pressurized or not. Lithium carbonates (Li_2CO_3) and potassium carbonates (K_2CO_3) are mixed in a roughly 60/40 ratio for cells operating at atmospheric pressure with lithium carbonates (Li_2CO_3) and sodium carbonates ($NaCO_3$) mixed in a roughly 60/40 ratio for cells that are internally pressurized.

The electrolyte is critical to how MCFCs operate but also to how they fail. It causes corrosion within the cell, determines the operating efficiency to a great extent, **evaporates** if temperatures are not correctly maintained, tends to move within the cell from positively charged areas such as the cathode to negatively charged areas such as the anode, and can even escape from the sealed cells through gasket material.

One other critical issue needs to be understood when dealing with these cells: because the electrolytes are not molten until they reach operating temperatures, the final fabrication step cannot be done until the cell is heated. The electrodes (anode and cathode), the electrolyte, the ceramic matrix the electrolyte is held in and any other stack components like bipolar plates and current collectors must be assembled and then heated until the electrolyte melts, usually around 450°C but that can vary depending on the carbonates used in the electrolyte and their ratio. When the electrolyte **melts** (turns from a solid to a liquid), it is absorbed into the porous, ceramic matrix and acts to seal the system internally as well. This can take a number of hours, depending on the size of the cell stack and must be done whenever the cell is turned off and the temperature drops. For that reason, high temperature fuel cells like the MCFC are usually run continuously once they are brought on line. It is one of the main reasons why these are usually used in larger installations, since turning them on and off will substantially lower their operating lives. Even in continuous operations the cells see considerable temperature variation from gas inlet to gas outlet, with the outlet sides seeing higher temperatures. This causes secondary problems within the cell, such as differing corrosion rates and expansion/contraction problems since the electrolytes will **expand** and **contract** as temperatures change. Usually, the pressure and temperature differentials within the cell are the control variables for these cells.

MCFCs provide the highest fuel efficiency operations, use the widest array of fuels, have limited environmental impact, and have substantial operating lifetimes (currently in excess of 30,000 hours). This is offset by their high initial capital costs since they are not well suited to high volume manufacturing and can in a sense be viewed as handmade due to the assembly and burn-in requirements. Their high temperature operational and maintenance constraints also limit their applications so that they are currently viewed as less viable than other fuel cells.

Molten Carbonate Fuel Cell Reactions

The reactions for the MCFC are different from the other fuel cells. As discussed above, they involve carbonate ions (CO_3^{2-}).

At the anode, the reaction is:

$$H_2 + CO_3^{2-} \leftrightarrow H_2O + CO_2 + 2e^- \qquad \text{Eq. 8a}$$

At the cathode, the reaction is:

$$2CO_2 + O_2 + 4e^- \leftrightarrow 2CO_3^{2-} \qquad \text{Eq. 9}$$

Unfortunately, these two equations do not match. Notice that there are two electrons in Eq. 8a but four electrons in Eq. 9 and one carbon dioxide (CO_2) in Eq. 8 but two carbon dioxides (CO_2) in Eq. 9. We have to match the number of molecules and electrons in the two equations, and to do that, we simply double everything in Eq. 8. It is critical to remember that in chemistry, especially in fuel cell chemistry, everything has to balance out. This is true not only in the chemical equations but also in the input and output channels. If you generate too much water and it builds up, the cell stops working. If you build up too much CO_2 or don't have enough CO_3^{2-}, then the cell stops working. For that reason, the cells themselves are important, but so are all the other systems that make up an entire fuel cell unit.

$$2H_2 + 2CO_3^{2-} \leftrightarrow 2H_2O + 2CO_2 + 4e^- \qquad \text{Eq. 8b}$$

It is important to realize that CO_2 must be supplied to the cathode as well as oxygen. It reacts there and that is how the carbonate ion (CO_3^{2-}) is resupplied to the electrolyte once the cell begins producing electricity. It must be done because in Eq. 8, the carbonate ion (CO_3^{2-}) is consumed in the reaction. For that reason, it must be supplied to the cell just as the fuel and the oxygen (O_2) must be supplied or the cell will stop producing electricity. If any of these supply systems fail, then once the carbonate ions (CO_3^{2-}), fuel, or oxygen (O_2) is used up, the fuel cell stops operating. The CO_2 is produced in Eq. 8 at the anode, so it can be used to supply the cathode, but this requires that the CO_2 gas can either transfer through the electrolyte to the cathode or be piped from the anode to the cathode. Electrolytes by definition allow ions to transfer easily but CO_2 is not an ion (it does not have either extra electrons or less electrons than it should to be neutrally charged), so in fact it will not transfer through the electrolyte. For that reason, most MCFCs take the gas generated at the anode and pipe it externally back to the cathode for use. This is done to eliminate the added complexity of designing an internal piping system that would be able withstand the high temperatures and corrosive atmosphere of the molten carbonate electrolyte.

Remember that in the AFC, CO_2 is considered bad and must be excluded from the cell to prevent poisoning. It is important to understand that what works in one type of fuel cell will not necessarily work in another and it is critical that the student understand the basic operation of all the cells.

For the MCFC, other fuels can be used as well. That means other reaction must be possible at both the anode and the cathode of another fuel is used. Most of the fuels mentioned like methane (CH_4) are **reformed** (converted into carbon and hydrogen) so that in the end, hydrogen is actually the fuel used. That is not always true though, for instance when carbon monoxide (CO) is used, since only carbon (C) and oxygen (O) and not hydrogen (H) are available.

For an MCFC using carbon monoxide as a fuel, the reaction at the anode (these equations will be balanced) is:

$$2CO + 2CO_3^{2-} \leftrightarrow 4CO_2 + 4e^- \qquad \text{Eq. 10}$$

The reaction at the cathode is:

$$2CO_2 + O_2 + 4e^- \leftrightarrow 2CO_3^{2-} \qquad \text{Eq. 9}$$

Notice that the reaction at the cathode is the same no matter if hydrogen or carbon monoxide (CO) is used as the fuel. It is only the reaction at the anode that changes. Notice too that the carbonate ion (CO_3^{2-}) is still used in the anode reaction even though water (H_2O) is not produced when carbon monoxide (CO) is used as the fuel. This is important, because it means that the design of the cell and the parts used do not require changing when a different fuel is used. This is one of the strengths of the MCFC.

Using other fuels other than pure hydrogen creates problems beyond water management or piping systems. Using fuels that are not **purified** or chemically created at the fuel cell, such as methane from landfills or fuels derived from coal introduces a number of different things into the gas stream that can cause problems in the fuel cell. These include fine **particulates** which can plug the system, sulfur and nitrogen compounds that can destroy the electrolyte, and metals which can deposit on the electrodes. Even gas reformation, where carbon (C) and hydrogen (H) compounds are broken apart, do not completely eliminate the hydrocarbons (hydrogen and carbon atoms bonded together into a molecule in various ratios). The remaining hydrocarbons then tend to break apart at the anode, coating it with carbon, which reduces and eventually eliminates the reactions at the anode. Remember as well that since CO_2 is generated at the anode and then piped to the cathode as a supply gas, any contaminant present at the anode from the fuel gas feed will also be present at the cathode.

The general form of the MCFC is the same as other fuel cells, as shown in Figure 4-4. The student will notice a surprising similarity to the drawings showing the various cells so far. This is done intentionally to drive home one of the critical points of fuel cells. Fuel cells are chemical reaction chambers designed to strip electrons from an atom or molecule to create electron flow and thus electricity. The part of the cells themselves differ very little, mostly in the way the atoms or molecules that have lost their electrons move from one part of the cell to another by using different types of electrolytes. Associated systems such as fuel delivery, cooling, and wiring are also very similar but for slightly different reasons, since those parts have to fit into already existing infrastructures such as power grids or natural gas delivery systems.

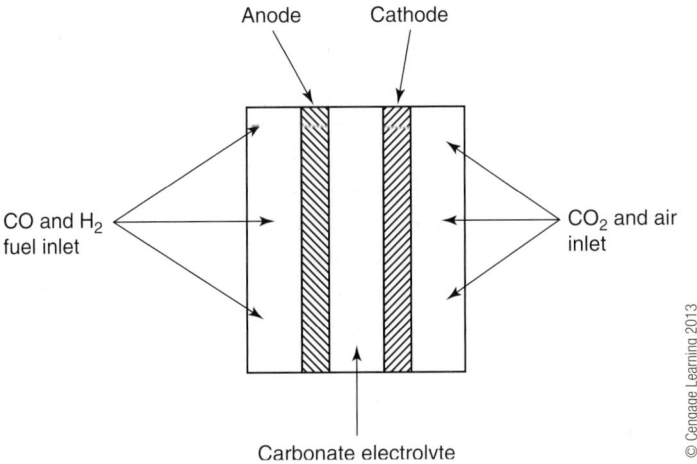

Figure 4-4 Molten carbonate fuel cell.

Since MCFC systems use high temperature to melt carbonates and then use a porous ceramic to contain the molten material, certain precautions must be taken. Ceramics are brittle so while the cells are being prepared for their first use, they must be treated accordingly. The amount of molten carbonate in the electrolyte layer is critical, yet until the first use is brought to temperature, the carbonate is solid and thus will be difficult to evenly distribute within the porous ceramic. In general the cells can operate between 10% carbonate filling and 60% but outside this range, the internal resistance of the cell increase dramatically. Too much molten carbonate in the ceramic matrix causes flooding just as in other cells, while too little means enough ions cannot move from cathode to anode and gas buildup occurs.

PHOSPHORIC ACID FUEL CELL (PAFC)

Phosphoric Acid Fuel Cells (PAFC) systems were the first modern cells to be available commercially. Even though other systems have had success in commercialization attempts during the first decade of the twenty-first century, at least a third of units being sold toward the end of that decade were still PAFC systems.

The systems generally operate at temperatures around 200°C (392°F) so they do not require special materials, but neither do they create enough heat to be able to capitalize on cogeneration. As in other low temperature cells then, the heat must be dealt with as a problem rather than used as an asset. Both liquid or air cooling systems are used In PAFC systems, with cooling loops generally used for several cells at a time rather than each cell having to be cooled. As in most cooling systems, liquid processes are more efficient but more complex and more expensive to make and maintain while air cooling is cheaper but requires much larger channels to effectively remove heat.

Since PAFC systems operate at lower temperatures, they require platinum catalysts to drive the reactions rather than cheaper non-noble metal catalyst like nickel. Most systems use very fine particles on porous carbon electrodes. The carbon is held with a polymer binder, usually PTFE, and serves to both disperse the fuel gas uniformly across the electrode as well as act as a conductor to transfer the electrons into the wiring system.

Smaller systems operate at atmospheric pressure but larger systems generating **megawatts** generally operate at higher pressure to increase efficiency of operations. There is a limit to the pressures theses systems can tolerate however, since above a certain level, the silicon carbide matrix holding the acid electrolyte will begin to break down. Usually, higher pressures require that the entire fuel cell stack be pressurized so it is held within a sealed unit. Nitrogen is the preferred pressurizing gas, in part to limit **corrosion** since phosphoric acid produces a corrosive gas. Individual cell voltages are also limited under pressurization for the same reason, since electrochemical reactions that cause corrosion can occur in the electrolyte if the voltage is too high.

The electrolyte used is very concentrated, generally being 100% **phosphoric acid** (H_3PO_4). Phosphoric acid is used because it is relatively cheap, is stable and will not **volatilize** (turn into a gas) at the temperatures used. Perhaps as importantly, the acid will not be contaminated by CO_2 nor will it react to form carbonates like AFC electrolytes will. The electrolyte acid will be used up over time and since fuel cells are almost always sealed units, no matter what type they are, a reservoir of acid must be provided. The electrolyte life must be matched to other components in the PAFC so that a relatively uniform operation exists. While supporting units can be designed to last for years because the individual fuel cell stacks can be replaced, it would not be economical to have a platinum anode designed to last three or four years when the phosphoric acid ran out in nine months. Usually, PAFC cell stack designs include a **reservoir plate**.

The acid is contained in a bed of **silicon carbide** (SiC) particles held together with a polymer binder, usually PTFE. This is a good place to talk about molecules. PTFE is a balanced molecule because in chemistry everything has to balance. It consists of two carbon atoms bonded together (ethylene) and each of the two carbon atoms has two fluorine atoms attached (tetra fluorine) so each molecule has two carbons and four fluorines. However, the chemical also has "poly" attached to the name, meaning that there are many of the six atom tetrafluoroethylene molecules tied together. The structure has to be balanced though, so the fluorine atoms must line up on opposite sides of the chain so the molecule is stable. This type of structure, where one molecule is repeated in very large chains is called a polymer, otherwise known as plastics. PTFE is a very common polymer used on nonstick cooking pans. The structure is shown in Figure 4-5 and notice how the fluorine molecules line up on either side of the carbon backbone.

The particles are small, usually less than 0.1 mm with the entire electrode being about 0.2 mm thick to limit **ohmic loss** across the cell caused by a solid that will not conduct ions being in the electrolyte. Small particles also help to keep the liquid acid in place and limit any gas pockets so that reactant gas cannot travel across the electrolyte. This is a problem with all liquid electrolytes that can cause damage not only to the cell itself but also limit the necessary reactions. As mentioned above, both pressure and voltage must be limited to keep the carbon and platinum from corroding at the interface between electrolyte and electrode.

Like other low temperature cells, the anode in particular is subject to poisoning if the fuel gas contains contaminants like carbon monoxide (CO) or sulfur. Usually, hydrogen is produced from hydrocarbon fuels such as natural gas or coal gas with contaminants removed before the fuel reaches the cell. A number of different processes are used to clean the fuel gas, with carbon monoxide being removed using chemical reactions while sulfur is removed using filter beds. **Palladium membranes** that pass only hydrogen gas are also used to clean hydrogen fuel gas.

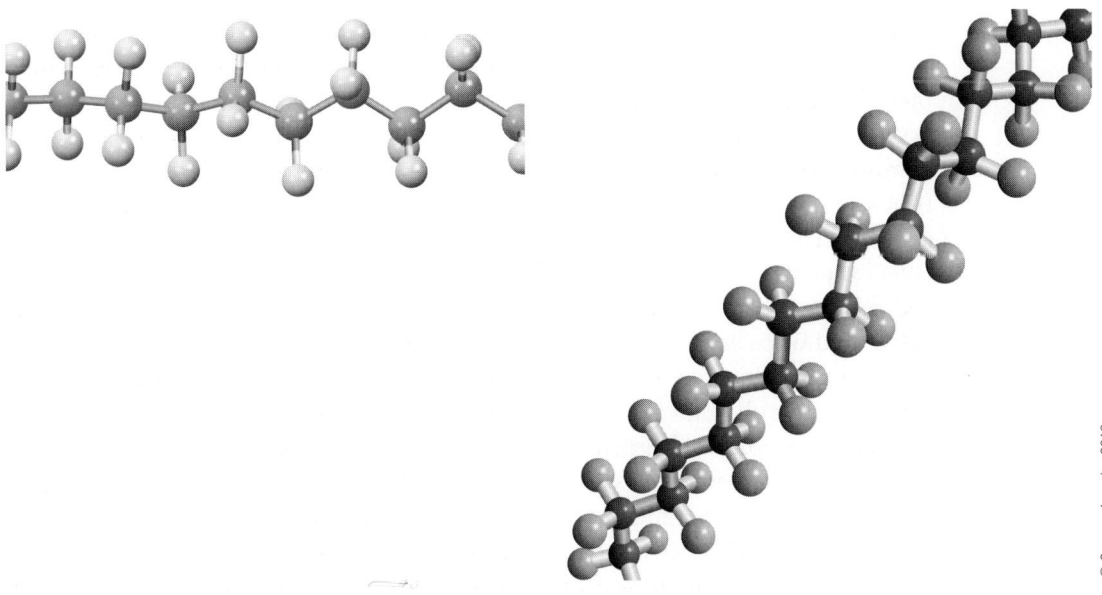

■ **Figure 4-5** PTFE molecule.

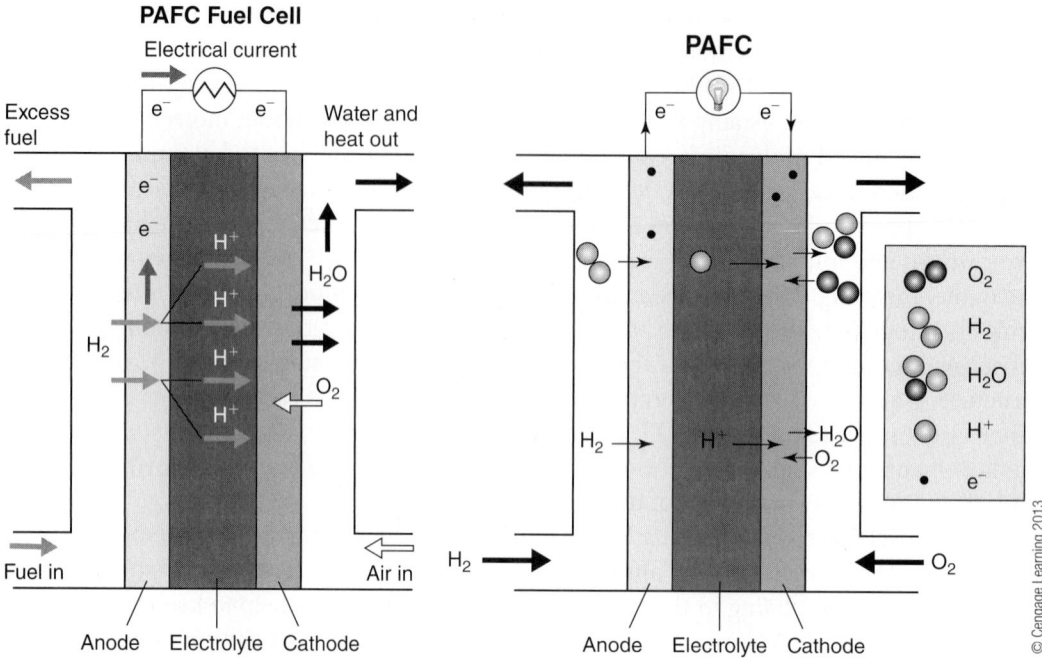

Figure 4-6 PAFC cell.

Unlike some cells though, the liquid electrolyte will move under load, causing alternating cycles at the electrodes of being flooded followed by loss of liquid as gas flows in, is reacted and then chemical transfer occurs between the reaction site and the electrolyte. This cycling will damage the electrodes over time, causing performance drops and ultimately cell failure.

System life for PAFC systems is currently around 40,000 hours of operation. PAFC cells are shown in Figure 4-6.

Phosphoric Acid Fuel Cell Reactions

Individual phosphoric acid fuel cells are much the same as the cells in SPFC or MCFC systems. Differences in material used or sealing systems are required by the different solids, gases, liquids, and pressures in the system as well as the associated balance of plant. As with all fuel cells, the temperature of operations, internal operating pressures and fuel gas determine to a great extent the system performance. PAFC systems are particularly sensitive to gas pressure and composition since the phosphoric acid (H_3PO_4) electrolyte will vaporize over time. Gas enters the cell in the form of fuel and the oxidant stream (O_2). Gas also leaves the cell. Over time the operating temperatures and pressures of the PAFC vaporize a portion of the electrolyte acid which in turn is taken out of the cell system by the gas stream. This has two significant impacts. The electrolyte is lost and once enough is gone, the cell operation will degrade and eventually stop. Because these cells do not contain electrolyte recycling stream like MCFC systems, enough acid must be set into the cell during production to account for loss to vaporization. In addition, the acid containing gas stream must be dealt with as it leaves the system since it will corrode interior parts and vent to atmosphere. Increasing temperature and pressure in fuel cells will increase performance but always at a cost. In PAFC systems, for instance, increasing temperature increases the vaporization of acid, which leads to larger reservoirs in the cell, which increases complexity and cost.

The chemical reactions used in PAFC systems are the same as in other fuel cells where hydrogen gas (H_2) is used as fuel.

At the anode, the reaction is:

$$2H_2 \leftrightarrow 4H^+ + 4\,\text{electrons}^- \qquad \text{Eq. 2}$$

At the cathode it is:

$$O_2 + 4\,\text{electrons}^- + 4H^+ \leftrightarrow 2H_2O \qquad \text{Eq. 3}$$

The protons (H^+) move from the anode to the cathode where water is created. Unlike a PEM cell, which has the same reactions occurring, water must be removed to keep the acid electrolyte from being diluted.

SOLID OXIDE FUEL CELL (SOFC)

Solid Oxide Fuel Cells (SOFC) are solid. Like SPFCs and some forms of AFCs, no liquid is involved in their operations, so that many of the operating difficulties with some fuel cells do not apply with solid oxides. SOFCs operate at high temperatures though, from 600°C to 1000°C, bringing up a different set of issues in fabrication and operation. In general, SOFC systems can operate at the highest temperatures of all fuel cells and are split into two temperature ranges, below 800°C and above that, a distinction mainly having to do with what materials can be used. Below 800°C, materials such as stainless steel can be used, while above that temperature special, high cost materials are required. As with all high temperature cells, start-up times are long and temperature cycling causes a variety of failures within a short number of cycles.

The cell itself usually consists of a sandwich of thin ceramic layers, one acting as an anode, the other acting as the electrolyte and the third as a cathode. Remember that anodes act to strip the electrons off a fuel gas while cathodes act to recombine that electron with other gases. The electrolyte acts as a **transport medium** for the charged ion resulting when the electron is stripped off. The electrolyte in SOFCs can also support the anode or cathode. This is a good place to reinforce the fact that individual fuel cells are usually small. Even a large solid oxide individual cell for instance may only be eight inches by eight inches and as thin as a piece of construction paper. It is important to realize that the thinner the individual fuel cell, the less voltage loss that will occur, meaning the more volts are available for the application. This is called ohmic loss and occurs when the electrons have to follow a path that is either very long (very long for an electron that is) or through a material that does not conduct electricity (or ions if the material is an electrolyte) very well. The electrons lose energy when they are forced to follow long paths or a path with high resistance, and that energy is then not available to do work as electricity. Very thin cells such as solid oxide cells limit ohmic loss and maximize current density, as do cells operating at over 800°C. Figure 4-7 shows a SOFC.

Ceramics are **inorganic**, **nonmetallic** substances that are made with heat. Inorganic substances are those that do not contain carbon. Nonmetallic is a bit harder to explain since in chemistry, most of the elements are metals. For ceramics, nonmetallic usually means there are oxygen atoms involved. For instance, SiO_2 (sand) is a ceramic since its principle component is a nonmetallic oxide molecule (silicon bonded to two oxygens). The usual electrolyte in SOFC is an oxygen

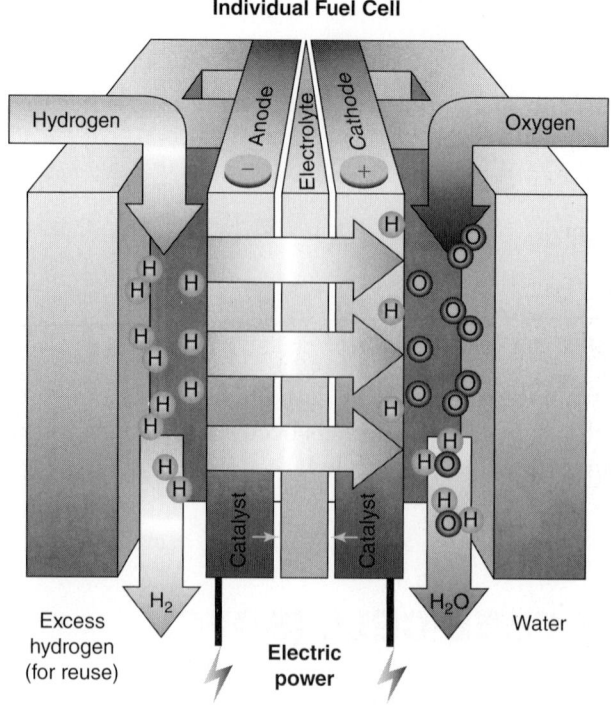
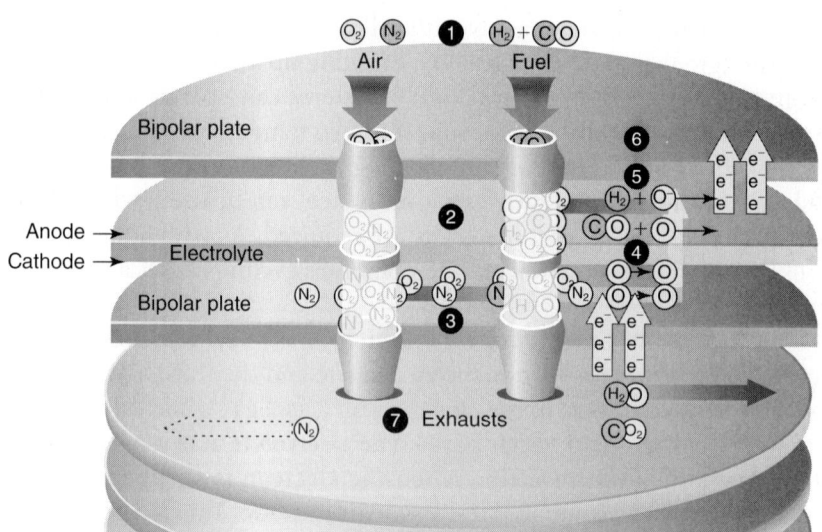
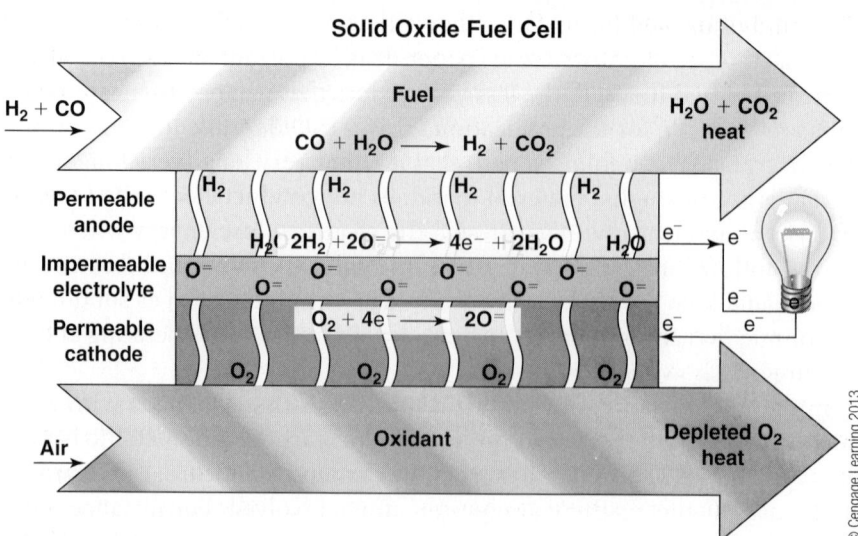

Figure 4-7 SOFC cell.

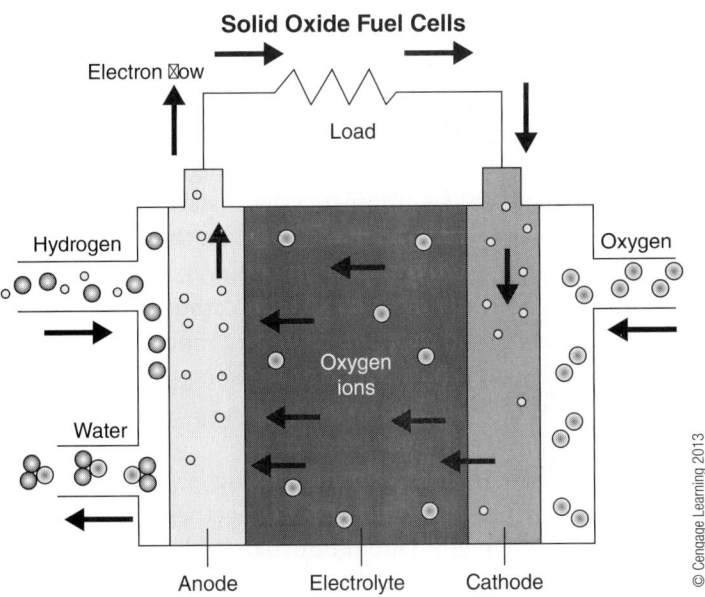

Figure 4-7 (Continued)

and **zirconium** molecule called **zirconia** (ZrO_2). You might have heard of cubic zirconia, sold as artificial diamonds in every mall in the world. The electrolytes in SOFCs are cousins of that, modified to tolerate the higher temperatures. These modifications usually consist of alloying the zirconia with another ceramic material to stabilize the overall structure. Currently, the most common stabilized zirconia is a mixture of **ytria** (Y_2O_3) and zirconia (ZrO_2).

The first SOFC was fabricated in the 1930s, but the use of zirconia in electric circuits goes back to the end of the nineteenth century, when Nernst discovered that if you get it hot enough, it will conduct ions. Recall that we are talking about a ceramic material and that most ceramic materials act as **insulators**, that is to say that electrons (electricity) will not travel through them. Generally, that also means ions will not travel through them either. In the case of zirconia, the ions will conduct from cathode to anode at the operating temperatures used, but electrons will not, at least until the temperature reaches 1500°C. The stabilized zirconia is used as the electrolyte, but the anode and the cathode usually share the zirconia structure with the electrolyte as well. Since the cell operates at high temperature, less expensive catalysts can be used. In most cases, the anode is nickel on a porous zirconia (20%–40% porosity) substrate while the cathode is a mixed alloy of **lanthanum, strontium,** and **manganes**e (strontium-**doped** lanthanum manganite) on a porous zirconia **substrate**. Due to the solid state nature of the SOFC, neither anode or cathode are easily contaminated by impurities in common fuel gases.

The common zirconia structure of SOFC systems minimizes the problems individual cells have when different materials are used. As materials get hotter, they generally get a bit larger, a process known as **thermal expansion**. When an anode is made of one material, an electrolyte in made of another material and a cathode made of a third type of material, then the three will expand at different rates and amounts as the temperature increases. Because ceramics are **brittle**, this could cause them to crack as the one with the largest thermal expansion pushes against the one with the smaller thermal expansion. Having all three similar means the individual cell will be relatively stable and reasonably strong over the large temperature range involved, room temperature to possibly 1000°C in isolated areas of a cell. Zirconia does not react chemically with the other things in a fuel cell, such as hydrogen, oxygen, nickel, lanthanum, and manganese, which limits cell corrosion and improves **operating life**.

Two general designs are used for SOFC systems, **tubular** and **planar**, although several others have been used as well, including corrugated configurations, banded structures, and what is called bell-and-spigot. Tubular systems start with a support tube, and then add a stabilized zirconia electrolyte, an anode on the outside and a cathode on the inside. Planar systems start with the thickest layer, often the stabilized zirconia plate, then add the anode on one side and the cathode on the other (self supported) or layer the three one at a time on the interconnect structure or other external structure (externally supported). Figure 4-8 shows a tubular configuration for a fuel cell, while most of the figures presented so far represent planar (flat) configurations.

No matter the system used, all are then sintered. **Sintering** is a high temperature process for metal and ceramics that flows the material together either without melting it (solid state sintering) or by melting it but only barely above the melting temperature (liquid phase sintering). It is usually done using small particles, many times with the particles suspended in a liquid to make handling easier and safer. Several processes are used to set the various layers either involving particles as described above or deposited from a gas or plasma phase. Vapor or **plasma deposition** processes are batch processes where the material to be added on are vaporized or run through very high temperature plasma and then either sprayed or electrically set using positive and negative charges. Slurry or **tape casting** processes set a slurry of liquid and ceramic particles over a single piece or on a moving line of material, then transfer this into a furnace where the liquid is carefully boiled off and finally the ceramic particles are sintered together at a high temperature. Some of these processes can set layers down as thin as 0.2 **microns** (0.0000079 inches). Keep in mind that the operating temperatures of some SOFC systems will cause internal sintering to occur over the life of the unit, in particular causing electrodes to degrade.

Current tubes are approximately 150 cm long, 2 centimeters in diameter and 1–2 millimeters thick, with the planar cells being even thinner. Tubes cost substantially more to produce, since vapor or plasma deposition is expensive while planar cells are cheaper, since tape casting can be done on an automated system. Tube cells have substantially higher ohmic losses since the paths are very long through and around the tube while planar cells have very low ohmic losses, thus the current density is higher for planar cells, making the cost of produced electricity cheaper. Tubular cells are substantially easier to seal than planar cells since ceramics are porous by nature and tend not to hold gases well. This property is useful for getting gas to the electrodes and then into the electrolyte, but not so useful when the gas has to be kept in the cell and out of the rest of the system. Sealing planar cells in some systems requires using a material that will almost melt at the operating temperature so it flows around the edges as a seal, much like MCFC systems require the high operating temperatures to seal themselves. This is further complicated by any internal pressure used in the cell, something many designs call for since higher cell pressures in SOFC systems increase performance.

All fuel cells must be tied together in order to produce a useable amount of electricity since the individual cells only produce limited amounts. In SOFC systems, the two types of systems used two different methods. In tubular systems, the gas can be supplied to and removed from each tube cell, whereas in the planar system, both anode and cathode gas have to flow through the entire structure. This is an issue for all fuel cells and will be explored in more depth in the next chapter.

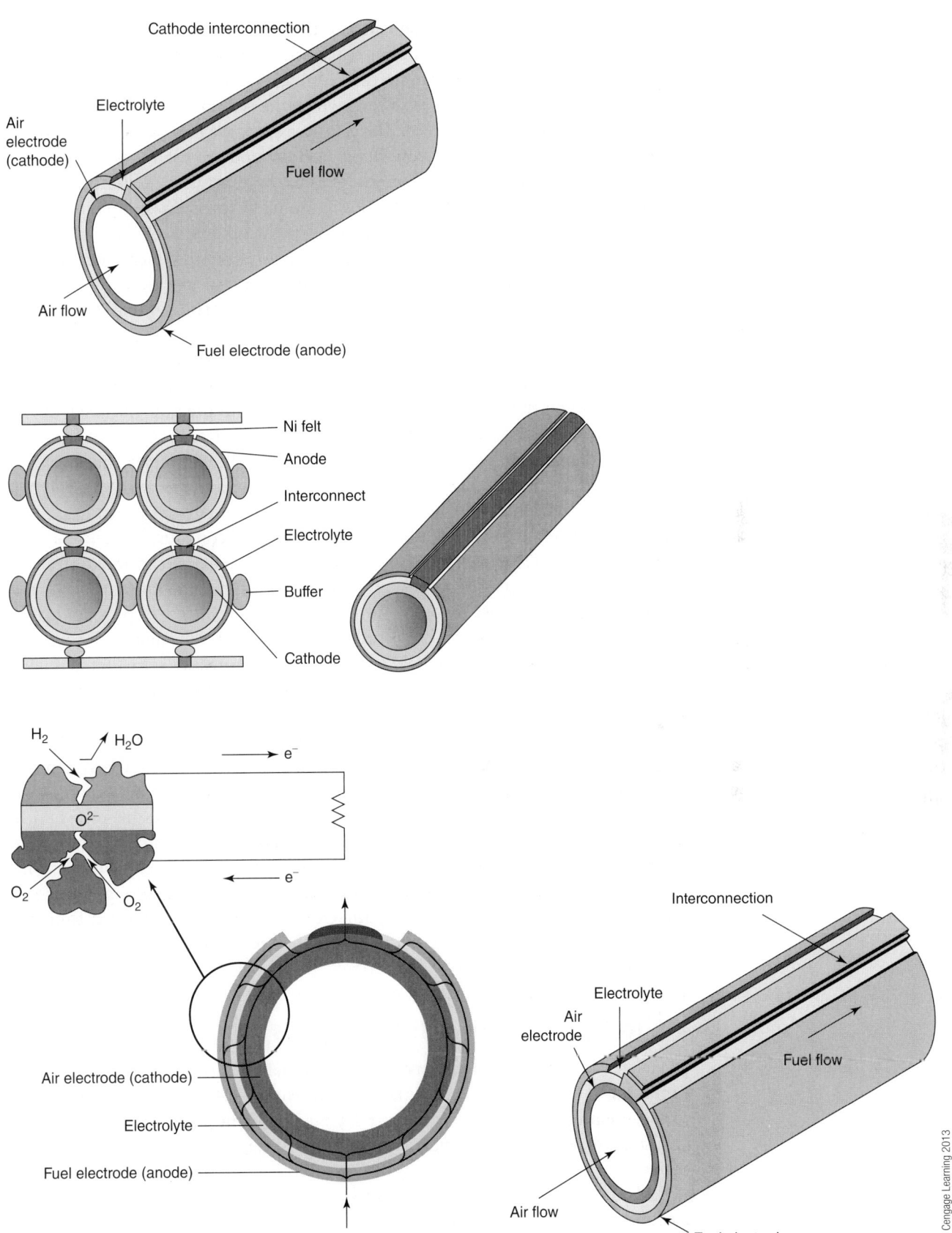

Figure 4-8 Tubular SOFC cell.

To introduce the subject, consider the two types of SOFC systems. In a tube, the fuel gas goes to the outside of the tube, reacts with the anode catalyst and then continues on through the electrolyte to the cathode, where it reacts with the air supplied inside the tube. In a planar system, where the cells are stacked like pancakes, it is more complicated. In between the cells in a stack are **bipolar** or **interconnect plates**. These plates generally conduct electricity and are used to direct the two gas supplies to the appropriate electrode. Many systems use a conducting metal with two sets of ribs on either side. The ribs going horizontally in the system will supply air to a cathode while the vertical ribs on the other side of the interconnect (bipolar) plate will supply fuel to the anode sitting next in the stack. Keep in mind that for SOFC systems, the high operating temperatures means these interconnects must be stable at high temperatures, cannot react with either gas flowing in the system, must have thermal expansions that are very close to the ceramic plates sandwiched between them and must be provide a gas tight seal between the actual cells. The design of associated equipment like interconnects depends more than anything else on the kind of fuel cell is being used and the how the stack of fuel cells is put together. The type of material used for associated items depends more on the pressure and temperature involved as well as the chemicals and mechanical stresses present in the system. Lower temperature systems will generally use metallic interconnects while higher temperature systems currently use **lanthanum chromite** ($LaCrO_3$). Figure 4-9 shows a planar SDOFC configuration.

Like other high temperature fuel cells, SOFC systems can make use of a number of different types of fuels. This is because higher temperature cells can internally reform fuel gas instead of requiring pure a supply of pure hydrogen as a fuel. SOFC temperatures do not require precious metal catalysts for internal reforming, relying on the nickel catalysts instead. Internal reformation using SOFC systems does require that the carbon in the fuel gas be kept from the anode or else it will deposit there and eventually block gas access. The most common way of dealing with this is with steam, requiring a separate subsystem. So efficient are SOFC systems at reforming fuel gases that some designs use the systems to produce hydrogen and then produce electricity and carbon monoxide (CO) as by-products. Combining several systems to reform natural gas, use waste heat in a combined heat and power co-generation cycles as well as provide CO and hydrogen as fuels to other fuel cells can result in very high efficiencies. Systems combining with internal combustion engines have even been designed, where gasoline is partially reformed by the SOFC to produce electricity, then the remaining hydrogen rich, lower hydrocarbon fuel is injected into the combustion engine for greater efficiency and lower emissions.

Solid Oxide Fuel Cell Reactions

The chemical reactions and forms of the individual cells used in SOFC systems are similar to other fuel cells. Conventionally, hydrogen gas (H_2) and carbon dioxide (CO_2) are used as fuels with methane (CH_4) used in some cases as well.

If hydrogen is used as a fuel, the reaction at the anode is:

$$2H_2 + 2O^{-2} \leftrightarrow 2H_2O + 4\, electrons^- \qquad Eq.\ 11$$

In this case, the oxygen ion is transferring from the cathode to the anode, again a different ion movement than in some other fuel cells, where ions travel from the anode to the cathode. The reaction at the cathode is:

$$O_2 + 4\, electrons^- \leftrightarrow 2O^{-2} \qquad Eq.\ 12$$

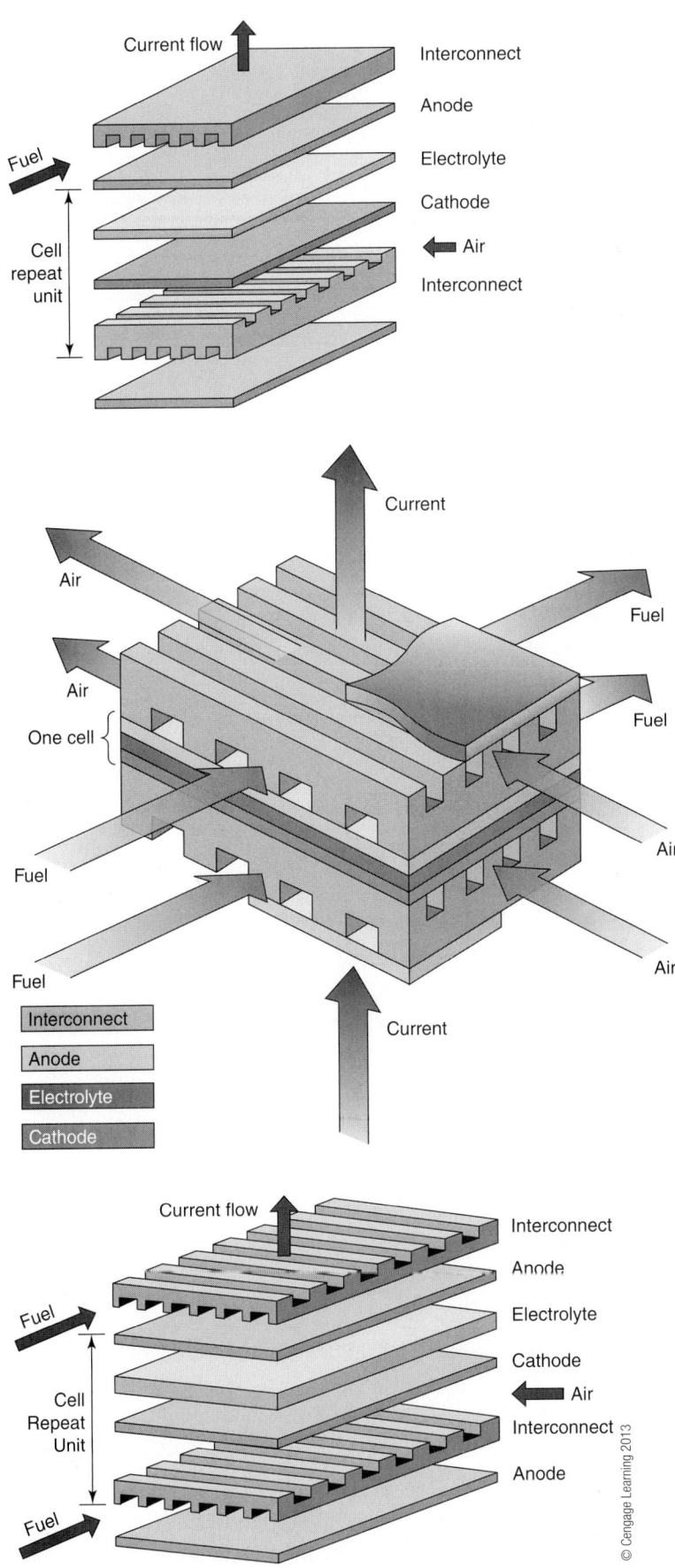

Figure 4-9 Planar SOFC cell.

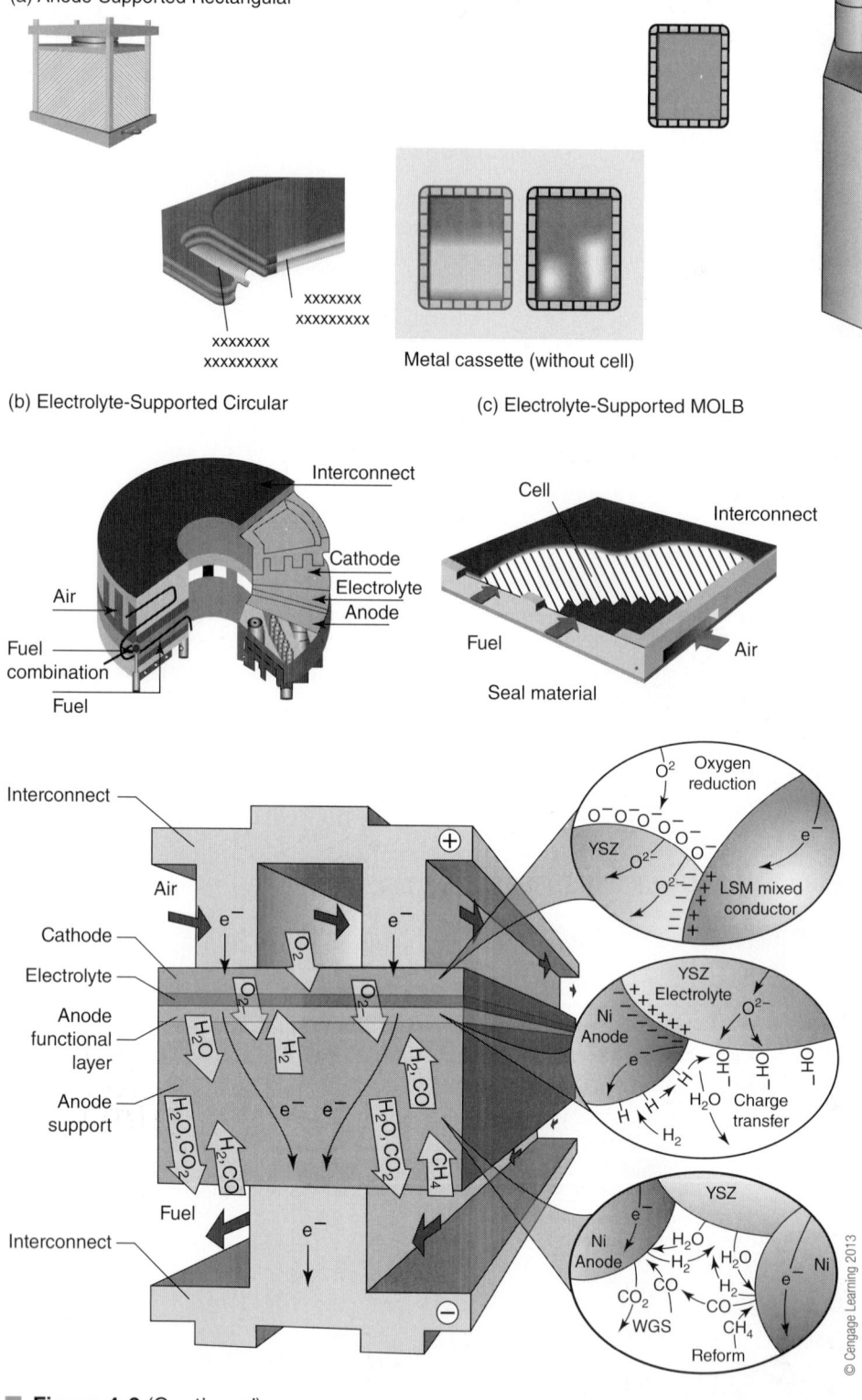

Figure 4-9 (Continued)

Notice the oxygen ion has a -2 charge, indicating it has two excess electrons. When hydrogen (H_2) is used as a fuel, the oxygen gas (O_2) molecule supplied to the cathode either as pure gas or as air (21% oxygen in air) must first be broken apart then extra electrons added to the two oxygen atoms. The solid oxide electrolyte must be able to pass the oxygen ions (O^{-2}) but as with other fuel cells, not pass other things, such as the electrons being produced at the anode.

Adding Eq. 11 and Eq. 12 gives the overall chemical equation for a SOFC using hydrogen as a fuel.

$$2H_2 + O_2 \leftrightarrow 2H_2O \qquad \text{Eq. 13}$$

In other words, SOFC systems can make electricity simply by making water. This is an important point, because the reverse can also be true. SOFC systems can make hydrogen and oxygen from water if electricity is supplied to the system. Solar energy or wind power thus can be used to create electricity when possible and some percentage of this electricity used to produce hydrogen and oxygen. When the wind stops blowing or the sun goes down, the stored hydrogen and oxygen gases are then used to create electricity and water to ensure a steady supply of electricity. Interest in combined systems of that sort is increasing and may well be the future of green power generation. Figure 4-10 shows a possible configuration for such a hybrid system.

If carbon monoxide (CO) is used as a fuel, the reaction at the anode is:

$$2CO + 2O^{-2} \leftrightarrow 2CO_2 + 4\,\text{electrons}^{-} \qquad \text{Eq. 14}$$

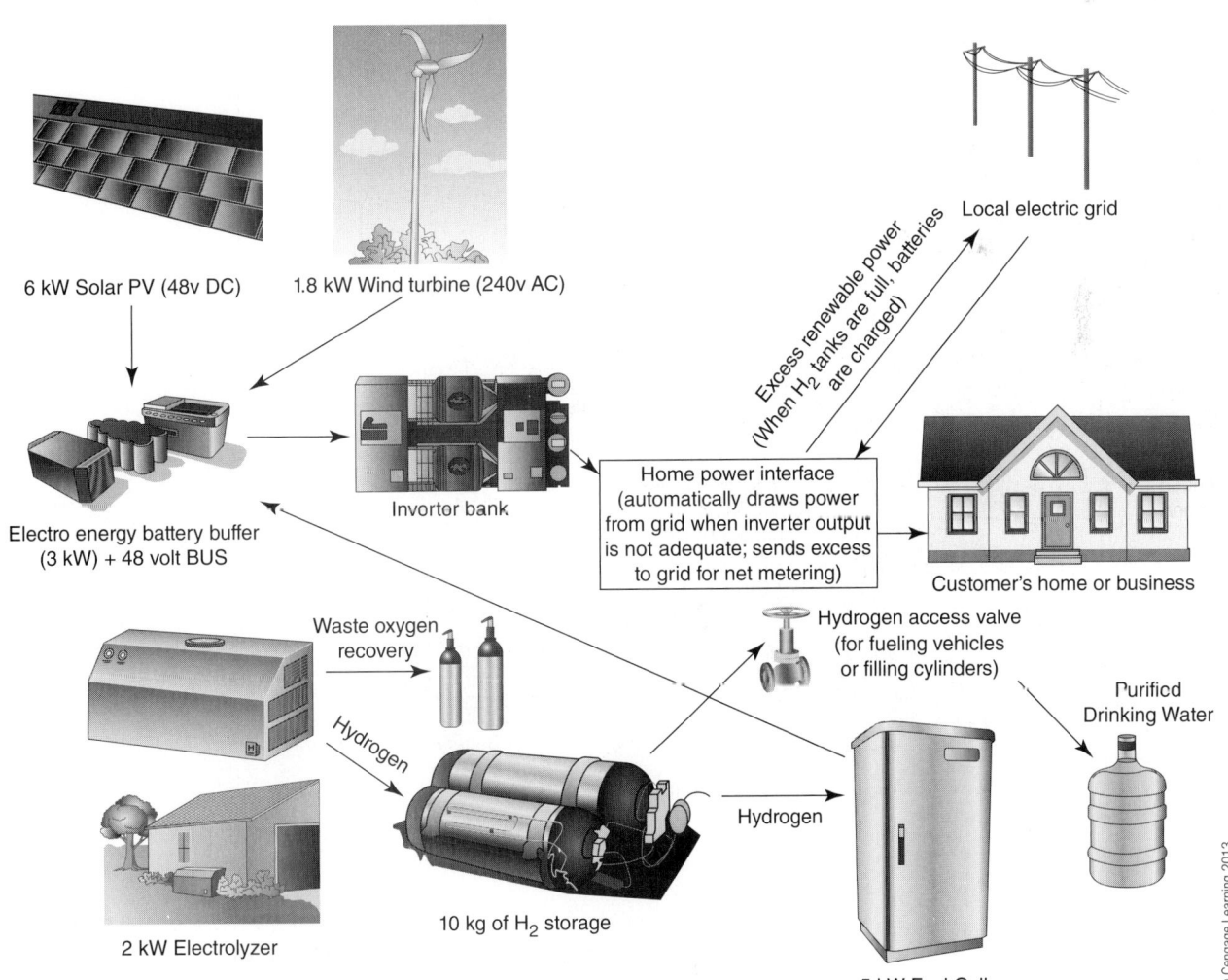

Figure 4-10 Hybrid systems.

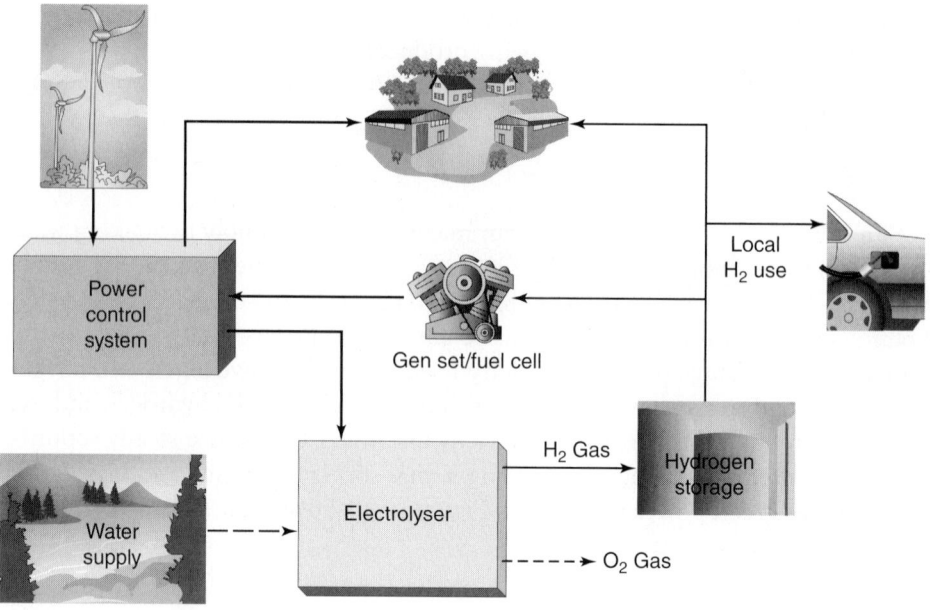

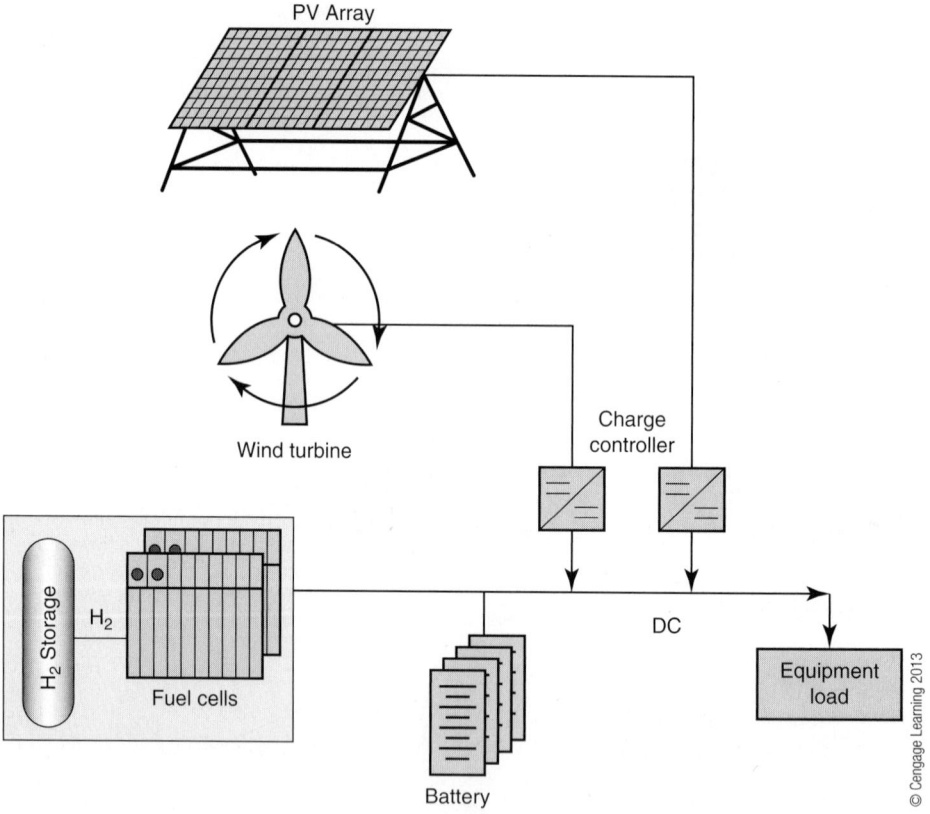

Figure 4-10 (Continued)

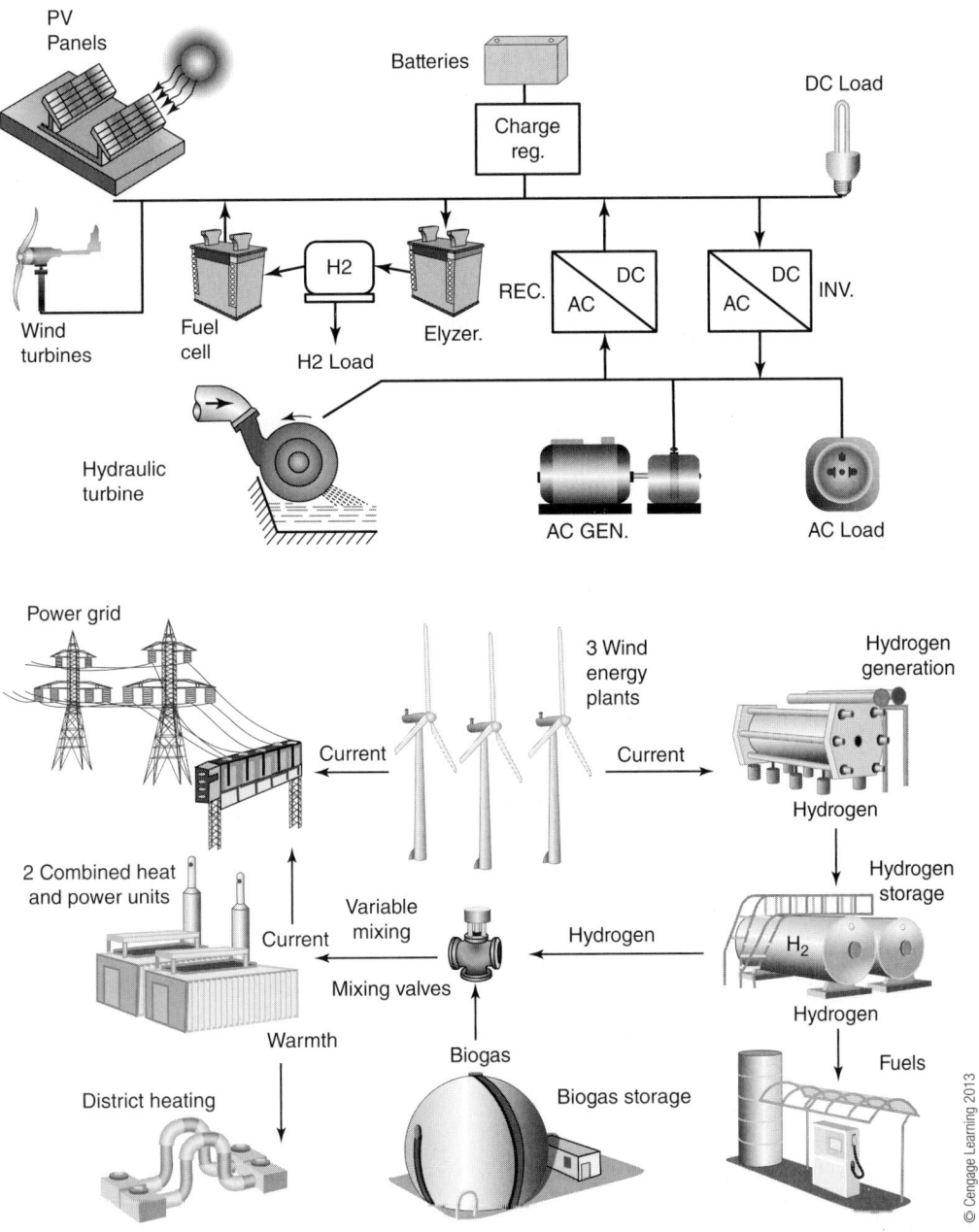

Figure 4-10 (Continued)

Again, the oxygen ion (O^{-2}) is passed from the cathode to the anode. The cathode reaction is the same as the one where a SOFC uses hydrogen (H_2) as a fuel.

$$O_2 + 4 \text{ electrons}^- \leftrightarrow 2O^{-2} \quad \text{Eq. 12}$$

This is a relatively simple system, where carbon dioxide (CO_2) does not have to be externally recycled like an AFC. CO will not poison the electrodes since it can be used as a fuel but other contaminants can cause trouble, in particular, **hydrogen sulfide** (H_2S), **hydrogen chloride** (HCl), and **ammonia** (NH_3). These three gases are issues with most of the high temperature fuel cells that internally reform fuel gases such as natural gas, methane, or coal gas, since they are contaminants present in such fuel gases. In general, hydrogen sulfide (H_2S) causes the most significant problems in high temperature cells followed closely by hydrogen chloride (HCl), with ammonia (NH_3) being considerably less of an issue.

Chrome can also cause problems in SOFC systems. Metals are not something that generally affect in fuel cells unless the systems are at high temperatures and pressures as SOFC systems are. Metals can move over time and in some SOFC designs, thin foils are used. If the chrome migrates under temperature and pressure, the corrosion resistance of the foil can be ruined and the cell eventually fails as a result.

Both temperature and pressure within the cell are important to SOFC systems, in large part governing internal ion conductivity and thus individual cell performance. SOFC systems are commonly used in **cogeneration** systems so the high pressures can be used to prepare the high temperature outlet gas as an input gas for a turbine. Since both turbine and fuel cell are operating at higher performances and no compression is needed for the gas going into the turbine, the whole system becomes more efficient.

KEY WORDS

Knowing the terminology used is critical when dealing with fuel cells. Following is a list of the important terms in this chapter, which are also in bold typeface within the chapter. It is recommended that students be required to submit definitions of some of these words as homework assignments in which they look the terms up in other books, articles, or on the Internet.

4-electron reactions
acid
alkaline
alkaline based electrolyte
Alkaline Fuel Cell
ammonia
anode
atmospheric pressures
atoms
balance
base
batteries
bipolar plates
brittle
by-product
carbonate ion
carbonates
catalysts
cathode
cell voltages
circuit
cogeneration
conduction
contract
corrosion
current
cycles
Direct Internal Reforming
doped
efficiencies
efficient
electrically neutral species
electrolyte
electron
endothermic
evaporates
exothermic
expand
generator
high temperature cells
high temperature fuel cells
hydrocarbons
hydrogen chloride
hydrogen sulfide
hydroxide ion
Indirect Internal Reforming
inorganic
insulators
interconnect plates
internal combustion engines
ion exchange membrane
kinetics
lanthanum
lanthanum chromite
loop
low temperature fuel cells
manganese
mature
megawatts
melts
microns
milliamps
molecules
molten carbonate
Molten Carbonate Fuel Cells
nonmetallic
non-noble metals
ohmic loss
operating life
operating temperatures
oxidation/reduction reaction
palladium membranes
particulates
parts per million
peak loading
phosphoric acid
Phosphoric Acid Fuel Cells
planar
plasma deposition
poisoned
polymer
pores
porous electrodes
potassium hydroxide
potassium ion
power generating
purified
reaction chamber
reactions
recirculated
reform
reformate
reformed
reservoir plate
scalability
silicon carbide
sintering
sodium borohydride
solid oxide
Solid Oxide Fuel Cells
solid polymer
Solid Polymer Fuel Cells
solution
strontium
substrate
tape casting
temperature sensitive
thermal expansion
transport medium
tubular
turbines
vapor
viscous
volatilize
water management
ytria
zirconia
zirconium

DISCUSSION QUESTIONS

1. Why would a low temperature fuel cell better handle on-off cycling than a high temperature fuel cell?

2. What are the five major types of fuel cells?

3. Why would a PAFC system usually operate at lower temperatures?

4. What is a hydroxide ion?

5. What is the name of a chemical reaction that absorbs heat while it is occurring?

6. What is 17 parts per million (ppm) expressed as a percentage?

7. In the reaction $2H_2 + O_2 \leftrightarrow 2H_2O$, how many molecules are reacting and how many molecules are being produced?

8. What is Direct Internal Reforming?

9. Why would a PAFC need a reservoir plate?

10. What is ohmic loss?

11. What is an inorganic substance?

12. What is the difference between a bipolar plate and an interconnect plate?

13. How many electrons are transferred between the reactants and the products in the reaction $H_2 + O^{-2} \leftrightarrow H_2O$?

CHAPTER 5

FUEL CELL MAIN COMPONENTS

objectives

This chapter will introduce the student to fuel cells as collections of parts. A discussion of the individual fuel cell electrolytes and electrodes will further differentiate the five types along with some differences encountered during fabrication. Some of the concepts critical to understanding the internal operations of both cells and cell stacks as well as some of the problems encountered during production and operations will be expanded upon.

INTRODUCTION

Fuel cells are extremely simple **engines**. They have really no moving parts to speak of; they do not rely on complex mixtures of gas, internal pressure, fuel ratios, **spark initiators** or follow-on generators to work. They are relatively small but **scalable**, and unlike most electricity generating schemes they do not rely on a combustion reaction that by its nature produces **pollutants**. They are expensive, but that is a relative issue. Automobiles, for instance, have virtually all of the same issues that drive up costs in fuel cells. They both operate at elevated temperatures and pressures requiring special materials. They both make use of expensive metals to make the reaction sequence happen: fuel cells at the anode and cathode, automobiles in electronics and catalytic converters. They both produce a product that is grossly expensive compared to competing technologies: fuel cells viewed as orders of magnitude too expensive compared to the cost of electricity from conventional plants and automobiles compared to the mass transit cost of moving individuals. Yet the automobile industry has little difficulty in selling systems of very high cost and limited life span where fuel cells are still viewed as unacceptably priced. At some point, it will become apparent that fuel cells will change global society just as mass produced automobiles or computers did. Students well versed in that technology may find themselves in high demand and with substantial opportunities available to them.

A single fuel cell produces around 0.7 volts when it is under a full load. Variation in this number exists from fuel cell to fuel cell but for the most part that is a good number to remember as a baseline. There are a few important chemical and electrical issues that govern how that amount of voltage can be produced and then how much can actually be used. Two of the most important concepts are how a chemical reaction begins and how electrons move (electricity) within any electronic device.

The reactions that occur at the **electrode** to produce an **electron** usually happen by themselves or require some form of added energy to make it happen. The **chemical reactions** fuel cells make use of do not happen without some form of help. That is why they all require **catalysts**. It is why cells operating at higher temperatures need catalysts that are less effective (cheaper) than catalysts in lower temperature cells. It is why only a few fuels actually work in fuel cells. The reasons behind this have to do with how things work. If you need to push a car on a flat road, you first have to get it going; once it is moving, the effort needed to keep it moving is less than the effort needed to get it going in the first place. Chemical reactions are much the same. The energy required to start the reaction is more than that needed to keep the reaction going once it starts. Catalysts in part bypass this **overpotential** to make things easier, in the same way that rocking a car makes it easier to get going. The formal term for this in chemistry is called the **activation energy**, and it means just that, the energy needed to activate a chemical reaction. Figure 5-1 shows a graphical representation of activation energy and the shortcut that is available when a catalyst is used.

Overcoming the activation energy requires putting a **chemical path** into the system, or to put it differently, by putting either some form of extra energy in or

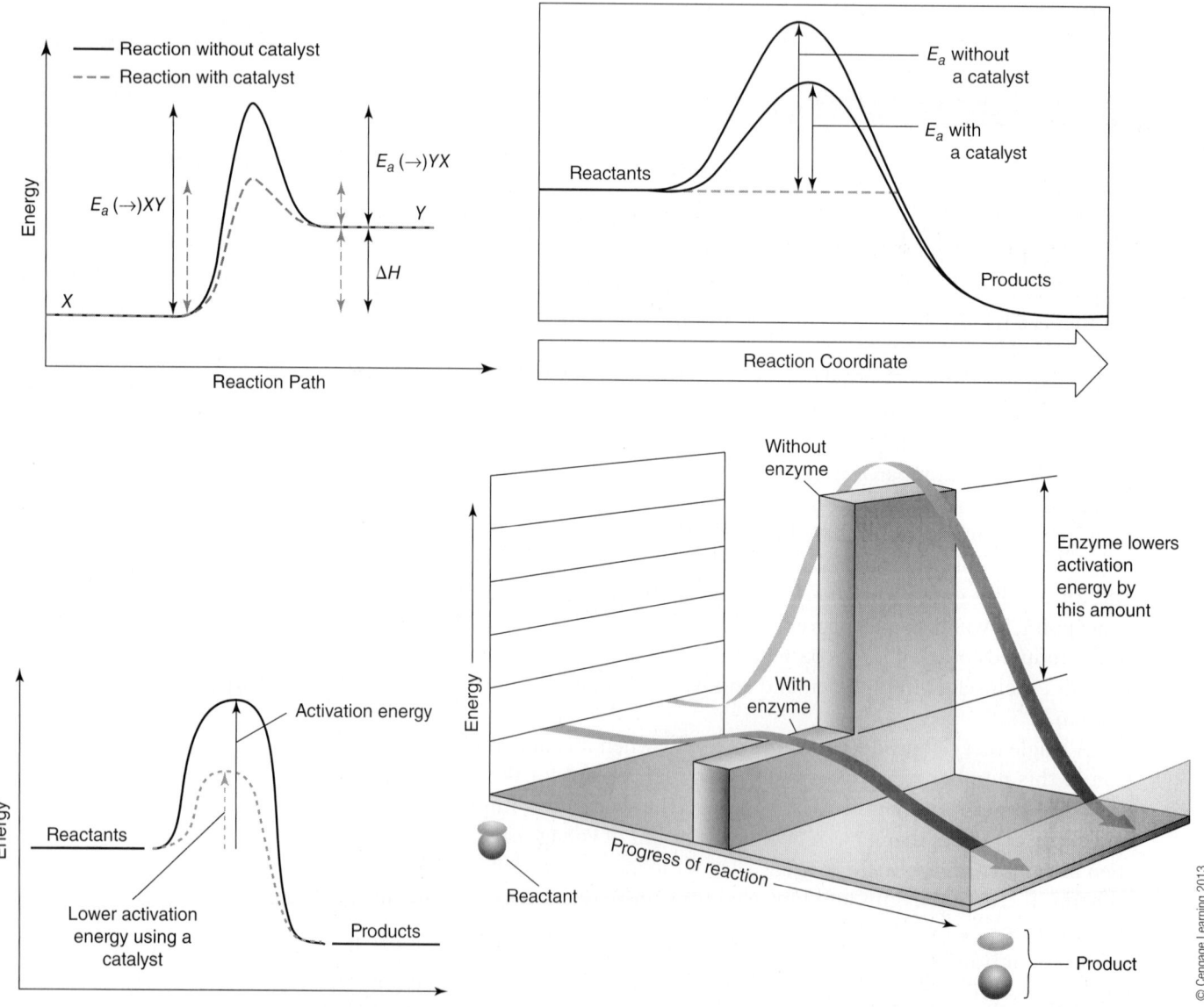

Figure 5-1 Activation energy and catalysts.

some kind of shortcut in. Automobiles do this by adding a spark into the reaction chamber with a spark plug to provide extra energy. Fuel cells do this with a catalyst to provide a chemical shortcut. Recall that for the most part, we want our energy to do **work** but both the activation energy and anything extra we put in is usually lost as heat. In other words, we pay to put it in, but we don't get anything for our money but heat. That is why **cogeneration** systems are particularly effective, because they can recover at least some part of that heat and then do something with it (perform work).

Once a reaction is going though, it will either keep going or it will stop. Many things can make a reaction stop, for instance, if the temperature drops or the pressure changes too much or one of the reactants runs out. Catalysts not only work to start a reaction, they help the reaction to keep going. Changing the temperature or pressure will also keep a reaction going as long as there are still enough **reactants**. Most cells will produce electricity better when the temperature or pressure is increased. Temperature and pressure are forms of energy and the more energy that is available, the more likely it is that chemical reactions will both start and keep going.

With the reaction going, electrons move into the wiring system and **ions** that have lost their electrons move into the electrolyte. For electricity to exist, a **circuit** has to be established. That circuit begins at an electrode where the split between electron and ion begins. The electron moves into some type of collector system, flows out to a wiring system, is used to do some type of work, flows back into the cell and to the other electrode, where it meets up with an ion. The cell voltage is determined by the chemical reactions, but the available voltage is determined by the system as a whole. All of the cell components resist the flow of electrons or ions to some extent. This combined resistance to electron or ion flow is called the **ohmic loss** or sometimes called the area-specific resistivity. This loss differs between the types of fuel cells, between different designs of the same type of fuel cell and even in the same fuel cell as it ages in service. This loss should ideally consume less than 0.1 volts per square centimeter of cell.

It is important to realize that electrons are all the same and so are ions. The electron that first split from a certain ion need not be recombined with that particular ion. Any electron will do for any ion.

Chemical reactions that take place with the help of catalysts also depend on how much catalyst is available. For fuel cells, this is even more important because as discussed in the first chapter, the electrode, the fuel, the catalyst, and the electrolyte must all be essentially in the same place for the cell to work. Just making the electrodes larger won't work, something Grove discovered in his original fuel cell. The catalyst surface area has to be very large and to do that, it is set as very fine particles over the entire electrode. This allows **gas** to disperse evenly, provides the maximum amount of catalyst surface for the gases to contact and provides a channel for the electrolyte.

In this chapter, we will look at some of the individual parts of the cells in a bit more detail so the student can better understand how they all work together so that chemical reactions occur in an orderly way and continue to do so in a way that provides electrons in a steady stream to the system in general. While important different concepts will be introduced for each type of fuel cell, the student should realize that all of these concepts are important to all of the other fuel cell types as well. It may be that certain things are more important to certain cell systems (like water in SPFC [PEM] versus water in SOFC), but all of these concepts have to be considered for all the fuel cell types.

FUEL CELL SYSTEMS IN GENERAL

We have mentioned the three main ways in which fuel cells differ from one another, the reactions at the electrodes, the electrolyte being used, and the operating temperatures used. In this chapter, we will discuss some of the other differences encountered in the various types of fuel cells as well as provide more detail about the main types themselves.

AFC

Since AFC systems run at low temperatures using hydrogen (H_2) as a fuel, many of their components are similar to SPFC (PEM) systems. The original AFC designs date from around 1900 and were developed as an alternative to Groves' acid electrolyte design. Current designs grew from Bacon's work in Great Britain and the refinement of that work that went into the American space program. The advent of **Teflon®** and then its related product **Nafion®** that enabled SPFC (PEM) systems made AFC systems less attractive to research, but industry has continued to develop the technology, albeit at a slower pace than during the advent of space exploration.

One of the main reasons why AFC systems fell out of favor was because their life span was a relatively short 5000 hours. This was acceptable in the main application of the time, spaceships, but became unacceptable when the technology began to move to more commercial applications where 30,000 to 50,000 hours was expected. In addition, AFC systems used in the space program in particular were seen to have low **power densities** (although they have high **efficiencies**) and experienced rapid poisoning of the alkaline electrolyte when in operation. In general, fuel cells are now judged primarily by how much power (electricity) they produced per unit of fuel consumed, how much power was produced per unit of volume taken up by the cell itself (power density), and by how much power they produced per unit of money needed to make and then run the cell. In the original space applications, they were not judged by those features so much as by their reliability and their broader contributions to the mission at hand such as producing water. In space applications, the AFC systems were acceptable, but in the more commercial application, they were much less acceptable. That is changing, however, as industrial producers move away from the cash cow of space and develop commercial systems that last longer, reduce overall cost, and perform reliably for extended periods of time.

That is not to say that AFC systems are at too much of a disadvantage when compared to other systems. When the static alkaline electrolyte used in space was replaced with a **recirculating electrolyte**, not only did the operating lifetime increase, but both water management and cooling schemes were greatly simplified, lending the AFC systems a distinct advantage. AFC systems can also reach **standard operating conditions** from a **cold start** in several minutes as well as operate consistently in a wide range of **ambient humidity**, things other cells struggle with. Alkaline cells can tolerate long periods of inactivity without major damage occurring to either electrodes or electrolyte. They can also use cheaper cathode catalysts or a lesser amount of expensive catalysts even at their low operating temperatures. This is because the **cathode oxidation reaction** occurs faster in a **basic solution** than it does in an acidic solution. In addition, the substantial experience industry had gained over the years of developing and producing systems for the space program provides a major advantage over other systems. Figure 5-2 shows a typical alkaline fuel cell configuration.

AFC systems have regained some of their appeal as Polymer Electrolyte and Solid Oxide systems have taken considerably longer to develop into mature products than was first anticipated. Probably the major issue currently facing the industry is **electrode stability** and life expectancy because the combination of corrosion from the alkaline electrolyte and mechanical wear tend combine to limit how long electrodes last.

MCFC

Molten Carbonate Fuel Cell (MCFC) systems operate at relatively high temperatures, typically ranging from 550°C to 700°C (most of the molten carbonate electrolytes now in use **freeze** below about 520°C). Like other fuel cells, they utilize somewhere less than 80% of the fuel delivered to the anode (it is better to have

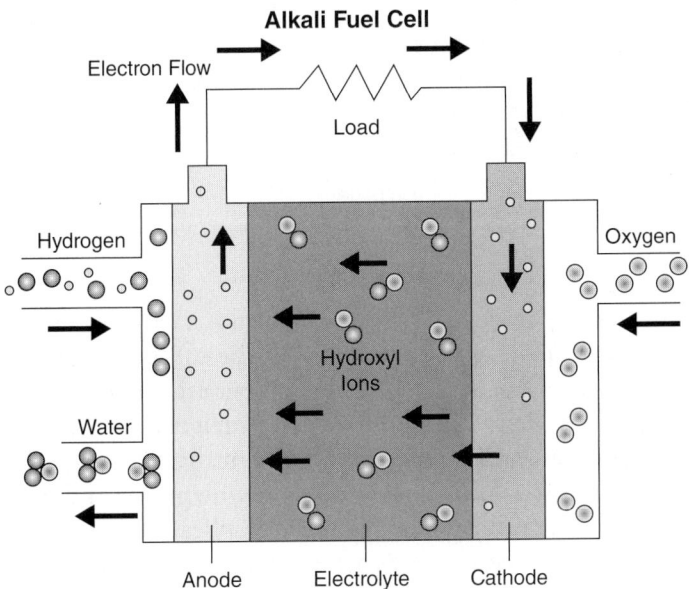

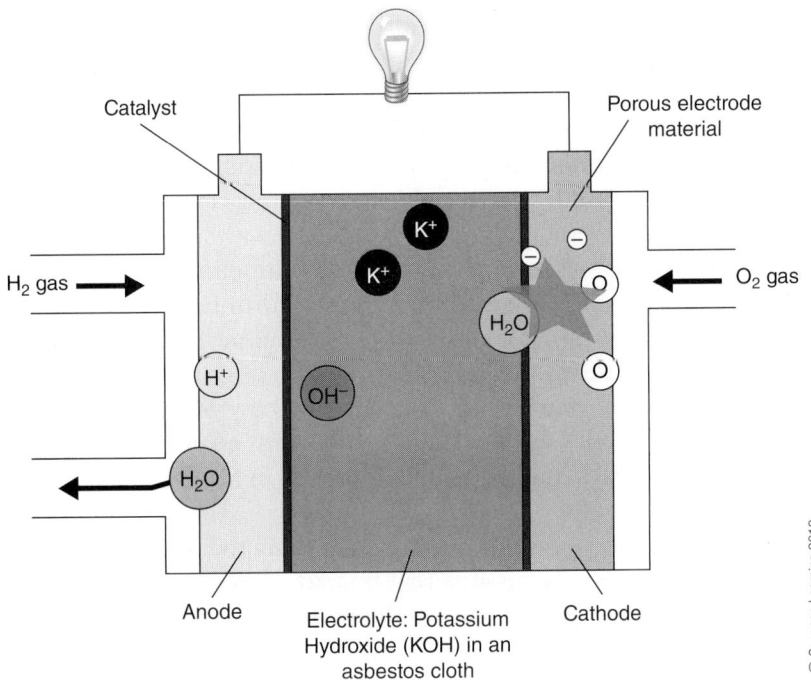

Figure 5-2 Alkaline Fuel Cell.

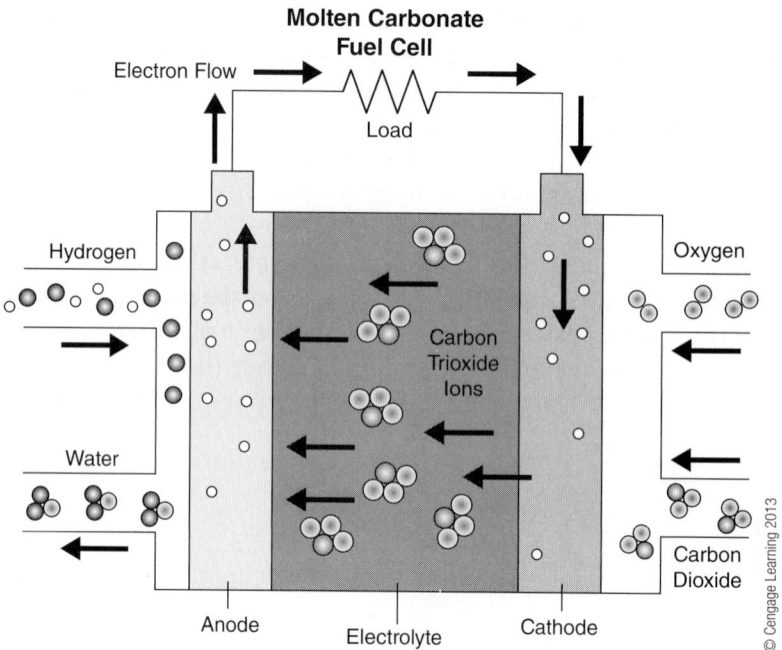

Figure 5-2 (Continued)

excess fuel and use all those expensive catalyst sites than to not have enough fuel and thus waste the money you spent) and slightly less at the cathode, with the individual cells delivering around 1 volt or less. The higher operating temperatures mean systems are usually larger, delivering up to several **megawatts** of power in some cases, although current systems tend to be sized in the one megawatt range more often. They are industrial systems used to provide **power and heat** to large buildings, campuses such as hospitals, or marine applications. Principle contaminants are **sulfur** and **chloride compounds**, with permissible levels in the 0.5 parts per million or **ppm** range for each (one molecule of the particular compound in one million total molecules, keeping in mind of course that a million molecules might be how many there are across the width of a very thin hair). Sulfur in particular is damaging and can build up in the system so in cases where **purging** is not used, input levels can be in the 0.01 ppm range. Unlike some other cells, MCFC systems are also damaged by **particulates** in the fuel gas feed as well as trace metals like **lead** (Pb), **cadmium** (Cd), **mercury** (Hg), and **tin** (Sn) since these can build up on the electrodes and react with the electrolyte. Like other fuel cells, these numbers can change with design and operating parameters but only within limits.

As mentioned, MCFC systems must be brought up to temperature in order to seal. This can cause a unique problem in that nonuniform heating during the initial start-up sequence can have a long-lasting effect on performance and on internal operations. If the difference in temperature is too large across the stacks, the **internal (ohmic) resistance** within the cells can vary by a considerable amount. A large loss of efficiency and output can result with no chance to recover until the system in brought back down. The initial start-up in MCFC systems is the single most important aspect of the system.

Once the MCFC systems are up and running, controlling the **internal pressure** in the system becomes critical. Having different pressures at the anode and cathode causes gas to build in the system, causes the molten electrolyte to move in response to the pressure and can cause **hot spots** to arise in the stack. This not only lowers system output but physically damages the stack as well as some of the associated systems. That is one of the reasons why MCFC systems operate better in applications where a steady current output is needed rather than being used where the **load** will vary much. This can complicate MCFC systems, especially if they are used in light industrial applications where the load demand varies considerably and in some cases can cease completely for hours or days, for instance during a shutdown or maintenance. In most cases, the MCFC system has to remain at a steady power output during those times and in many cases feeds electricity back to the **grid** when it is not powering the local site.

Like other cells, the basic design of the MCFC is rather simple. An anode, the liquid electrolyte held in a suitable matrix, a cathode, and an interconnect (bipolar plate) make up the individual cells and stacks are tied together to produce the necessary output. Figure 5-3 shows a typical MCFC configuration.

Like other high temperature cells, **internal reformation** of fuel gas is possible for these systems, precious metal catalysts may not be required, and combined heat

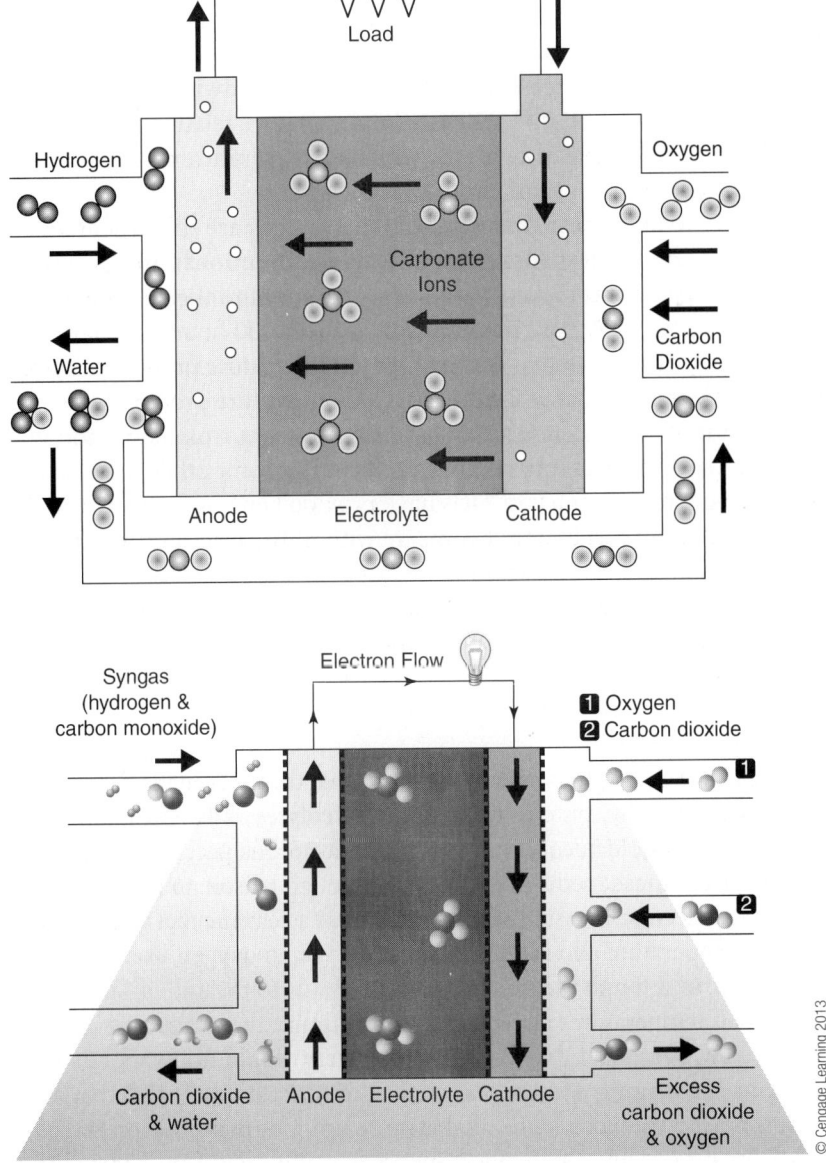

■ **Figure 5-3** Molten Carbonate Fuel Cell.

and power (CHP) systems are used to increase overall efficiency. The tradeoffs are also the same as other high temperature systems, higher materials cost, thermal expansion and contraction problems, corrosion, and the inability to be used in situations where cycling on and off is needed.

PAFC

Phosphoric Acid Fuel Cells (PAFC) came about following the success of the American space program in the 1950s and 1960s. Research funding for both small and large fuel cell systems was widely available while the moon determined the space program, and to some extent, foreign policy. PAFC systems were developed along with **steam reforming** of natural gas to provide fuel for fuel cells. Acid based systems such as SPFC (PEM) and PAFC developed partly in response to pure hydrogen being unavailable as a fuel outside the space program but the general availability of **hydrocarbon** fuels, which included trace contaminants other cells had difficulty dealing with. In particular, the acid based systems were tolerant of **carbon dioxide** (CO_2), one of the major contaminants of AFC systems then used in space applications.

PAFC systems are relatively advanced as far as being commercially available. They are used mainly in larger, **stationary installations** of up to 20 megawatts and have enjoyed reasonable success in many locations worldwide. Because they operate at lower temperatures (150°C–200°C), the systems are relatively simple and stable but must use precious metal catalysts, which are a significant percentage of their overall costs, in some cases as much as 20%.

Although PAFC systems are commercially available, they are still expensive. In general, they produce electricity more efficiently than conventional **turbine based generators**, but the capital and fuel costs add a large amount to their overall price. Systems in operations typically have lifetimes well over 40,000 hours and have run continuously for up to 100,000 hours, putting them in the mature product category. It is important to understand that while fuel cells are not mature products, they are reasonably well understood and reliable. What has kept them from being adopted over the years is not so much that they do not work as well as some other technologies but rather because hydrocarbon based fuels have provided energy to the world for over a century at a cost fuel cells cannot compete with. If the price of conventional power generation changes due to economic, environmental or cultural reasons, fuel cell systems may well enter the marketplace at a pace more like cell phones did at the end of the twentieth century than automobiles did at the end of the nineteenth.

SOFC

Solid Oxide Fuel Cell (SOFC) systems are unusual in that the fuel cell itself is entirely solid. AFC, MCFC, and PAFC systems use **liquid electrolytes** and SPFC (PEM) systems make use of water molecules within the solid but porous polymer electrolyte. All of these systems use these methods for a simple enough reason: to make the passage of ions between electrodes easier. SOFC systems use a ceramic that when heated to a high enough temperature allows specific ions (O^{-2} or an oxygen atom with two extra electrons) to pass through the solid from the cathode to the anode. SOFC systems operate at high temperatures not so much to maximize the production of electrons (electricity) but because if they did not, the system would not work at all. As mentioned before, the **ceramic electrolyte** will not conduct unless it is at temperature. Figure 5-4 shows a view of a SOFC that is more up close than previous images, with the molecules drawn out and the open spaces that allow gas flow evident. This type of structure is more like what all fuel cell assemblies look like on a small scale.

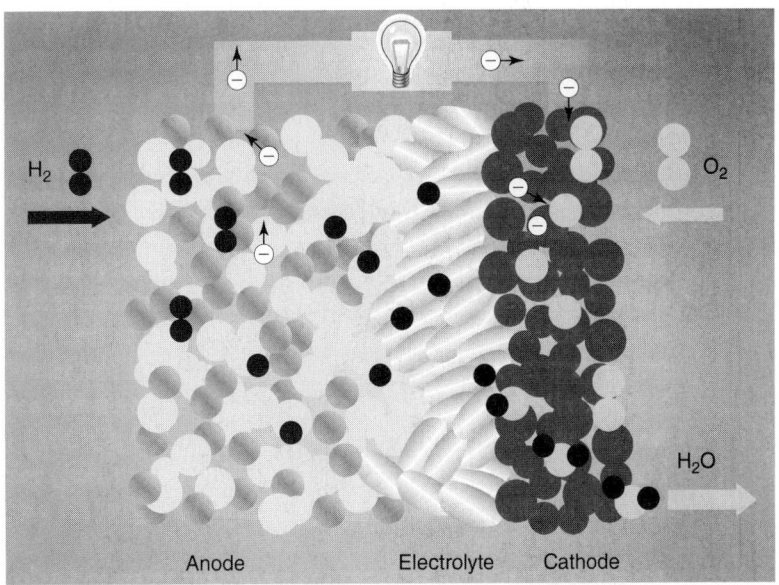

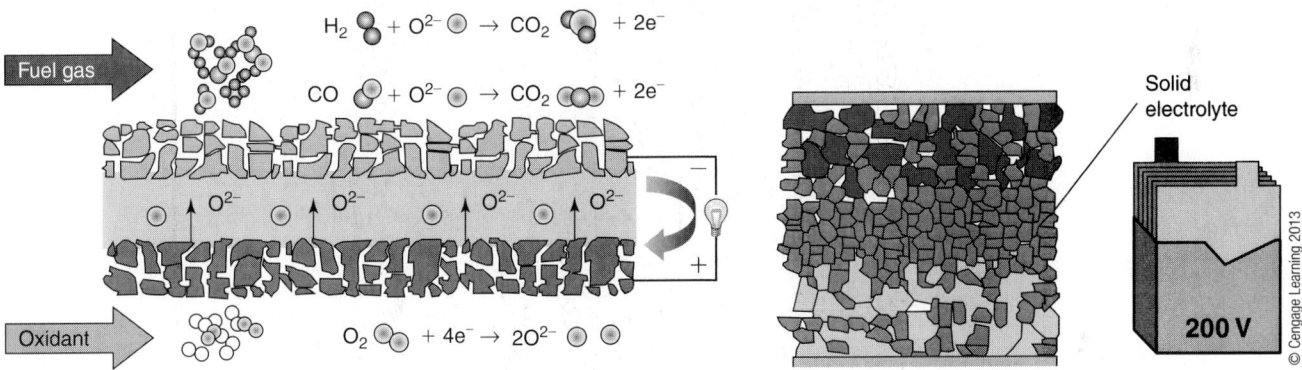

■ **Figure 5-4** Solid Oxide Fuel Cell.

Because the cells are solid, they can be fabricated relatively cheaply using the same type of techniques used in industries from fine china to printing T-shirts. The two major types of SOFC systems, **planar** (flat) and **tubular**, use different methods of fabrication. Tubular designs generally **extrude** one of the three central parts (anode, cathode, or electrolyte), then add the other two layers perhaps by dipping the tube in a **slurry** or **spray casting** the other layers. Figure 5-5 shows a typical method for high temperature spray casting.

The most common methods generally use the cathode as the underlying layer since it is thicker than the anode, then add the electrolyte layer, and finally the anode layer. Planar designs may use a modified version of this where the extrusion is flat instead of a tube. **Ceramics** are relatively weak structures until they are sintered at high temperatures to bind the molecules together, so **staged sintering methods** may be used. Extruded tubes would be sintered to provide an underlying layer that can support the ceramic slurries that are added one at a time afterwards. These are generally referred to as supported structures, so a **cathode supported** Solid Oxide Fuel Cell would first form the cathode, sinter it to provide a strong support, and then add the rest of the cell. Anode, cathode, and electrolyte supported cells have all been produced, and even interconnect (bipolar) supported cells have been used. Figure 5-6 shows a typical tape (slurry) casting method.

As mentioned in previous chapters, these cells are very thin because the ceramics have difficulty conducting the ions fast enough once past a certain thickness.

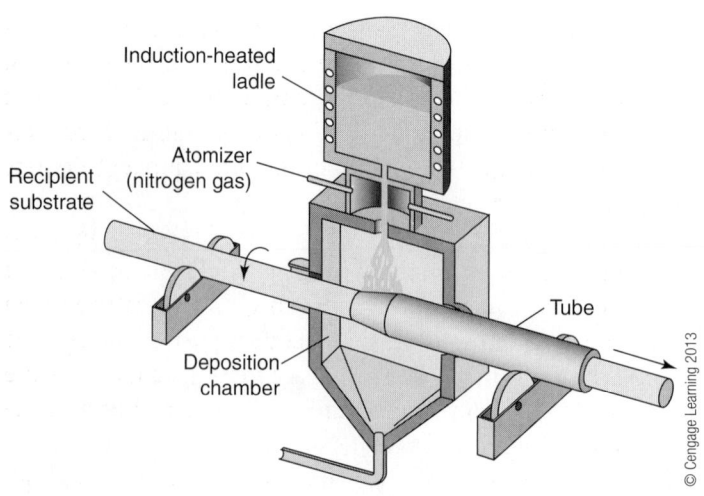

Figure 5-5 Spray casting.

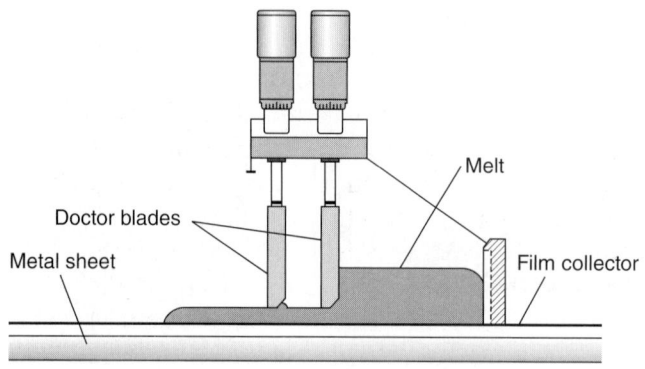

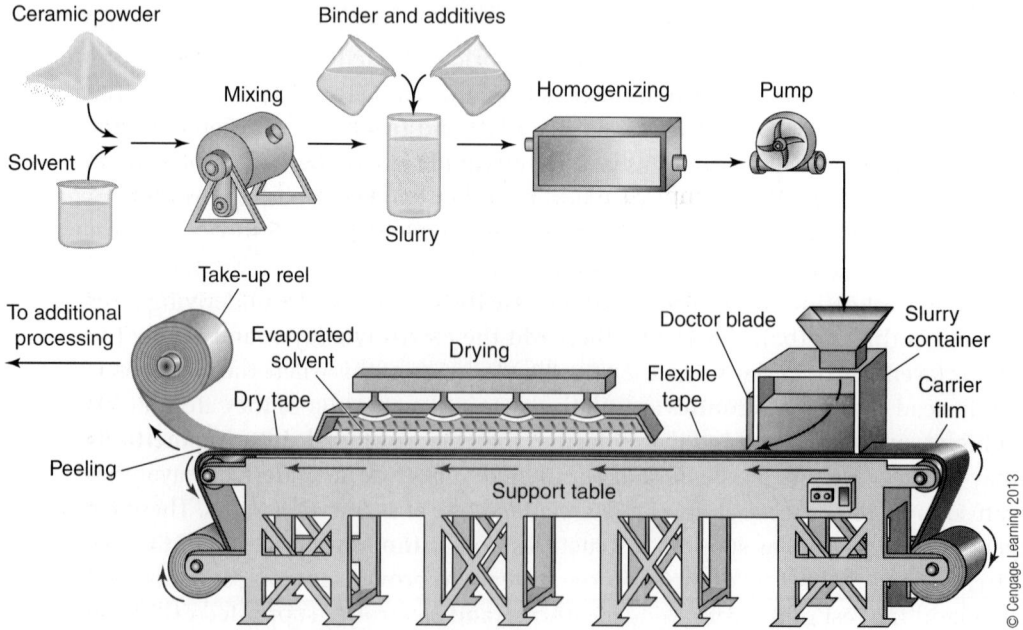

Figure 5-6 Tape (slurry) casting.

The optimum thicknesses vary with the proprietary ceramic mixtures, but a sandwich of electrolyte and electrodes will be less than several hundred micrometers. The thicker the sandwich is, the higher the operating temperature will have to be to obtain the sort of conducting speed required. SOFC systems split the ohmic resistance at about 45% occurring in the cathode, 25% in the interconnect (bipolar) plates, 18% in the anode, and 12% in the electrolyte.

SOFC systems run at around 1000°C or less (by a few hundred degrees) depending on the design used. This causes problems in design and construction that lower temperature operations do not see, ranging from **high temperature creep** (the atoms in solid materials move as if they were in a **viscous liquid**) to the ceramic parts of the cell reactivating the sintering processes so the material changes while in service. Chromium in particular is a problem in SOFC systems because the atoms can migrate through the solid material and it can vaporize in localized areas if the right pressure and temperature occur. Perhaps the most critical issue in high temperature operations is in making sure the **thermal expansions coefficients** match for all the parts. Thermal expansion occurs when something heats up (most things expand when heated and shrink when cooled with water being the most notable exception). The basic reason behind this is because atoms do not ever stay still but rather vibrate constantly since they are so small that even the slightest change in energy (from temperature or pressure or other atoms) can affect them. The more energy they are subjected to (such as raising the temperature), the more they vibrate. The more they vibrate the more volume they occupy; since they are all tied together by atomic bonds, they cannot go too far when they vibrate or the bonds will break (this is how melting and finally vaporization occurs). Occupying more volume is another way of saying they are bigger than they were before, which is called thermal expansion. This is a very important issue in many things, from jet engines to fuel cells, because if several things are confined in a space and the temperature goes up, they will all grow. In a SOFC stack, the interconnect (bipolar) plates confine the solid electrodes and electrolyte. If the electrodes and electrolyte grow more than the interconnects as the temperature raises, they will push against the interconnects. Because they are ceramics, they are not particularly strong and will crack or even explode under extreme circumstances. All high temperature fuel cells must match materials to keep thermal expansion from destroying the cell.

SOFC systems also have particular gas flow designs that differ from other fuel cells. Since the cells are solid and operate at high temperatures, they must seal not only when the operating temperatures are reached but also as the cell warms up. Seals are one of the critical parts of these cells and glass is typically used (many times Pyrex®), since glass structures do not really melt but rather flow like lava in scenes of Hawaiian volcanoes. They flow as they warm up and eventually seal in place. If the cell has to be turned off and it cools, the seals are set and as long as care is taken, they retain reasonably good sealing capabilities until they are reheated to fully seal again.

Gas flow in SOFC is much like other cells, coming in from opposing sides of the cell to feed the anode and cathode in a **planar design**. **Tubular designs** (the interiors can be round or ovals or triangles or any other shape) require different flow patterns. Generally, the fuel is fed to the exterior of the tube and the oxidant (air) is fed to the interior of the tube because that is where the anodes and cathode are. The solid oxide electrolyte separates the two electrodes as a layer between the inner cathode and outer anode. This usually means that in a tube bundle (they are not stacked) the fuel is fed to a central chamber containing all the tubes, and so that chamber has to be airtight. The oxidant (air) is fed to each tube using plenums (manifolds). The reactions happen on the inner and outer tube surfaces with one end of the system sealed. As with most fuel cells, excess fuel carries the final products out of the main sealed chamber.

Fuel cells use a variety of **flow patterns**, particularly when interconnects or bipolar plates in a stack of fuel cells are flowing gas to an anode on one side and a cathode on the reverse side. The **flow channels** in the plates can feed one electrode using vertical passages while the channels on the opposite side of the plate can use horizontal flow channels, as shown in Figure 5-7. They can also both be either

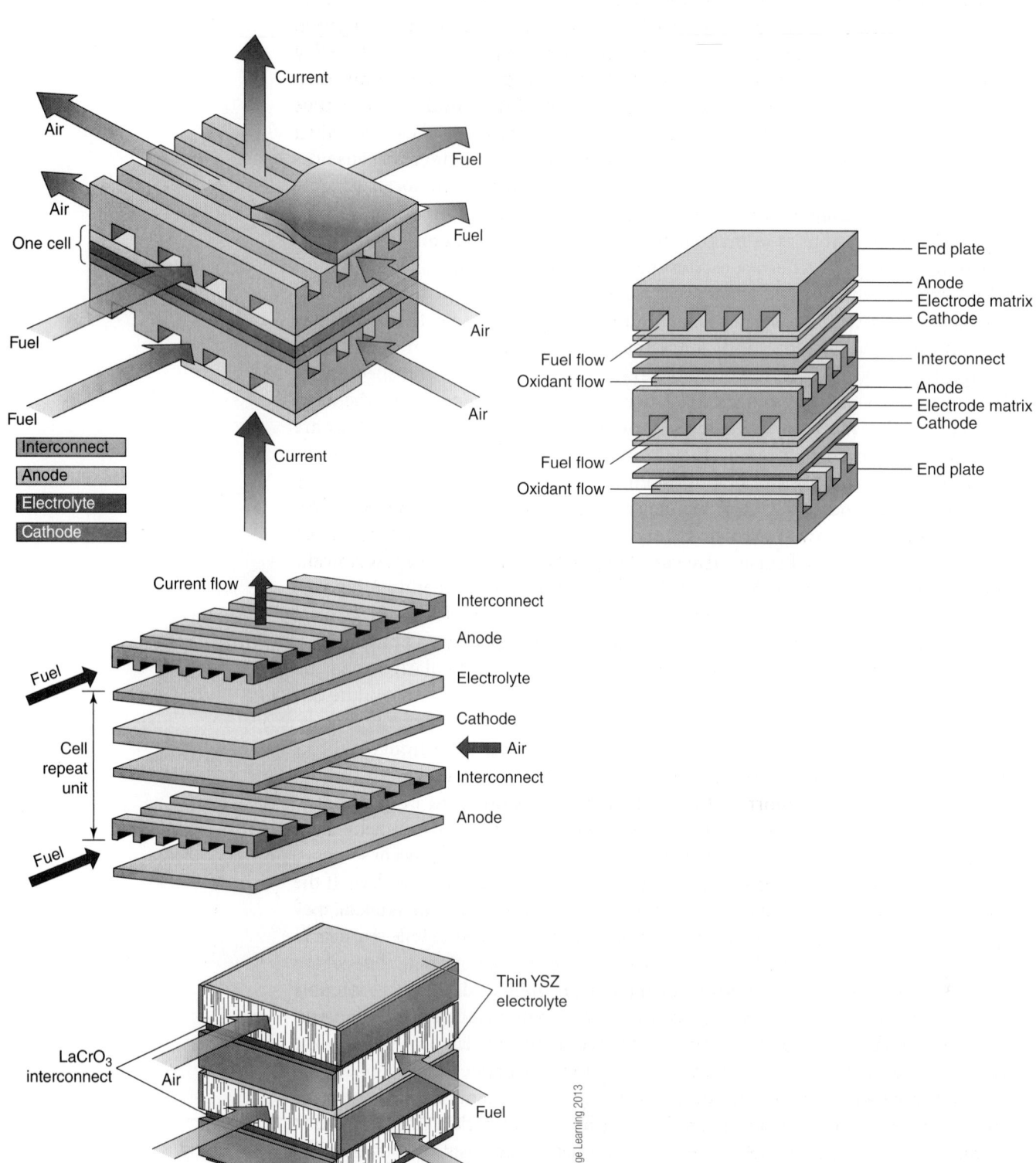

■ **Figure 5-7** Solid Oxide Fuel Cell flow channels.

horizontal or vertical channels. Gas itself can flow the same direction on both sides of the **interconnect (bipolar) plates** or flow countercurrent on the two sides or flow one from top to bottom while the other is flowing from right to left. Designs are yet matured enough to have the same basic characteristics, so virtually any combination of channels or flow direction might be seen as designers attempt to balance feed rates, cooling requirements, and humidification requirements as well as a number of other issues. Even the shapes of the channels can differ, being rounded in a corrugated design or rectangular in a planar design. The only common issue is generally that the gas must reach the electrodes in uniform volumes and pressures to avoid surging the reactants (some catalysts sites have reactions occurring while others do not). This creates **hot spots**, which in turn causes substantial problems in the cell as a whole, and is a concern in all fuel cells.

The flow of gas within the cell is important, but so too is the gas itself. Because SOFC systems operate at such high temperatures and can use both hydrogen (H_2) and carbon monoxide (CO) as a fuel, they are less prone to contamination and poisoning than some other fuel cell types but molecules like ammonia (NH_3), hydrogen chloride (HCl), and hydrogen sulfide (H_2S) will cause problems. All fuel cells make use of gas cleaning systems whether they use a hydrogen feed (external reformation of fuel) or whether the cell itself creates the fuel used at the anode from a feed gas (internal reformation).

ELECTROLYTES

Fuel cells have names and as discussed, those names come from the different electrolytes between anode and cathode. One way or another, the ions (atoms which have lost electrons) have to move between electrodes but the electrons cannot, Electrons have to be created at the anode and then travel through the wiring system to the cathode so that electricity is created. Fuel cells use pretty standard reactions at the anode and cathode with some variation, but the way ions pass in the electrolytes is different for each fuel cell.

SPFC (PEM)

We will start with SPFC systems keeping in mind that much of what is important to them is just as important to the other types of systems as well. Since the main difference between fuel cells is the electrolyte in use, that topic will begin the discussion.

The design starts with the solid polymer (the polymer electrode membrane). As mentioned, the most common polymer in use is a modified **Teflon®** known as **Nafion®** manufactured by DuPont. The membrane is made from **polyethylene** polymer stock. **Ethylene** is two carbons bonded together, each with two hydrogen atoms bonded to each carbon as well. A large number of these (poly) are then bonded together, forming a long **chain** of hundreds and even millions of carbon atoms. Every carbon in the chain is bonded to two other carbon atoms (one to the left and one to the right) to form the chain. Each carbon is also bonded with one hydrogen atom above and one below (left and right and above and below are relative terms but they used here to give the student an idea of atomic arrangements). The carbon atoms each have four bonds, a hallmark of carbon atoms. The polyethylene can then be changed by removing the hydrogen atoms bonded to each carbon atom in the chain and replacing them with fluorine atoms. This replacement produces what is called **Polytetrafluoroethylene** (PTFE), more commonly called Teflon®. The Nafion® is then made from Teflon® by

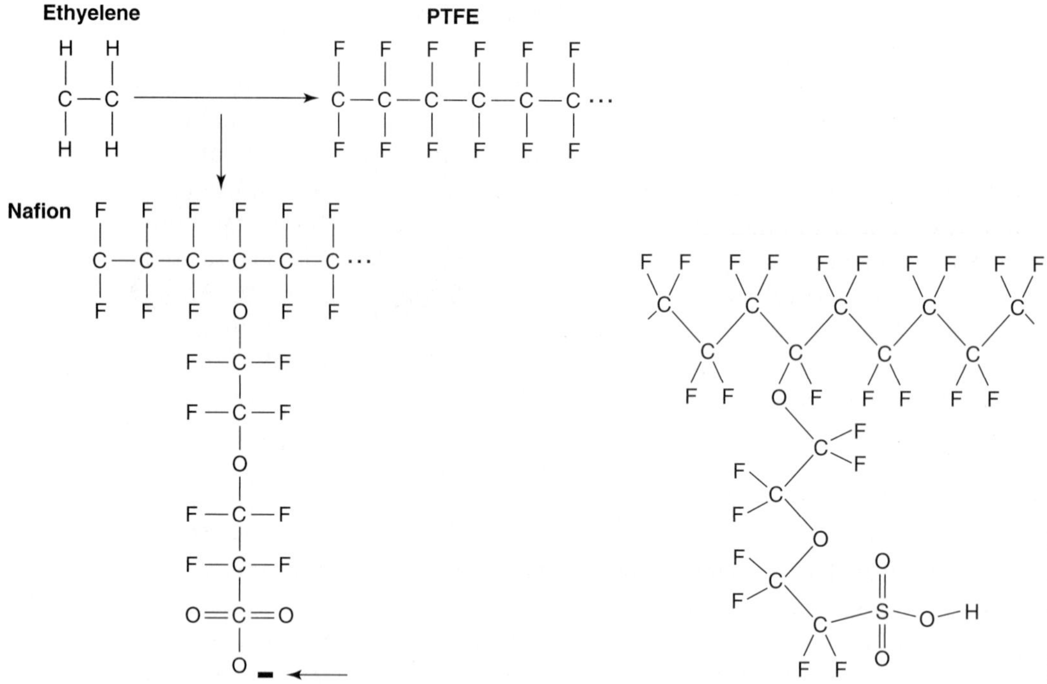

Figure 5-8 Ethylene to Nafion®.

adding what is called a **pendant molecule** with carbon, oxygen, and fluorine atoms in the middle of the PTFE chain. The various molecules are shown in Figure 5-8.

Notice the arrow in the bottom right of the figure. The Nafion® molecule is negatively charged, which is to say it has one excess electron in the very large polymer molecule. This means that for it to be stable, it must have a positively charged atom or molecule (called a **counter-ion**) nearby to ensure an overall neutral charge. This is one of the reasons why it passes the **proton** (H^+) so well in the PEM cells. The proton (H^+) goes into solution and causes a build-up of **positively charged particles** which in turn causes those particles to try to move away from each other and go to a place that has less positively charged particles, or perhaps more **negatively charged particles** (the effect is the same). What happens is much like marbles moving down a tube. Push a new one in one end of the tube, and an old one pops out the other end of the tube. The Nafion® polymer is like the tube, with the negatively charged pendant ends being the wall. As new protons (H^+) enter, they force the line to move, pushing the far end proton (H^+) to where the electrons and the oxygen are. The electrons and oxygen are at the cathode and they react to form water. Those protons (H^+) are then used up in the creation of water, causing a never-ending need for more protons to balance the negatively charged electrons arriving at the cathode through the wiring system.

That is the simple view. For the more complex view, we will start with the proton (H^+). Hydrogen protons do not actually exist when they are in water. Instead, the hydrogen atom (without its one electron) hooks up with a water molecule to form H_3O^+ (**hydronium ion**), seen in Figure 5-9 (note that the full molecule is positively charged).

When this structure forms, it is called the **protonation of water** (because of course it is adding a proton to a water molecule). This happens because the water molecule itself is what is called a **dipole**, where its electric charge is out of balance and the electrons are more likely to be near the oxygen atom in the molecule than near the hydrogen atoms in the molecule (resulting in a molecule that bends the

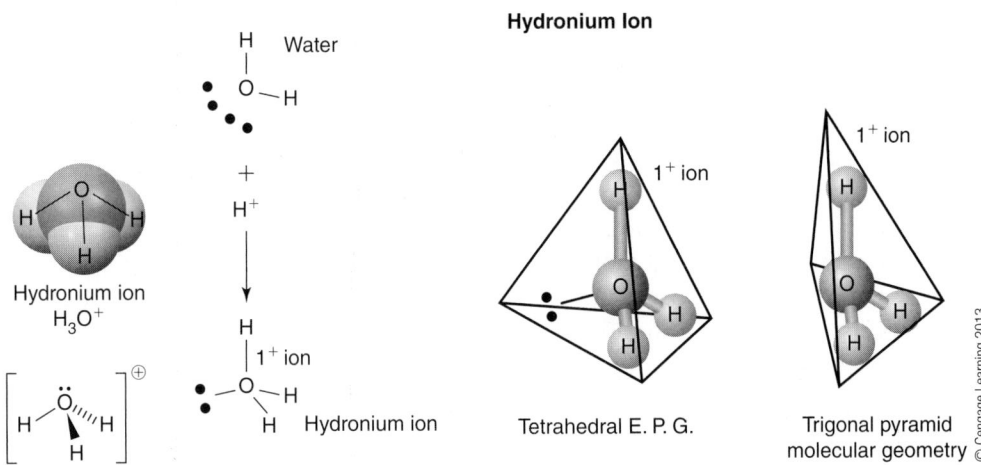

Figure 5-9 The hydronium ion.

two end hydrogen atoms down from the central oxygen atom so the ends are a bit positive and the center a bit negative). The positively charged proton (H^+) can then come in and form a weak bond with the (kind of) negatively charged oxygen part of the water molecule. This is important because the hydronium ion (H_3O^+) behaves much like the water molecules around it when it moves from the anode to the cathode. Remember the positively charged ions like hydronium move away from the anode and to the cathode. The flow of the hydronium ion (H_3O^+) moves the proton (H^+) required for the fuel cell to work, but it also tends to drag the water molecules with it. This is an important concept in any fuel cell where water is present; it is called **electro-osmotic drag**. It occurs because water is a dipole and when the charged hydronium ion (H_3O^+) moves, the dipole water molecules follow the charged ion like a line of kids in follow-the-leader. This is much like iron powder moving along with a magnet. In general, each hydronium ion (H_3O^+) can drag a number of water molecules with it, contributing to the flow of water within the membrane.

The membrane itself is a bit more complex as well because there are many things in the world that work, we just are not too sure exactly how or in some cases even why. Nafion® is one of those things. Nafion® is a version of a common polymer used in many products. The backbone of the polymer is polyethylene, two carbon atoms and four hydrogen atoms combining to form a molecule and then thousands or millions of those molecules bonded together in a long line. Polyethylene is one of the most common polymers (plastics) in use today. When the fluorine is added, the resulting PTFE is durable and resists being attacked by other chemicals because both the carbon to carbon chemical bond and the carbon to fluorine chemical bonds are very stable. **Chemical attack** is when an element (or another molecule) comes in contact with a molecule. If the contacting chemical is more likely to form a chemical bond with one of the other elements in the original molecule than an atom already in the molecule then chemical attack occurs and atoms switch places. In that case, all the species involved are changed in one way or another. Because the carbon to carbon and carbon to fluorine chemical bonds are stable, few other chemical elements can disrupt them. The PTFE portion of Nafion® is **hydrophobic**, so it repels water, and thus keeps water moving within the main structure of the polymer electrode. When the side (pendant) groups (**perfluorovinyl ethers** in case you are wondering) are added, they serve two functions, the first being to provide a charged area and the second to provide a **hydrophilic** (attracts water) area. The combination of the hydrophilic area in the side group area and the hydrophobic areas around the backbone provide a channeling arrangement so water can pass but must stay where it can carry the proton along with it. When the water with the hydronium ion (H_3O^+) comes near the end of the charged pendant molecule, the proton (H^+) can transfer to the negatively charged pendant. The proton (H^+) moves across the polymer electrode in a series of jumps from water to a charged pendant and back to water and to the next charged pendant. For that reason, the ability to conduct protons (and thus the amount and consistency of current flow) is proportional to the water content in the polymer membrane and to the number of pendant molecules.

The internal structure of the Nafion® seems simple enough, but is not. The tangle of backbone/pendant Nafion® molecules, which are very large as far as molecules are concerned, keep the water moving and also serves to block the movement of larger ions while allowing the smaller ions like protons (H^+) to move. The ways in which the Nafion® accomplishes these things remain in doubt however. The way the channels are arranged so they can pass water when one part of the large molecule repels water while another part attracts water is still being argued. There are several models, but the one most likely at the moment has the hydrophobic backbone arranged consistently in a lengthwise manner in order for the water to stay as far away from it as is possible when all the backbones are taken into account. The hydrophilic pendants will also twist the backbones so that they are arranged within the water. Keep in mind that this only works if there is the correct amount of water in the polymer membrane, neither too much nor too little. The hydrophilic pendants then form very small channels, probably in the form of rods but this is still in question. Because water is a dipole, formed when the electrons shift toward the oxygen in the H_2O molecule, the water will align itself with the positively charged hydrogen ends of the dipole toward the negatively charged end of the hydrophilic pendant molecule, making the channel even more pronounced. When a hydronium ion (H_3O^+) moves into this, it disrupts the arrangement since it is very positively charged while the water dipole molecule is less positively charged. This movement probably helps to pump the water down the channel, with the water flipping position as a hydronium ion (H_3O^+) comes near, then the hydronium ion (H_3O^+) giving up its proton (H^+) and going back to being just water (H_2O), which in turn flips to align its dipole. If too many hydronium ions (H_3O^+) build up, the charge distribution will be disturbed in the system. This will cause flow as well as more protons (H^+) moving to the pendant molecules as the charge balance tries to come back into balance. That in turn causes more flipping of water molecules, and the whole dance starts again. Perhaps you begin to see just why some of these processes, while working fine in reality, are hard to explain in theory. Water must also be balanced within the cell as a whole, further complicating the matter as will be discussed in a following section.

There are other polymers being investigated to replace Nafion® but none yet currently available in bulk. These include polymer-zeolite nanocomposite, other perfluorinated polymers but with sulfonic acid pendant groups that are closer together, sulfonated polyphosphazene-based membranes, trifluorostyrene and substituted trifluorostyrene copolymeric compositions, and phosphoric acid-doped poly(bisbenzoxazole) high temperature ion-conducting membranes. Plastics might seem pretty simple when you twist open a bottle of soda, but they are not simple and their names can give students nightmares. Since the membranes are the heart of Solid Polymer (PEM) Fuel Cells, considerable work is being done to come up with options for manufacturers. Given the nature of the still emerging industry, it seems reasonable to assume that the student may encounter a number of different models using different membranes.

AFC

AFC systems can operate at temperatures ranging from around 100°C to 250°C depending on the concentration of the **alkaline electrolyte** used. Acid and base (alkaline) are terms meaning several different things in chemistry. In this case, when we use alkaline, we mean water with excess hydroxide ions (OH^-). The higher the concentration of the alkaline (base), the higher the temperature the cell can operate at so that an electrolyte that is 85% potassium hydroxide (KOH) and 15% water will run at over 200°C while the 100°C range systems generally use from 35% to 50% potassium hydroxide (KOH) in water. The electrolyte is contained within a matrix such as asbestos to keep it in place.

Alkaline is another word for base, as in acid versus base. Recall that a base in this case is considered to be the compliment to an acid in water. An acid breaks up to add more hydrogen to the water (forming the hydronium molecule of H_3O^+), while a base (or alkali) adds the OH^- molecule to the water. They are complimentary in a water system since one water molecule (H_2O) can break up to add one free hydrogen proton (H^+) and one free hydroxide molecule (OH^-). If an extra hydrogen proton (H^+) is added, then the balance swings toward hydrogen and the water becomes acidic, while if an extra hydroxide (OH^-) molecule is added, the balance swings toward the basic (alkali) side. If the water molecule breaks up and adds one of each, then they cancel each other out. As mentioned, the electrolyte used in AFC systems is **potassium hydroxide** (KOH), which goes into solution with water as K^+ and OH^-, adding more of the hydroxide to the solution to make in more basic (alkaline); the K^+ does not alter the acid/base characteristics (pH). Remember also that the water molecule is a dipole, so the positive potassium atom will align with the center of the water molecule where the oxygen is a bit negatively charged while the hydroxide ion will align with the two ends where the hydrogen is a bit positively charged. This forms a nice stable arrangement in water. While there are other definition of acids and bases (alkali), the basic definition of excess hydrogen or hydroxide generally works well in describing fuel cells.

In space, the AFC electrolyte was sealed, just as electrolytes are sealed in other fuel cells. This worked well enough in space, where operating lives could be in the several thousand hours, but as mentioned above, for terrestrial commercial applications, substantially longer operating lifetimes were needed. The lifetime was limited in part because the potassium hydroxide (KOH) electrolyte reacts with any carbon dioxide (CO_2) in the air used at the cathode to form **carbonates** (CO_3), as mentioned in previous chapters.

$$2KOH + CO_2 \leftrightarrow K_2CO_3 + H_2O \qquad \text{Eq. 7}$$

In space, this was not an issue because they did not use air; they used pure oxygen which contained no carbon dioxide (CO_2). That is too expensive for commercial uses however, and something had to be done. There are few options for cleaning gases of **trace amounts** of other gases, and in fact this is one of the ongoing issues with commercialization of all types of fuel cells. Some of the contaminants causing trouble in fuel cells require less than one molecule out of one hundred thousand molecules be present (ten **parts per million**). One can visualize how difficult it is to achieve levels like this if you think of trying to find one light blue marble in a dump truck having one hundred thousand deep blue marbles and to do this day in and day out over the 50,000 hour lifetime of a fuel cell. AFC system developers decided that since the potassium hydroxide (KOH) electrolyte was a liquid at a reasonable temperature, it was easier to just pump the liquid out once carbonates formed and replace it with new liquid. Since KOH liquid is inexpensive, it is a relatively cheap fix to the problem as well.

AFC systems then are the only ones that replace the electrolyte on a regular basis without having to change the electrodes out. The actual electrical producing parts of the fuel cells themselves, the anode, electrolyte and cathode (the **membrane electrode assembly**) may only last for tens of thousands of hours, but that does not mean the unit is scrapped along with the spent membrane electrode assembly like a battery is thrown away. Most fuel cell systems are modular in design so the fuel cell assembly can be removed from the associated (**balance of plant**) items and replaced, and then the system brought back on line. In AFC systems, the electrolyte is constantly removed from the system, leaving the anode and cathode in place and operating. Once the anode and cathode degrade, they too can be replaced just as in other fuel cell systems.

This does require careful design. Remember that for fuel cells to work a **three-phase area** must exist at both the anode and the cathode. The catalyst, the required chemical reactants as well as the electrons and the ions must all be together in order for electricity to be produced. Recycling an electrolyte liquid must be done in such a way that the porous gas diffusion layer supplying either fuel or oxygen is not blocked by excess liquid flow or lack of liquid, the electrolyte is at the catalyst site for long enough that the ion moves into the electrolyte and finally that the electrolyte does not move out of the system before the ion can get from one electrode to another.

The circulating electrolyte simplifies the general design and limits cost in other ways as well. Because the system essentially goes dormant when no current draw occurs, even the AFC systems with limited operational life span can last long enough for many commercial uses. In a car, for instance, the engine does not last for 10 years, it lasts only several thousand hours since it does not run most of the time. The car lasts for 10 years and the engine does not wear out in that time because it is not actually running. Replace the internal combustion engine with an AFC system, and little difference would be seen. Put a high temperature fuel cell in a car, and the time to power up the cell is unacceptable for people having to get to work on time. The recirculating electrolyte can even assist in start-up operations since it can be heated outside of the cell and used to shorten the time-to-operating-temperature even further. The critical internal water amounts can also be controlled using the recirculating electrolyte as can the fuel cell cooling cycles, eliminating both of those secondary system requirements and lowering the initial capital cost. Remember from Chapter 4 that an alkaline cell produces more water at the anode than is consumed at the cathode so water regulation is an issue. By taking out the electrolyte, the water can be removed in an associated **dehumidification** circuit. The hydroxide (OH^-) concentration within the cell can be controlled and since that is a critical issue (recall from Chapter 4 that the anode reaction has hydrogen and hydroxide forming water), constant concentrations will stabilize and improve system operations. Since even low temperature operations tend to occur at temperatures above the boiling point of water, gas (steam) can develop inside the cell as well as unreacted fuel gas and air building up to some extent. Any gas bubbles interfere with the three-phase requirement of a fuel cell and recirculating the electrolyte removes this gas. Finally, a circulating electrolyte acts as a very efficient sealing system.

Because the AFC systems operate at lower temperatures, the materials used for the system are determined by the chemical properties of the electrolyte rather than the operating temperatures of the fuel cell. Potassium hydroxide (KOH) is corrosive, especially in the higher percentages used in higher operating temperatures but the cost of materials needed to deal with such corrosives are reasonable as opposed to high temperature alloys. Potassium hydroxide (KOH) also has a low **surface tension**, meaning that it will flow over most surfaces relatively easily rather than beading up and stopping. This means it can leak out of even very small cracks or holes that other liquids will not flow through, causing a number of problems ranging from safety issues for personnel to short circuiting of the unit, so that sealing schemes for piping and pumping are generally more comprehensive than some other fuel cells require. Extra care must also be taken at the cathode where hydroxide ions are produced (see Equation 6 in Chapter 4) since that area will have more of these ions and thus be more corrosive as well as having different flow characteristic than the electrolyte at the anode, where the hydroxide ions are being consumed.

One final issue with liquid electrolytes should be mentioned. When a fuel cell stack is used with a recirculating electrolyte, the liquid travels a path that in essence joins all the stacks together into a conducting circuit. Since the electrolyte is specifically chosen to conduct the hydroxide ion (OH^-) and all the stacks are connected, the ions will tend to go toward whichever place that has the lowest concentration of ions. That should be from the cathode across one cell to the complimentary anode, but circumstances will arise where that is not true, and then the ions travel along with the electrolyte rather than from cathode to anode. This has a serious impact on stack performance. The longest possible paths between cells in the stacks and between stacks are generally designed in to try and limit this behavior and the paths are made as narrow as possible to keep the ions moving within a single cell rather than travel between cells. Stacks can also be wired first in **series** and then in **parallel** to limit the internal voltages and available paths. A 12 stack system can then be 6 stacks wired in series and then the two stacks of 6 wired in parallel to make the 12 stack system.

MCFC

MCFC systems use a different type of control to keep the molten carbonate electrolyte material from the electrodes. The high operating temperatures and corrosive characteristics of the carbonate results in a situation where few if any materials will perform well for extended periods of time in the way Nafion® does in a PEM cell or PTFE does in a PAFC. Instead of a controlling material being used to maintain the electrolyte balance within the cell, MCFC cells use variable porosity. As mentioned in previous chapters, the electrolyte is contained in a porous, ceramic matrix, usually made of fibers. The ceramic most commonly used is $LiOAlO_2$, a mix of lithium, oxygen, and aluminium. This ceramic material serves to contain the liquid without adding any control features, such as those found in Nafion®. Instead, the electrolyte ceramic is sintered so that it contains very fine pores while the electrodes are sintered so that they contain much larger pores. The sizes of the pores used as well as their ratios are generally considered proprietary. The smaller pores contain the electrolyte and when they open up to the larger pores, liquid stays within the smaller pore, helped by capillary action and the pressures within the cell itself. MCFC also use a complimentary physical principle to keep the liquid electrolyte where it belongs by using catalysts at the anode that the electrolyte will not **wet**, that is to say that the liquid will not contact the solid to any great extent or for very long, moving off it as quickly as possible. Nickel is one such metal and using it allows the cell gas to contact the catalyst since the electrolyte avoids it.

This unique issue means MCFC systems experience problems other cells do not. The carbonates tend to migrate during operation in part because of the electric charge and ultimately, one end of a stack will have excess electrolyte while the other end will not have enough, shutting the stack down. The smaller pores experience corrosion, especially in the area where the pore sizes change and can be plugged by contaminants among other things. These problems cause the molten carbonate to shift within its ceramic matrix, affecting the internal resistance of the cell. This is a problem in MCFC systems since the electrolyte is the single largest contributor to the ohmic resistance, up to 70% in some products. Changes in electrolyte distribution during operation usually increase resistance, lowering power output.

Since the resistance of the electrolyte increases with thickness, one of the major efforts in MCFC research has been to decrease the thickness. Initial methods used pressed fibreboards that were up to 0.2 cm in thickness but newer methods using ceramic **tape casting** processes now produce electrolytes of less than 0.25 mm in thickness. Keep in mind that thinner is not always better, since thinner electrolytes will degrade faster and contribute to an increased rate of carbonate evaporation, one of the principal causes of cell failure in MCFC systems.

The usual methods to produce the electrolyte is to mix the $LiOAlO_2$ in a solvent along with various binders and flow agents to make a slurry that will spread but not run when cast on a smooth, moving surface. Particular care must be taken to ensure the ceramic powders used in the slurry are of uniform size. A blade moves excess material off to set the final thickness, and the binder and other agents are then baked out. The green ceramic pieces of anode, cathode, and electrolyte (the electrodes can be made using similar methods) are then assembled into a stack. These methods of manufacturing have reduced the ohmic loss through the electrolyte by a substantial amount and now allow fuel cells of up to a square meter to be manufactured. As mentioned before, stack fabrication is not completed until the MCFC system is heated for its initial start-up. At that time, not only are the seals set, but the final electrolyte finishing is done, with the molten carbonates distributing through the electrolyte structure.

PAFC

Phosphoric acid (H_3PO_4) was chosen not only because acid systems tolerate contaminants reasonably well, but because it is stable at the operating temperatures and pressures used in low temperature fuel cells. It will not boil or break down, does not react with the other chemicals in the system (CO_2 does not poison it like in AFC systems), and will pass protons (H^+) between electrodes. Remember that acids generally consist of a mixture of water, water with an extra hydrogen atom attached (the hydronium ion), and the negatively charged molecule that remains once the acid has lost the proton to water. Thus, hydrochloric acid will have a chloride ion remaining in the water once the HCl molecule has broken up. Most acids free up protons and thus qualify as fuel cell electrolytes where an ion must pass between electrodes, but none of the other acids are stable enough under the particular circumstances present in fuel cell operations. One of the most important parts to remember about liquid based fuel cells like PAFC or MCFC systems is that the electrolyte must remain a liquid. A PAFC cell will freeze if the internal temperature of the cell falls below about 42°C (108°F). Going from a liquid to a solid during the freezing process in most cases (but not all with water being a notable exception) causes a decrease in volume. The electrolyte material will pull back, causing gas to move in and then push back when it melts back into a liquid, causing internal stress and interrupting ion movement in areas where the gas pockets are. More importantly, the reactions stop. Remember that chemical reactions have activation energies they must overcome in order to get going. Stopping and starting these reactions require energy as does reheating the cell to melt the electrolyte and get to the operating temperature. This damages the cell and consumes energy.

Originally, the acid was diluted in water because the material inside the cell could not stand up to anything more concentrated. Materials research done in the late 1960s and early 1970s produced a holding matrix capable of withstanding more and more concentrated acids so that pure phosphoric acids are used as electrolytes, increasing the efficiency of the cells. The acid is now held in a matrix of

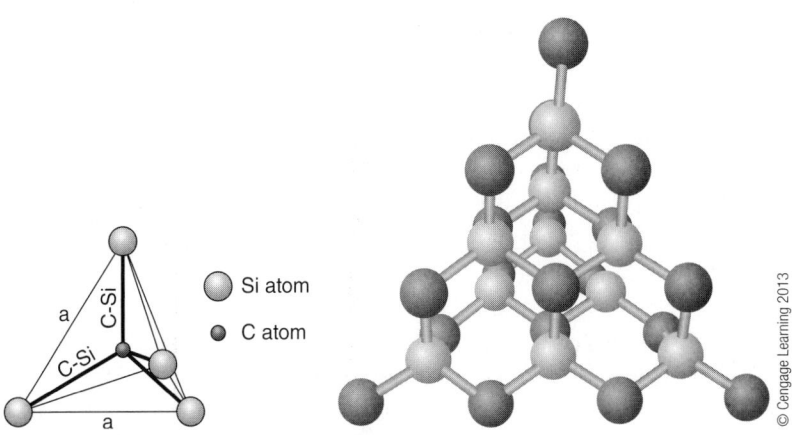

Figure 5-10 Silicon carbide.

silicon carbide (SiC) bound by a PTFE polymer. These are made of small particles (1 micron) bonded together with the PTFE and are made to be strong enough to hold together under the compressive forces on each cell when the stacks are assembled. The small particles are used in part to produce internal pores small enough to keep the liquid in place and thus prevent gas crossing over from one electrode to another, since anodes do not tolerate cathode gasses well nor do cathodes deal well with anode gasses. The SiC matrix is limited in how much internal pressure it can handle though, so PAFC systems run either at atmospheric pressure or a small internal pressure (less than 200 mbar). Figure 5-10 shows the SiC molecule. By mixing this structure in with PTFE, the less rigid PTFE is supported under comprehension by the SiC diamond.

Acid is also used up during operations. Unlike an AFC, electrolyte is not pumped through the system to replenish the supply and clear out contaminants. This is mostly because pumping pure phosphoric acid requires pumps and lines capable of withstanding the acid. Current PAFC systems generally use an electrolyte reservoir plate that holds enough in reserve to last the lifetime of the stack. Keep in mind that fuel cells are very thin. PAFC systems generally have electrolytes that are 0.1 mm to 0.2 mm thick so the reservoirs do not have to be very large. As in all fuel cells, the thin electrolytes keep resistance between anode and cathode small, increasing efficiency.

SOFC

SOFC electrolytes are ceramics that conduct ions at elevated temperatures. There are a number of different ceramics currently used or considered for use in the industry. The most common are **stabilized zirconia** structures, where alloying elements are set into the atomic matrix to keep the main zirconia structure from being degraded at the temperatures and pressures used in the cells. The term alloying atom is used to give the student a better idea of what it being done and the scale being used. In metals, alloying elements are added in large percentages as in stainless steel where as much as 40% of a metal is an alloying element. In applications like electronics and fuel cells, these percentages may be considerably lower, and the atoms must occupy particular positions in the atomic structure of the material; a method known as **doping**. These alloying atoms doped into the structure can also be used to improve the ionic conductivity of the electrolyte. Yttrium and scandium are two such alloying elements used. Other materials such as ceria with gadolinium alloying atoms or lanthanum gallate with strontium alloying atoms doped in are also know to work in the solid oxide cells. Figure 5-11 shows a yttrium stabilized zirconia molecule.

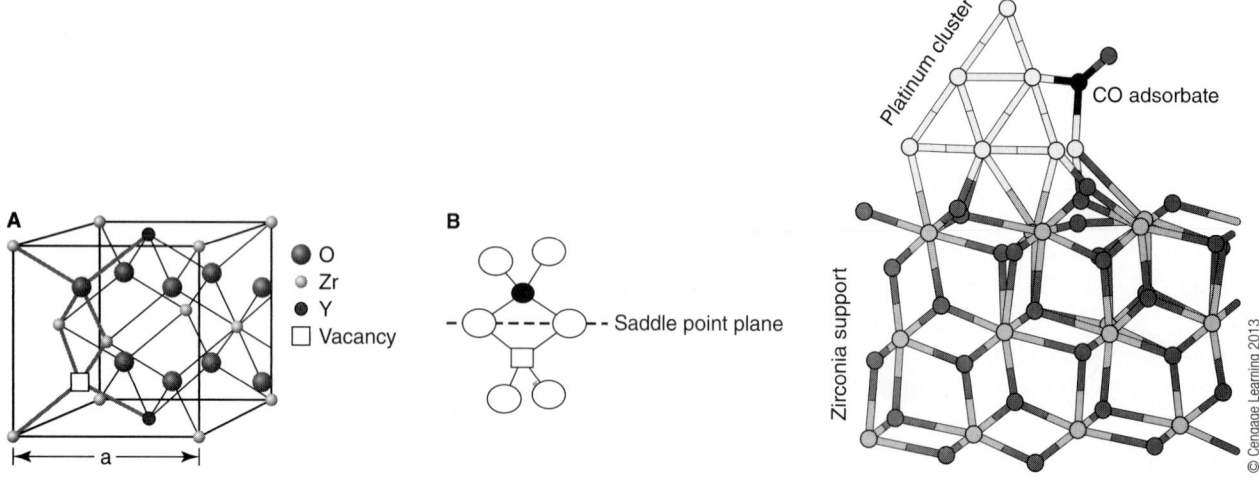

Figure 5-11 Yttrium stabilized zirconia molecule.

It is not so important to know the names of these elements or the differences between them as it is to understand that the atomic structures of ceramics are very complicated and that a considerable range of properties can be obtained by substituting certain atoms into the structures. While it is beyond the scope of the book to go into any great detail, doping is relatively simple in nature. An element that is close in size and structure to zirconia is introduced in a relatively small amount. In this case, the element ytrria takes the place of less than 10% of the zircon atoms in the zirconia structure where the Zr^{4+} (zircon that has lost four negative electrons so it is positively charged +4) atoms are replaced by Y^{3+} atoms. To balance the zirconia ceramic molecule (ceramics in this case meaning a metallic atom of Zr or Y combined with a nonmetallic atom of oxygen), two oxygen atoms must give up the four electrons Zr needs to balance back to zero (neutral charge). Oxygen easily gives up two and Zr easily takes four because of where they are in the periodic table. Yttria cannot take four electrons because of where it sits in the periodic table, it can only take three, and so two yttria atoms in the structure take six electrons from three oxygen atoms. When yttria replaces zircon in the structure then, for every two Zr replaced (had used eight electrons from four oxygen atoms to balance the charge), the two Y atoms that take their place (now only six electrons from three oxygen atoms are needed to balance the charges) force one oxygen atom out of the structure so that the whole package stays electrically neutral.

This type of doping produces what is called a **hole** and other oxygen atoms can then move into and out of these holes, explaining why zirconia can conduct oxygen ions. Moving oxygen ions does require high levels of energy be available to move everything around and that energy comes from the high temperature. Zirconia is not the only ceramic that conducts ions at high temperature so others will also conduct electrons to some extent as well. If the wrong circumstance develops within a cell, the anode and cathode ceramics can also conduct ions and that causes a short circuit in the cell.

The SOFC electrolytes themselves then are different than an AFC polymer where the hydroxide ions (OH^-) can move because the electrolyte is in fact composed of such ions and to pass them is somewhat like adding identical marbles into the top of a funnel while taking the same kind of marble out of the bottom. SOFC electrolytes are more like polymer electrolytes where ions move in a channel structure. In oxide cells oxygen ions (O^{-2} or oxygen with two extra electrons) are passed from one site to the next as they transfer from the cathode to anode. Ceramics are oxide materials (ceramics by definition contain oxygen atoms in the main ceramic structures) to start with, so they pass oxygen in a coordinated movement across the electrolyte in a manner somewhat like a bucket brigade. Buckets can pass from hand to hand because hands can hold buckets, but if it was tried with feet or heads passing the buckets it would not work very well. Nor will it work without buckets, since passing water using a hammer (which hands can also hold) won't work because hammers won't hold water. The holes and the ions moving through the network of holes have to match up. The oxygen atoms must pass in a consistent and predictable manner within the very complicated arrangement of atoms that make up a ceramic, and that is why only a few ceramics will work. Most ceramics will not pass oxygen ions in that manner.

There are thin film solid oxide cells and thick film cells. Thick films are those operating at higher temperatures of around 1000°C, whereas thin film cells operate at a couple of hundred degrees less. Thick films are along the order of 15 millimeters where thin films can be 10 times thinner (10–15 microns is the current limit to thickness using conventional, cheaper fabrication methods). There is a difference in how these cells are produced, with thick films being cheaper to produce but requiring better matches between electrolyte and electrode so chemical reactions and thermal expansion and contraction issues do not destroy the cell during processing. In general, thick films are used in systems where the systems are brought to operating temperatures slowly or perhaps operated nearly continually, whereas thinner films will better tolerate some operational (load) cycling and temperature changes.

One of the issues with SOFC systems then is in assuring that the electrolyte and the electrodes will match up when they are processed to form a solid structure. Remember that these are sintered at very high temperatures, some in excess of 1400°C, they are very thin (some layers are less than 15 microns thick) and the three are all different materials. When the materials are assembled, they are generally slurries of less than 50% solid material with the solid ceramic particles in the slurry having to be smaller than the ultimate thickness of the layer to be deposited and of very high quality. Five or ten micron particles are very small, very hard to produce, and very difficult to work with. Particles smaller than five microns are particularly difficult to work with.

One of the defining factors for oxide cells is the thickness of the three layer sandwich (anode-electrolyte-cathode). For the cell to create enough electricity, the total thickness of the electrolyte is determined by the material to be used and the subsequent temperature that will allow for enough ions to pass through that thickness. Then, processes that allow such thicknesses must be available and be able to produce a sandwich that is strong enough to handle cells stack assembly processes. All fuel cells have electrolyte issues, whether they be molten carbonate cells, where the cells are useless until they are first heated up to a temperature sufficient to melt the carbonate or solid oxide cells, where the brittle nature and thermal expansion differences of ceramics require specialized equipment to fabricate.

ELECTRODES

While electrolytes lend their name to the individual fuel cells, it is the electrodes that produce electricity because that is where the electrons journey start and end. What determines the nature of the electrode though is the chemical reaction that must occur there rather than how atoms or molecules move in space.

SPFC (PEM)

While the membranes act as transport mediums for the protons, PEM cell electrodes must perform many more functions. The electrodes must break electrons away from the fuel gas (hydrogen). They must act as electrical conductors for the electrons. They must allow the ions that are left once the fuel molecule has lost an electron access to the membrane that transports them to the cathode. They must provide the **surface area** needed for enough chemical reactions to occur that a usable current flow is created. Finally, they must be inexpensive enough to make the product cost reasonable. The point of the electrode is to provide a place where the gas fuel, the catalyst, the electron and ion pathways, and the water all come together in a regular and repeatable manner so that a steady flow of electricity occurs.

The **electrode assembly** usually consists of four primary parts, a porous membrane on the outer edge that distributes an even gas flow to the catalyst, a porous membrane that acts to support the catalyst, the catalyst, and finally the anode or cathode material itself. All these parts must also be electrically conductive so current can flow. Recall from Chapter 1 that two hydrogen gas molecules ($2H_2$) must be split into four atoms of hydrogen (4H), which then must be stripped of electrons to form four hydrogen ions ($4H^+$) and four electrons ($4e^-$). This cannot happen unless the hydrogen gas molecules can get to the catalysts on the anode. A **gas diffusion membrane** takes the incoming gas stream and distributes it so that an even supply of hydrogen gas molecules is provided to all the catalyst sites on the electrode. An even, steady supply of gas to an evenly dispersed supply of catalyst sites means an even, steady supply of electrons (electricity), but it also means that most if not all of the very expensive catalysts sites are being used to their capacity. In most cases, the gas diffusion layer and the SPFC (PEM) electrode itself are not separate parts, but rather made of a mix of carbon and PTFE mixed with the catalyst, usually platinum. The platinum particles set on the electrode are approximately 3.5 nanometers in size. This method has several advantages, lowering the cost of production, providing immediate catalyst sites for the hydrogen reactions, providing hydrophobic areas that ensure the hydrogen gas always has pathway into the electrode and providing nonhydrophobic areas where water can move. Using carbon in the mix ensures an **electrically conductive pathway** out of the anode electrode and into the wiring system leading to the cathode.

There are two general methods to manufacture the assembly, applying the electrode/catalyst layer to either the conducting backing material or the ion conducting polymer membrane. In one case, the PTFE, carbon, and catalyst mixture is fixed on carbon paper or cloth (0.2 to 0.5-mm thick) using proprietary methods and then this is hot pressed onto the ion conducting polymer membrane. Some manufactures set the electrode directly into the polymer membrane, producing the membrane, then rolling, spraying, or printing the carbon/PTFE/platinum catalyst mix onto both sides (anode and cathode). The catalyst containing layer is generally about 30 microns thick. In both methods, the anodes and cathodes are usually the same so that the electrode/membranes are uniform in size and structure. In the second case, carbon cloth or paper is then fixed to both sides of electrode/membrane assembly (outer edges) to provide a gas diffusion layer as well as a conducting layer. The carbon backing paper or cloth supports the catalyst, provides an electrically conductive layer and provides a path for both gas and water. This set is called a **Membrane Electrode Assembly** (MEA). Figure 5-12 shows a typical MEA.

The fuel cell (anode, electrolyte, and cathode) itself is sandwiched between two **bipolar** (bipolar because they conduct both forward and backward) **plates**. The plates are generally high purity graphite, although metals such as aluminum or stainless steel have also been used. Both of these metals are expensive, but other

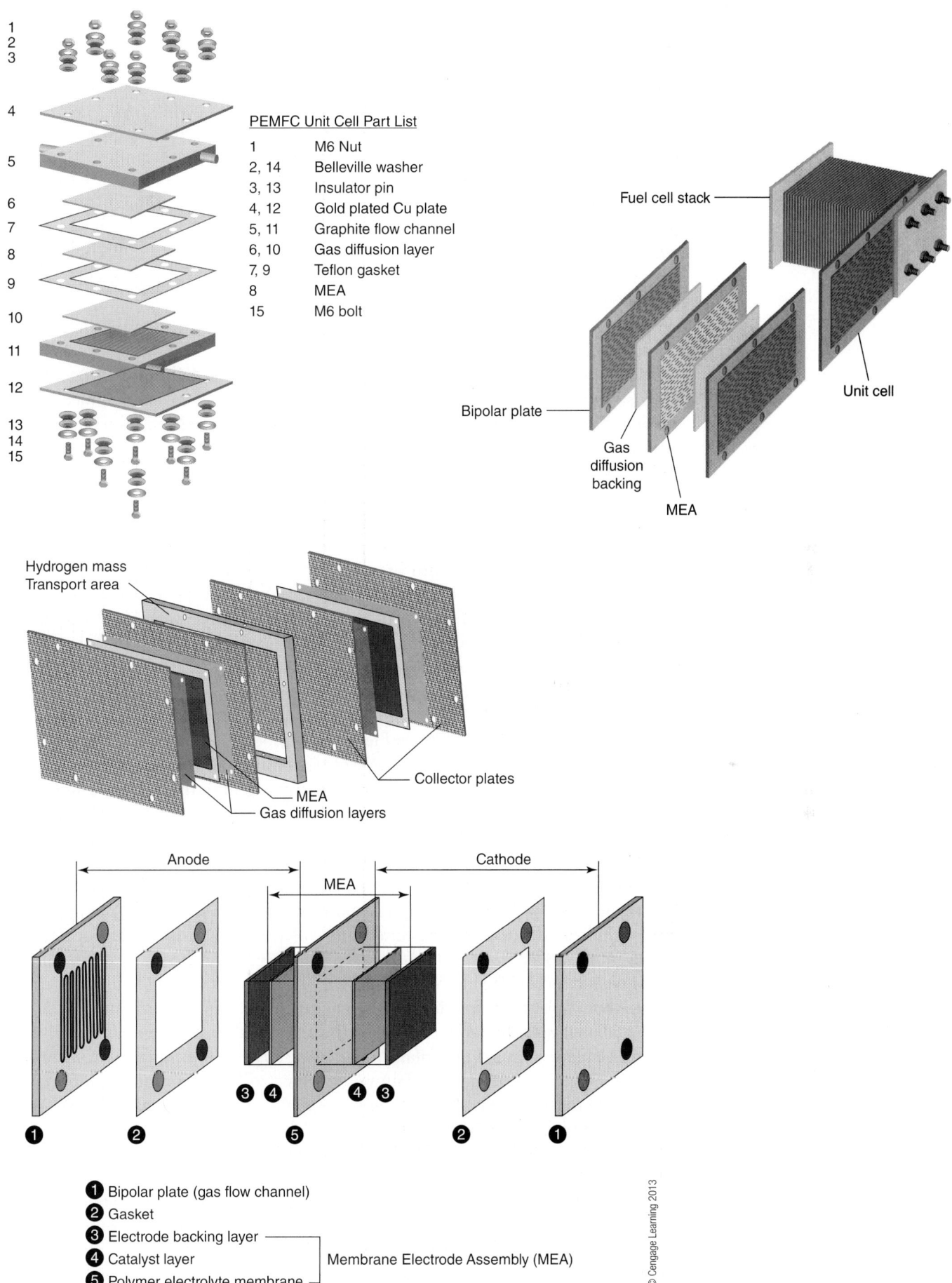

Figure 5-12 Membrane Electrode Assembly.

metals tend to corrode within the fuel cell stack and in some applications, the cost is acceptable. To this sandwich is added a number of other items, such as gaskets and spacers and more MEA sets and so on to build the **fuel cell stack**. The various parts of this assembly must be in intimate contact and be of uniform structure so that the complex movement of gas, electrons, ions, and water occurs at a steady state during operations to produce a constant flow of electricity. All of these things have to be kept in place as well. The air (oxidant) stream cannot mix with the fuel stream, and it is the bipolar plate that provides this seal. Cooling water has to flow and electrolyte water has to be kept at the right level so as not to either flood or dry out the cell, and the bipolar plate helps to regulate this flow as well.

The bipolar plate is thus critical to the operation of fuel cells. Remember that any individual fuel cell is relatively useless except as a laboratory curiosity because of the limited current and voltage produced. It is the bipolar plate and the associated wiring that allows for stacks of individual cells to be assembled into a product that produces useable amounts of electricity. The plate must collect the electrons produced and transfer them to the internal wiring running from anode to cathode so it requires a material with a minimum conductivity. It must distribute the fuel gas evenly to the anode gas diffusion layer and distribute air (or oxygen) evenly to the cathode gas diffusion layer so it requires a material that the gases do not diffuse through. In larger systems, it also has to carry coolants to maintain the working temperature within the cell stack without having the coolant mix with the two gases so it requires a material that is resistant to corrosion. It must seal the individual cells well enough to prevent gas, coolant, or water bleed-through and contain any pressures within the system, so it requires a material that is strong enough to withstand the forces within the cell. It acts as a structural member, so it requires material strong enough to withstand the outside forces acting on the cell. It must be light enough to minimize weight. It must be relatively thin to limit the size of the stack. It has to be compatible with the production methods of the MEA. It has to be made of material that will not poison the cell by migrating into the electrode. And of course, it cannot be too expensive to limit the overall cost of the fuel cell. Figure 5-13 shows the gas flow channels typically seen in bipolar plates.

Platinum is the preferred catalyst for both the anode and cathode in SPFC (PEM) cells when using straight hydrogen gas. If the gas is not pure, as happens when the hydrogen is made from methane, natural gas, or some other hydrocarbon (containing both hydrogen and carbon) source, then an alloy of platinum is used for the anodes, usually one containing ruthenium. Platinum is an expensive metal, roughly $1655 per ounce ($50/gram) in 2010. Over the many years SPFC (PEM) fuel cells have been produced, major efforts have been made to minimize the amount of platinum used because it is so expensive.

Catalyst usage varies depending on the manufacture, with the United Sates Department of Energy setting a **platinum loading target** of 0.4 milligram (1/1000 of a gram) of platinum per square centimeter of electrode/membrane size for automotive use to limit the cost for automobile applications. Original SPFC (PEM) cells used in the mid-1960s typically contained about 28 mg/cm^2 of platinum. Production units available around 2008 reported from 0.6 to 5 mg/cm^2 of platinum while research units report as low as 0.005 to 0.12 mg/cm^2, so the target loadings are now within reach. For alloy electrodes, the best current production loadings are in the vicinity of 0.25 mg/cm^2 of platinum and 0.12 mg/cm^2 of ruthenium. The catalyst bearing layer is approximately 10 μm thick.

Current platinum loading levels typically result in power production in the 0.5 to 1 amp/cm^2 range per MEA (the original Gemini cells operated at 0.037 A/cm^2 @ 0.78 V and used 32 MEA's in the stack, as shown in Figure 5-14) with experimental designs

CHAPTER 5 • FUEL CELL MAIN COMPONENTS

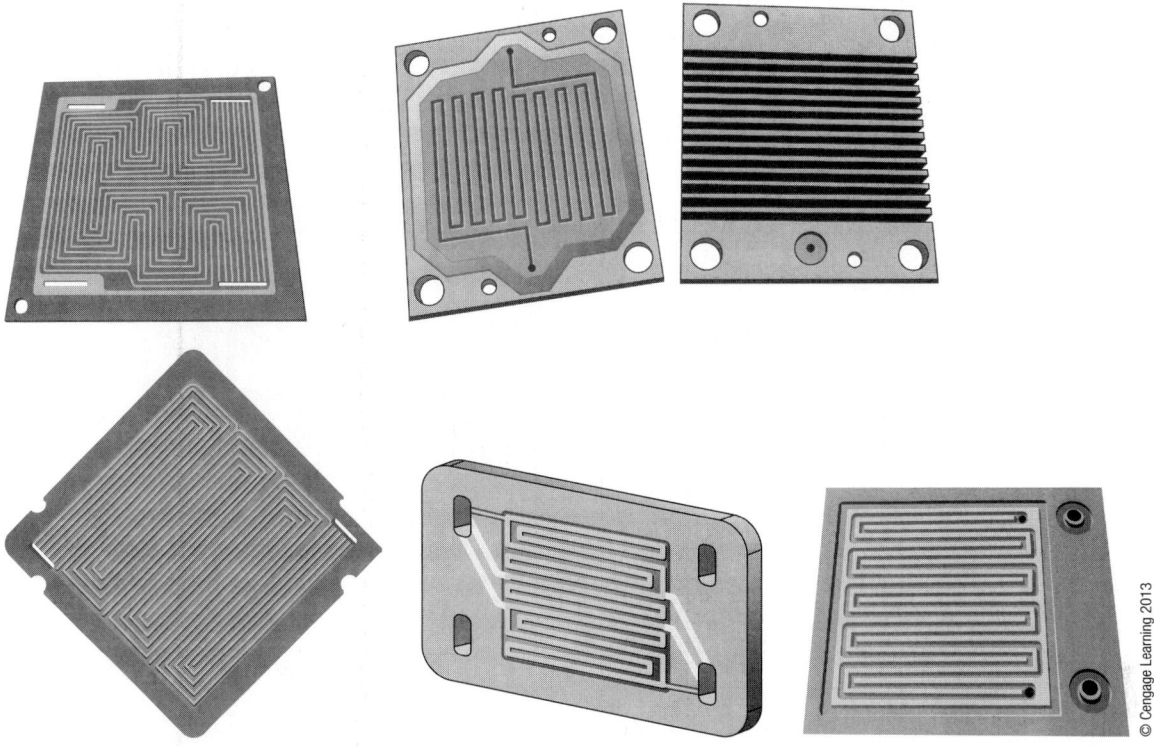

Figure 5-13 Bipolar (interconnect) plate.

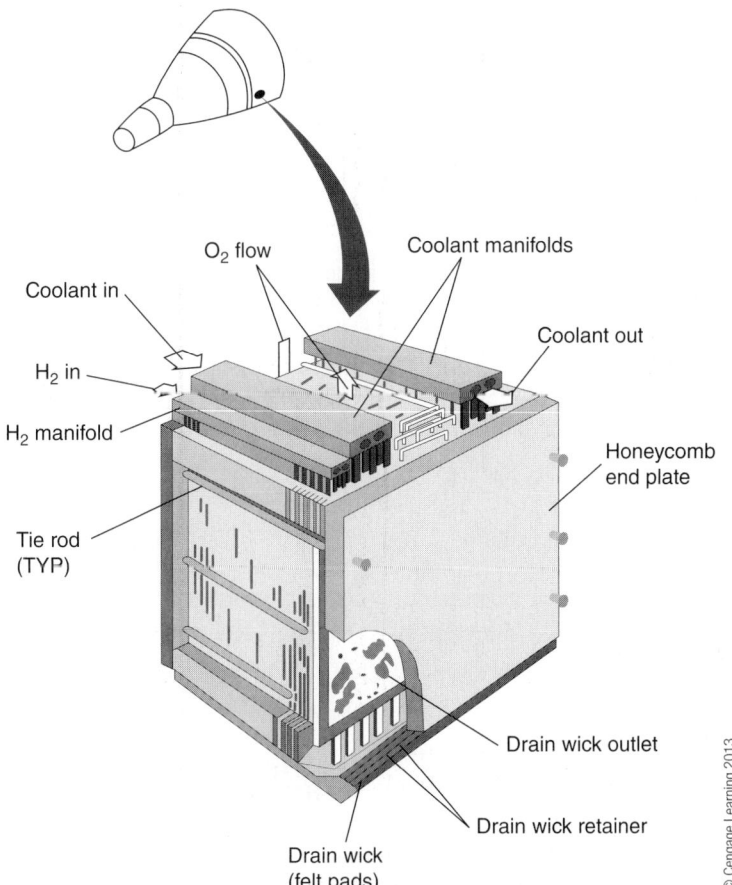

Figure 5-14 The Gemini fuel cell.

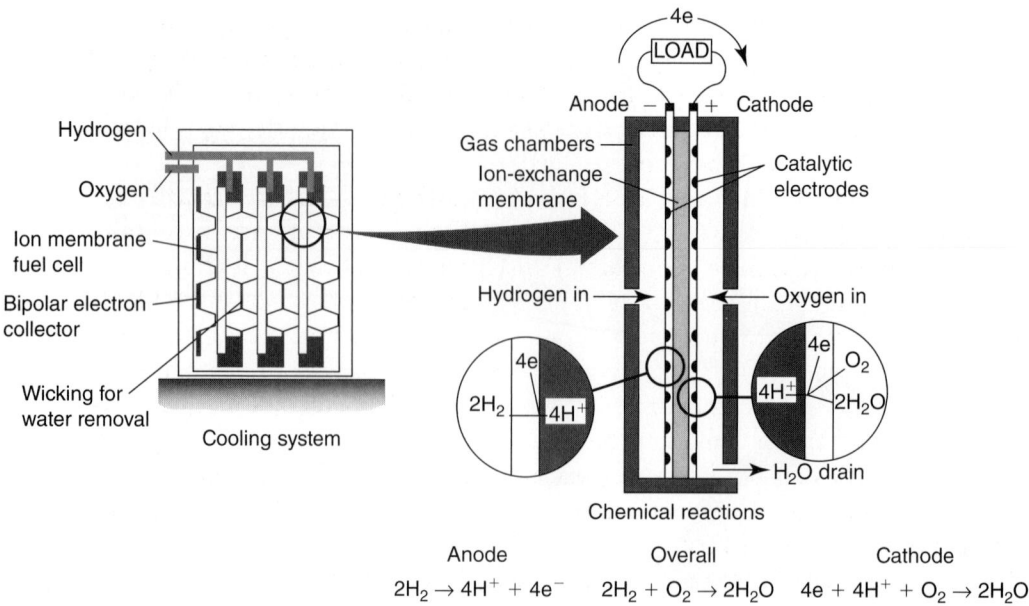
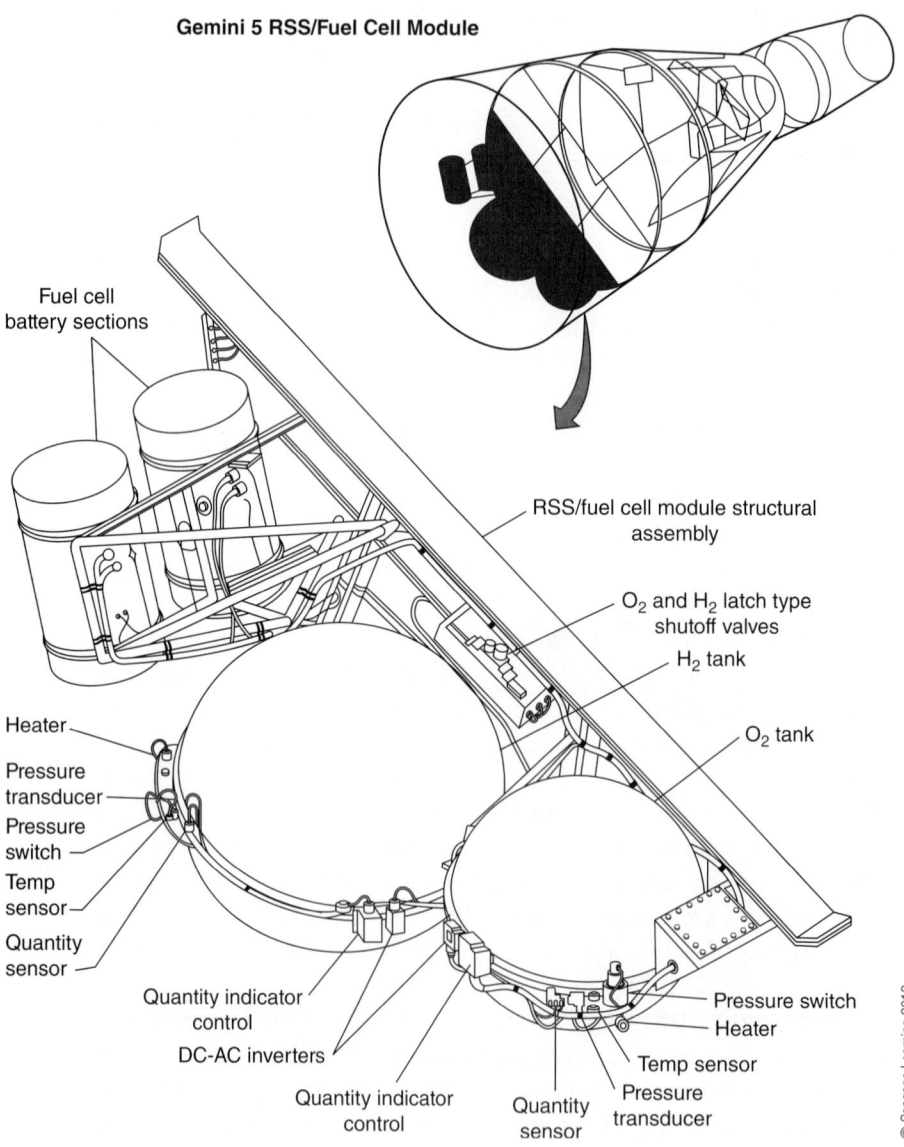

Figure 5-14 (Continued)

achieving as high as 4 amp/cm^2. These numbers equate to a catalyst cost of around $10–$20 per kW of energy produced, with this number varying based on the price of platinum or ruthenium. Current SPFS (PEM) cells have life spans of over 20,000 operating hours with the current begin produced by the individual fuel cells degrading at a rate of 0.1%–0.7% per 1000 hours of use.

The SPFC (PEM) fuel cell stacks operate at lower temperatures, usually below the boiling point of water (100°C or 212°F) because that enables them to make use of less expensive material in construction and because it allows for a quick start-up from room temperature. Unfortunately, it also causes problems if fuels are used that are less than very clean. The preferred fuel is hydrogen gas, but this is not widely available while other fuels such as methanol or even diesel are. The hydrogen contained in these fuels can be taken out using a separate system and then supplied to the cell, but such separations are rarely perfect. These **reformed fuels**, as they are called, contain trace amounts of such gases as carbon monoxide (CO), sulfur, and carbon dioxide (CO_2). These contaminants **poison** the catalyst sites, generally by reacting at the surface and staying there as surface molecules to block it so hydrogen cannot reach the catalyst, reducing the cell performance and eventually shutting it down. To minimize this type of poisoning, the cell operating temperature must be in excess of 120°C, quite a bit higher than the low temperature systems in use. Higher temperatures and high operating pressures offer other improvements in performances for the fuel cell as well, mainly because the higher temperatures lower the resistance of the electrolyte so the ions can pass easier and increase the rate at which the anode and cathode reactions occur.

Again, a tradeoff must be made in the system design, where the operating temperature must be matched to the fuel gas. Reformed gas is more convenient in the existing market, where hydrogen is not readily available. Unfortunately, reformed fuels generally have more than 1% contaminants in the stream, whereas for CO, 10 parts ppm or 0.001% will effect catalyst activity. The choice then is to operate the cell at the higher temperature or use a secondary system to further clean the reformed gas, neither of which are cheap options. Platinum and ruthenium alloys are also being developed that better tolerate contaminants, but these are more expensive. Each design choice made impacts the rest of the fuel cell, and the number of available options seen is a hallmark of a market still emerging. Once a mature industry arrives, the variation technicians see, and have to deal with, should stabilize.

While the fuel gas causes problems at the anode, the cathode must deal with another set of problems, mainly arising from its input. Most systems use air as the oxygen supply (oxidant). High **current densities** improve the cathode performance, since both electrons and ions respond to the higher densities (corresponding to more ions and electrons per unit area) by reacting faster. Unfortunately, higher current densities cause corresponding problems at the cathode. In particular, the ability of gas (the air) to move into the porous cathode is limited by the current flow and a layer of leftover nitrogen from the air builds up at the feed side of the electrode (air is about 78% nitrogen). These combine to block the entrance side of the cathode. These problems get worse when operating pressures become too great as well.

Both anode and cathode also have to balance what is called the **overpotential** or the loss of efficiency due to resistance within the cell. In electrical circuits, resistance is usually ohms and measures the loss of energy in moving electrons through a metal (electrons are very small particles but it still takes some energy to move them). The resistance of steel wires is greater than copper which is to say that electrons find it easier to travel in copper than in steel. The same is true on fuel cells but there are a few more components. There is basic resistance of electron flow through the various parts of the system, but at the anode, that resistance also includes the initial difficulty in driving the electron off the hydrogen atom at the catalyst site, followed by difficulty in transferring the electron from the catalyst site to the wiring system. There is also what are called **mass transfer resistances**, where the hydrogen gas molecules have to move through the gas channels in the bipolar plate, wait their turn at the catalyst site and then move to a spot where the reaction can occur. Since this is a physical movement of gas all along, it requires moving mass, even though the mass of a hydrogen (H_2) atom is small, there are billions of billions of them, and they all have to line up for their chance to react. This mass transfer is very complicated. It starts at the gas inlet, perhaps goes to a purification step, moves to a gas diffusion area where it has to divide into enough smaller streams to feed the entire cell uniformly, flows on to the electrode, loses its electrons and then shifts into the electrolyte to flow to the cathode, where it meets another gas stream flowing in the opposite directions with its own set of resistances to flow. This is complicated by the presence of water, by the presence of impurities and by any gas pressures within the cell. All of these things reduce the overall efficiency of the cell, perhaps by a very small amount, and go by the general name of **ohmic resistance**.

AFC

Fuel cell electrodes differ in a number of ways. In general, the differences have to do with the temperature at which the cells operate while similarities have to do with the way chemical reactants, catalysts and electron/ion conduction all come together at the same point.

Any discussion of electrodes should begin with the issue of cost and that starts with the catalysts used in electrodes. The predominant material used in fuel cells as catalysts is platinum. In low to medium temperature cells like AFC, platinum is not strictly necessary to start and maintain the chemical reactions so much as it is necessary to make these reactions all happen at the same rate. If the anode reaction in AFC is consuming hydrogen and hydroxide to make water faster than the cathode can produce hydroxide, then in short order the cell will shut down from lack of hydroxide. The reactions must match each other as well as the fuel and oxygen streams entering the cell or the cell stops working.

One of the most important issues with AFC electrodes is the reaction at the cathode, discussed in Chapter 4.

$$O_2 + 4e^- + 2H_2O \leftrightarrow 4OH^- \qquad \text{Eq. 6}$$

For reasons beyond the scope of this book, this **chemical reduction reaction** is more efficient in a basic (containing OH^+ as opposed to acidic which contains H^-) solution. The more basic the solution is (the more OH^+ in solution), the more efficient the reaction is as well. This has a substantial impact on the cathode then, because unlike most fuel cells, in particular those operating at lower temperatures, AFC systems can use either precious metal catalysts or cheaper nickel based catalysts. This gives AFC systems designers the ability to manipulate the system efficiency and power density to match applications and price targets.

Electrodes are designed and built with several **required elements**. Gas must be able to diffuse through the electrode to reach the catalyst sites. Catalysts must be widely dispersed to maximize the three-phase requirements of fuel cell reactions. The electrolyte (liquid potassium hydroxide in the case of AFC systems) must be in contact with the catalysts. Some way to transfer both electrons (electricity) and ions (OH^+ in the case of AFC) to the reaction sites on the catalyst must be in place.

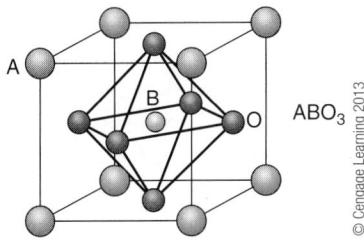

Figure 5-15 The perovskite structure.

The combination of materials used in an electrode must withstand gas, electrolyte, temperature and pressure.

Typical AFC system electrodes used in space had a porous carbon base combined with an interior nickel mesh. The gas distributes through the porous carbon while the current (electrons) are collected and transferred in the nickel mesh. The catalyst is added as very small particles. A number of different catalysts have been used for AFC applications, because of how long these systems have been in use and because substantial research was funded during development of systems used in space. Although platinum is the basic catalyst material in all fuel cells, palladium, copper, silver, tin, and gold combinations as well as ceramic based molecules such as **spinels** and **perovskites** (structure seen in Figure 5-15 where A and B are different atoms bonded to oxygen atoms) have all been used as well, particularly in AFC systems where the alkaline electrolyte allows for a wider choice. Less expensive catalysts though are not as stable in the alkaline electrolyte and corrode faster than the more expensive precious noble metals like platinum.

There are several different ways to make electrodes. Because AFC systems are low temperature and the alkaline electrolyte promotes the cathode reaction, they can use what are called **Raney metals** to combine the gas diffusion layer, the current collection and the catalyst sites. To make these, two metals are mixed, with one metal not being **reactive** in alkaline solutions (usually nickel or silver) and one metal being reactive in alkaline solutions (usually aluminum). The metal is put into an alkaline solution and the reactive part is removed in a chemical reaction. Catalysts are then put on the remaining metal (which is now porous) to make the combined electrode assembly.

Electrodes can be cast and sintered using ceramic technologies or rolled like paper. A simplified version of the method has a porous layer so that the fuel gas can pass to the catalyst sites and then a carbon layer that has the catalyst already in place (the **electrocatalyst layer**). Another carbon layer mixed with a hydrophobic polymer, such as polyethylene, Teflon® (Polytetrafluoroethylene), or even paraffin is set next. A different porous layer goes down in order to pass the electrolyte to the catalyst with the hydrophobic layer in between keeping the liquid away from the gas side. These layers can be set as thin fabrics or as slurries (mixtures of liquids and particles). A wire mesh collection layer can also be added to minimize resistance to electron movement in the cell but it has to be on the catalyst layer side in order to provide a conduction path. The piece is then sintered at high temperatures to give the strength and porosity needed. Electrode structures can have several layers set down in this fashion with some pressed or set at different angles to increase strength, reaction area and improve longevity. The piece produced is less than 0.5 mm in thickness.

It is important to realize that there are two type of porosity in electrodes. Gas porosity is needed to get the fuel or oxygen to the catalyst surface and these pores are very small. To get the electrolyte to the catalysts site, much larger pores are needed, since liquids do not travel as easily as gasses do. The size of the pore has significant effects on the electrode structure. If the liquid pores are too big, electrolyte can move into the gas diffusion layer and block the smaller pores so fuel or oxygen cannot flow. If the gas pores are too small, gas flow is too slow. Too many pores, large or small, can result in insufficient sintering (pores are empty spaces and too many of them mean there aren't enough solid parts touching for sintering to work), making the internal structure of the electrode too weak to handle the movement of gas and liquid. The structure will then break in service. There are also two types of gas being moved across the electrodes, a fuel gas and an oxidizing gas. If the fuel gas is hydrogen (H_2), then it moves through even the smallest of pores, and in fact will diffuse through solid metals if given enough time. Oxygen (O_2) on the other hand is a considerably larger gas molecule and does not move nearly as fast through even a porous structure. This problem is made worse if the oxidizing gas is air, a mixed gas mostly of nitrogen (N_2) and oxygen. Maintaining the flow of hydrogen and oxygen at even amounts so the reactions happen in time is not simple.

Carbon is used because it is a strong and conductive material while the PTFE controls porosity. PTFE is not conductive though, and certain designs do not allow for cells to be tied together in a conventional manner, with electrode, electrolyte and cathode sandwiched between bipolar plates (interconnects) because the conduction paths are not continuous. Monopolar designs requiring no bipolar plates (interconnects) are in service. They are generally cheaper than bipolar designs since the plates are relatively expensive and the resulting cell stack is thinner and easier to fabricate. The **current collector** used becomes critical since it is the only way to pass current out of the individual cells. Because the current collectors are then subjected to substantial interaction with the electrolyte, they must be tolerant of alkaline solutions. Nickel or nickel alloy meshes are used, and some are even gold coated to improve longevity. The current transfers out of the cell and in monopolar designs each cell is isolated so that individual cells can be monitored and in some designs removed from service if needed. Because most fuel cell designs use bipolar plates, they cannot isolate individual cells but only the stacks.

The electrodes fail for a variety of reasons. One such cause has long been thought to be carbonate formation within the porous electrode structures. The carbonates form as crystals and then block the pore structures to shut the cell down. This phenomenon may be as much a function of the internal pore structure as of the chemical reactions, since some researchers report finding carbonates and others do not. Very fine pores such as those found in Raney electrodes may act to promote such crystal formation while a less complex and more open structure may limit such formation. The alkali electrolyte will over time also destroy the electrode structure as well as many of the matrices holding the liquid electrolytes in place within the cell. Recirculating electrolytes help to minimize this type of destruction but not eliminate it, especially at higher temperatures or if the cells are even partially pressurized to improve performance. **Electrode weeping** is another issue, where in spite of a hydrophobic layer between the electrolyte and the electrode, water enters the porous gas diffusion layer blocking flow. There are several ways that fuel cells weep. Fine cracks can appear in the electrode if care is not taken during fabrication, especially when sintering is done. The hydrophobic layer can be damaged in service, in particular if the cell begins to reverse itself to break water into hydrogen and oxygen instead of combing hydrogen and hydroxide to form water. Gas bubbles will build up in the pores and eventually break the assembly apart. Hydrogen fueled cells, especially where the cell (and thus the gas) is pressurized can produce water within electrode pores when the cold hydrogen gas meets the warmer electrode material dropping the **vapor pressure** of the cell and **condensing** water. Contaminant gasses in the fuel or oxygen stream can also degrade the catalysts sites over time, reducing or even stopping the chemical reaction itself. Nickel meshes can also be oxidized during service increasing their resistance and eventually shutting down current flow.

MCFC

Like SOFC systems, MCFC systems are capable of **internal reforming**. This is an important part of high temperature fuel cells, since it helps to balance the high cost of materials as compared to low temperature systems (which can use cheaper materials but require an **external reformer**). When internal reformation occurs, steam is generally added to the fuel gas mixture in order to drive all the necessary chemical reaction, so the gas delivery system to the anode is more complex than lower temperature cells. At the anode, a series of reactions can take place depending on whether the fuel used is hydrogen or carbon monoxide as simple fuels or more complex hydrocarbon fuel gas that must first be broken down to form the two simple fuels that actually run the cell. Keep in mind that internal reformation produces the hydrogen or carbon monoxide for the cell and while it can be done at the anode, a separate electrode can also be used to reform an initial fuel such as natural gas which then feeds the anode. The two anode reactions, internal reformation of fuel and fuel cell production of electricity are very different. Fuel reformation, for instance, is an **endothermic** reaction, in that it must have a source of heat while the fuel cell reaction is **exothermic** and produces heat. Having the reactions separate in space has advantages but also disadvantages. Having one reaction creating heat while another is absorbing heat with both going on at essentially the same place is difficult to control, but one can also be used to supply heat to the other. Some of these fueling schemes can get complicated, since not only is new fuel introduced, but any fuels not used in the cell can also be recycled back as input. Fuel cells rarely use all the hydrogen or carbon monoxide fed to the anode so the plumbing of gas and cooling streams is critically important. Excess fuel is used for a number of reasons, among them the desire to make constant use of the expensive catalysts and to move the products of the chemical reactions out so make room for more reactions (otherwise pressure builds and eventually the reactions stop).

MCFC anodes are nickel alloys made in such a way as to be porous, since gas must be able to pass through the anode structure. Anodes are made using metal powders. These can be pressed into a shape and held under pressure while being sintered at temperatures less than the melting point. The time and temperatures used depend on how porous a structure is needed, and in MCFC systems, porosity ranges from 55% to 75%. Anodes can also be tape cast like electrolytes, a particularly useful method when thin layers are needed. Anodes are usually less than 1 mm in thickness and can be up to one square meter in size.

Most of the alloys contain chrome or aluminum, used primarily to stop the material from further sintering when the cell operates at temperature. Sintering can occur at temperatures above about 25% of the melting point for some materials, provided enough time is allowed. MCFC systems can run at 700°C or even a bit higher, and that temperature is high enough that nickel will sinter in a matter of days while the fuel cell will be in service for months. If something is not done (especially when internal reforming is used) with the main anode, then in short order, the porous anode sinters together, essentially forming a clump although the melting point itself is never reached. Clumps of metal are not porous and so the cell will be ruined. That is one of the main reason why secondary reformers are used rather than reforming being done at the main anode; the operating temperature of the reformer can be lower than the operating temperature of the main cell stack. Another problem with internal reforming being done at the main anode is that some of the **alloying element**, like chrome, will react chemically with some of the elements in the ceramic holding the molten carbonate electrolyte, like lithium. To counter that, materials engineers add aluminum but then aluminum may react with something else. Because sealing the stacks require a substantial amount of torque be applied to keep the individual cells from leaking gas or fluids from one to another, anodes will also move the alloying elements in what is called **high temperature creep**, transferring atoms (in particular chrome) at temperature out of the anode and into the electrolyte where it can deposit and cause short circuits within the cell. Matching alloys, elements and operating temperatures in any high temperature operation is difficult and complicated, and it is even more so in fuel cells because of the chemical reactions either happening or that can happen in the **reducing** and **oxidizing atmospheres** at the anodes and cathodes.

Internal reforming on the main cell anode does have one very important feature that makes the more complicated plumbing, materials design and reaction balancing worth the effort. Using the main anode to reform fuel gas will result in a more **efficient conversion** of the hydrocarbon fuel into fuel gasses for the cell. The reasons behind this are beyond the scope of this book, but in essence, the rules of chemistry force more of the chemical reactants to react than normally would because the hydrogen fuel is being consumed and the reaction tries to make more of it to compensate for the loss. If internal reformation at the main anode is used though, the anode has to contain more nickel to accommodate both the reformation reactions and the main fuel cell reactions. Although designs vary, many direct internal reformation (DIR-MCFC) set the reformation nickel first, then pass the cell fuel gas to the main reaction nickel deeper in the anode structure, in essence creating a two layer anode. This approach is called **indirect internal reformation**, where reforming fuel gas with the main anode is called **direct internal reforming**.

Unlike other fuel cells, the anode in molten carbonate cells also does additional duty as a reservoir for the molten electrolyte. It can reform fuel gas, promote the main fuel cell reaction and act as a reservoir; truly a multitasking piece of equipment. Remember from earlier chapters that there are strict rules governing whether atoms and molecules react, whether that reaction will give off heat or absorb heat and how long the reactions take. The MCFC anode can act as a reservoir because the main fuel cell reaction at the anode (either a hydrogen proton or a carbon monoxide ion reacting with the carbonate ion) happens very quickly while the reaction at the cathode (carbon dioxide and oxygen gas reacting to form carbonate ions) takes considerably longer. Since the reactions happen faster, less nickel catalyst is needed, but rather than reduce the amount of nickel, which is relatively cheap, designers and engineers make the anode larger even though it does not have to be. As mentioned previously, one of the problems with MCFC systems is that at the high temperatures used, the molten electrolyte can vaporize. Unlike AFC systems, which can pump the molten alkaline electrolyte in and out of the cell, MCFC systems are closed because sealing the ceramic parts at the high temperatures is so difficult to do. Excess electrolyte is then stored in the anode and when the liquid in the electrolyte vaporizes, there is some to take its place, extending the life of the fuel cell using a very cheap and effective method.

As with all fuel cells, electrodes can be **poisoned by contaminants** in the fuel or oxygen input streams. Sulfur in particular causes major problem with nickel electrodes, coating the nickel surface so the hydrogen cannot access the catalyst sites and interfering with reactions that must occur for the cell to operate. Sulfur can also react with the oxygen at the cathode to form sulfur dioxide (SO_2) which in turn reacts with the carbonate in the electrolyte to destroy it. The amount of sulfur MCFC systems can tolerate is very small, less than 0.01 ppm or 0.000001% but there are methods to purge the cell so the tolerance can be increase to approximately 0.5 ppm (0.00005%).

The MCFC cathodes are generally nickel oxide (NiO). Where the nickel anodes have trouble with structural stability, the cathodes break up and the nickel moves into the electrolyte, traveling away from the cathode and toward the anode. This is because when the nickel breaks away from the NiO, it becomes **positively** (losing electrons) **charged** and the oxygen becomes **negatively charged** (gaining electrons). Since the cathode is positively charged, the nickel is repelled into the electrolyte where it tries to get to the anode, which is negatively charged (anything positively charged moves toward anything negatively charged). Many times, the nickel comes out of the liquid molten carbonate solution as nickel metal and forms lumps in the ceramic matrix holding the liquid, causing short circuits within the cell. In particular, the nickel can react with the carbon dioxide present in the cell if there is too high of a carbon dioxide gas pressure in the cell. The oxide will break up and form a carbonate ion with the reaction:

$$NiO + CO_2 \rightarrow Ni^{2+} + CO_3^{2-} \qquad \text{Eq. 15}$$

Generally, this effect is minimized by not letting the cell become pressurized and making the electrolyte as thick as possible. Most MCFC systems operate in the basic (as opposed to acidic) range as well to keep nickel oxide from being reduced at the cathode. Balancing acids and bases is a critical part of the chemical operations of fuel cells, especially for high temperature or liquid based electrolyte systems. Keep in mind that fuel cells are all **acid/base** systems of one kind or another, and must be balanced to operate efficiently, or in some cases at all.

PAFC

As in all fuel cells, the electrodes in PAFC systems must promote the necessary equations, pass electrons into the wiring systems, and ions into the electrolyte. They must allow even gas distribution from the input side through the internal support structure and to the area where gas, catalyst, and electrolyte meet. They cannot degrade in the electrolyte (concentrated phosphoric acid after all) or be chemically altered by reactions with anything in the cell—solid, liquid, or gas. Since the PAFC electrode faces a gas atmosphere on one side and a liquid acid on the other, it must keep the liquid in its place but at the same time still allow both gas and liquid to meet where the catalysts particles are. It must act to dispel the water created in the PAFC reactions. It must be strong enough to handle the compressive loads of the stack. It must not be so expensive that no one can afford to buy the system.

The original electrodes used when PAFC systems were being develop consisted of **platinum black** but the cost soon ended that structure. Current systems use platinum or platinum alloy particles (about two nanometers in diameter and roughly 10% by weight) on a fine carbon powder with these supported on carbon (graphite) paper. There are a number of variations having to do with making one side hydrophobic to drive the water away from the layer and keep the liquid acid electrolyte from flooding the gas diffusion layers or keeping the anode and cathode gasses away from each other or providing suitable electron channeling. Balancing the liquid and gas phases is generally done using PTFE, which is a very common item in the fuel cell industry. Carbon paper can be **impregnated** with the PTFE using relatively simple means after the catalyst and carbon particles are placed on the paper. The PTFE maintains the gas/water interface, keeping the liquid electrolyte on one side of the carbon paper and the gas on the other. The carbon paper supports the platinum/carbon particles and provides a conducting path for electrons.

In general, more platinum is needed at the cathode (as much as five times as much) than the anode since the cathode reactions are less favorable than those at the anode and to make the rates of reactions match between the two, a little extra help is needed. The cathode also contributes the most to the **ohmic loss**, as with most cells there being a higher loss using air than pure oxygen. Like other cells, carbon monoxide will lower **efficiency** at the anode as well so that greater currents are produced using high purity gasses. In general, though, the ohmic loss of PAFC systems is smaller than other cells.

Although efficiency increases with temperature and pressure, many PAFC systems run at **atmospheric pressure** and relatively low temperatures to increase the life of the cell, since the liquid phosphoric acid **evaporates** more quickly at higher temperatures.

SOFC

Electrodes are where the money is made in fuel cells because that is where electricity is made. SOFC electrodes are different from some other fuel cell electrodes in two important respects: they are ceramic structures and they do not make use of precious metal catalysts. Both of these result from the high temperatures used in SOFC systems. For the zirconia solid oxide electrolyte to **conduct ions**, it has to be maintained at a relatively high temperature (in the neighborhood of 1000°C). That temperature in turn requires that the two electrodes be not only capable of tolerating such temperatures but behave at those temperatures like the solid oxide electrolyte. Since the two electrodes structures as well as the electrolyte structure are also sintered together in the fabrication process at temperatures of approximately 1300°C, they must also be able to survive that temperature together as well. That limits the possible combinations considerably.

The most common SOFC anodes now consist of the same stabilized zirconia (stabilized with yttria and sometimes ceria) as the electrolyte. Stabilizing the structures is done in ways like holes are created as discussed in a previous section. Alloying elements do one of three things: they can take the place of Zr in the matrix (**substitutional alloying**), they can fit in between the Zr elements already in the matrix (**interstitial alloying**) or they can form secondary molecules within the matrix (carbon does this in steel forming iron carbide). This can either destabilize the structure, for instance, if it is considerably larger atom is alloyed in or stabilize the structure, for instance when an alloying element forms more (or stronger) bonds to the surrounding atoms so it is harder for them to come apart. Doping elements into the zirconia to stabilize them and then using the same base substrate for anode, electrolyte, and cathode eliminates thermal expansion issues, lessens the chance that high temperature creep will change the structure, and reduces the number of chemical reactions that may damage the material.

Nickel is added to the YSZ (yttria stabilized zirconia) to act as the required catalyst. A number of different catalysts have been used over the years in SOFC anodes, ranging from platinum and gold to iron and aluminum, but nickel now seems to be the preferred metal because of the advantages it offers. The nickel catalyst on YSZ is stable at high temperatures as well as being a good catalyst, adds some mechanical strength to the ceramic, and does not change the thermal expansion characteristics so much that the cells are damaged at operating temperatures or during cycling. In addition, by placing the catalyst on what was essentially one edge of the electrolyte, the resulting structure maintains the ion conduction pathways needed but changes the nickel bearing electrode side just enough to produce electron conducting pathways. Both sides keep the porous structure needed to pass the fuel gas.

The nickel in the anode has an added bonus in SOFC systems in that it will also catalyze the reformation of fuel gas to produce hydrogen for the cell. Like other high temperature cells, SOFC systems can break up hydrocarbons (carbon atoms and hydrogen atoms bonded together) with natural gas mainly used in the SOFC systems. The anode thus not only uses CO and H_2 as fuel, but produces those fuels itself (internal reformation) during operation. This is a considerable cost savings but in general only smaller cells make use of this feature as larger, industrial systems tend to use the anode exclusively for electron production and use external reformers to obtain fuel to limit **fouling** of the catalyst sites.

In early SOFC systems, precious metals were used in the cathodes because there was no other real choice. They did not work particularly well due to the thermal and chemical mismatch between metal and ceramic, but at least they worked well enough for some period of time. In the 1970s, ceramic structures were developed that not only could act as a catalyst for the oxygen reaction but were well matched to the stabilized zirconia ceramic structure of the electrolyte during fabrication and subsequent use. $LaMnO_3$ (one lanthanum atom, one manganese atom, and three oxygen atoms bonded together) doped with SrO (one strontium atom bound to one oxygen atom) proved to be particularly effective, and this is the cathode catalyst material most commonly used in SOFCs.

The **interconnects** (**bipolar plates**) used to tie planar systems are another important part of the SOFC even though they are not strictly part of the individual cell itself. Since individual fuel cells do not produce enough electricity to be of much use, multiple cells are tied together in stacks. Since SOFCs operate at high temperatures, the interconnects (bipolar plates) used must be able to tolerate the temperatures and have properties that match the ceramic fuel cell structures well enough so that the cells are not destroyed at operating temperatures or during the change in temperature seen in service. High temperature steels such as Inconel® may be a registered trademark are suitable for the high temperature (900°C–1000°C) service but do not match other properties well enough to keep the cells intact during service. Most interconnects use chrome as a base, usually in a **cermet** (a mixture of ceramic and metals) structure. Many of these are proprietary but all are stable at high temperatures and have thermal properties close enough to the stabilized zirconia used in the electrode/electrolyte assembly. Unfortunately, chrome can transition into a gas at the higher temperatures and move to the electrolyte interface with the electrodes where it redeposits, blocking the required three-phase region. Coatings can be used to stabilize the chrome or the operating temperature can be lowered to minimize this problem, but it continues to affect SOFC systems to some extent. Below 900°C conventional high temperature metals can be used, eliminating this problem.

Most SOFC systems also make use of current collectors. Usually, a **nickel felt** (long whiskers of nickel sintered together to form a fabric) is set between anode and interconnect to transfer electrons better within the cell. These are then tied together in the cell stack to provide a complete electrical pathway.

KEY WORDS

Knowing the terminology used is critical when dealing with fuel cells. Following is a list of the important terms in this chapter, which are also in bold typeface within the chapter. It is recommended that students are required to submit definitions of some of these words as homework assignments in which they look the terms up in other books, articles, or on the Internet.

acid/base
acidic solution
activation energy
alkaline electrolyte
alloying element
ambient humidity
anode supported SOFC
atmospheric pressure
balance of plant
basic solution
bipolar plates
cadmium
carbon dioxide
carbonates
catalysts
cathode oxidation reaction
cathode supported SOFC
ceramic electrolyte
ceramics
cermet
chain
chemical attack
chemical path
chemical reactions
chemical reduction reaction
chloride compounds
circuit
cogeneration
cold start
combined heat and power (CHP)
condensing
conduct ions
counter-ion
current collector
current densities
dehumidification
dipole
direct internal reforming
doping
efficiencies
efficiency
efficient conversion
electrically conductive pathway
electrocatalyst layer
electrode
electrode assembly
electrode stability
electrode weeping
electrolyte supported SOFC
electron
electro-osmotic drag
endothermic
engines
ethylene
evaporates
exothermic
external reformer
extrude
flow channels
flow patterns
fouling
freeze
fuel cell stack
gas
gas diffusion membrane
grid
high temperature creep
hole
hot spots
hydrocarbon
hydronium ion
hydrophilic
hydrophobic
hydroxide ions
impregnated
indirect internal reformation
interconnect supported SOFC
interconnects
internal pressure
internal reformation
internal reforming
internal resistance
interstitial alloying
ions
lead
liquid electrolytes
load
mass transfer resistances
mature products
megawatts
Membrane Electrode Assembly
mercury
Nafion®
negatively charged
negatively charged particles
nickel felt
ohmic loss
ohmic resistance
overpotential
oxidizing atmospheres
parallel
particulates
pendant molecule
perfluorovinyl ethers
perovskites
planar
planar design
platinum black

platinum loading target	reactants	substitutional alloying
poison	reactive	sulfur compounds
poisoned by contaminants	recirculating electrolyte	surface area
pollutants	reducing atmospheres	surface tension
polyethylene	reformed fuels	tape casting
polytetra-fluoroethylene	required elements	Teflon®
positively charged	scalable	thermal expansions coefficients
positively charged particles	series	three-phase area
potassium hydroxide	slurry	tin
power and heat	spark initiators	trace amounts
power densities	spinels	tubular
ppm	spray casting	tubular designs
proton	stabilized zirconia	turbine based generators
protonation of water	staged sintering methods	vapor pressure
purging	standard operating conditions	viscous liquid
Raney metals	stationary installations	wet
	steam reforming	work

DISCUSSION QUESTIONS

1. Where is the reaction chamber of an internal combustion engine?

2. Rank following cells according to how long it takes the system to come to full power: SOFC, SPFC(PEM), and AFC (shortest to longest).

3. What is the difference between the electrolyte in AFC systems and that in PAFC systems?

4. Look up the coefficient of thermal expansion for zirconia and for 316L stainless steel.

5. Will an electron move toward a cathode or toward an anode?

6. Will water move toward the Polytetrafluoroethylene (PTFE) in Nafion® or toward it?

7. In an AFC system, will adding more of the KOH electrolyte make the cell more acidic or more basic?

8. Name the parts in an SOFC membrane electrode assembly.

9. If you put a drop of liquid potassium hydroxide (KOH) on a countertop, what would it do?

10. If you had a powder made of a material that melted at 1000°C and wanted to sinter it to form a solid, would you use a temperature over the melting temperature or under it?

11. If you used pure oxygen instead of air in a fuel cell, how much less gas would you be able to use?

12. Name three required design elements that might be used when designing an electrode.

13. What is the difference between indirect internal reformation and direct internal reforming?

14. How many atoms are in one molecule of $LaMnO_3$?

CHAPTER 6

MANAGING FUEL CELL COMPONENTS

objectives

This chapter introduces the student to the heat, water and pressure management done in fuel cells. Fuel cells create and use both heat and water so managing them and the resulting operating pressures within the cell stacks must be done to ensure continued operations. Some of the fundamental methods used to deal with water in fuel cells are presented for the various types of cell. Heat is generated by fuel cells during the chemicals reactions that take place to produce electricity and that heat is used rather than lost to increase system efficiency. This chapter introduces some of the methods used to take advantage of the heat being generated. The methods of dealing with the gas and the pressures within the cell are also introduced.

INTRODUCTION

Fuel cells themselves are not very complicated. Gas goes in, contacts a catalyst, changes via a chemical reaction, transports the various parts (including electrons), recombines to form a different gas, and then exits. Making sure everything works together is more complicated than the process of making electricity is. The two most important control issues in fuel cells are water and heat. Water has to be balanced precisely and heat has to be controlled carefully, otherwise the cells will stop working. Managing the two systems of heat and water enable the cells to operate and without them, no matter how advanced the cell itself, the system will fail.

WATER MANAGEMENT

Water management is one of the most important issues facing fuel cells because water is critical to fuel cells. In some systems like SPFC (PEM), the cells will not operate without water nor will they operate with too much water. Water must be available in the correct amount, not only as liquid, but also in the gases present within the cell. Water is needed in reforming the input fuels. Water is used to cool cells. Water holds the protons in PAFC cells and the hydroxide ions in AFC cells. The operations of fuel cells depends to a great extent on the systems which ensure the timely and correct availability of water.

SPFC (PEM)

Water is needed in the polymer membrane so that the hydronium ion (H_3O^+) can migrate to the cathode to combine with the electrons and oxygen to form water. Because the anode material must be in direct contact with both the wiring needed to flow electrons to produce electricity and the water needed to transport the hydronium ion (H_3O^+), water is also critical to its performance. To a great extent, the ability of the membrane to conduct ions such as hydronium (H_3O^+) is governed by the level of water **saturation**, since saturation lowers the internal resistance to the ion flow and **dehydration** within the system tends to detach the membrane from the electrode, blocking transfer from the anode into the ion conducting membrane.

Water is generated in the fuel cell at the cathode as part of the set of chemical reactions producing electricity. This water is part of the membrane system, in that it is available to flow back into the system to maintain the water balance. Remember that the membranes are very thin, so the water will diffuse back through it and to the anode. This back-diffusion occurs when too much water builds up on the cathode side, and works in the same manner as when too many electrons occur at the anode, causing them to flow toward the area at the cathode where there are too few electrons. Too much water at the cathode then causes too much water at the anode (where the fuel is split into electrons and ions) which disrupts the fuel gas flow to the anode and the catalysts by eventually blocking fuel gas access to the anode and catalyst assembly entirely as the gas will not flow through water. Figure 6-1 shows the water is present in the membrane as discreet molecules, and as such can be easily disrupted.

The physical blocking of fuel gas by water is one issue. Drying the membrane so the protons (H^+) cannot move off the anode/catalyst assembly and into the water to make the hydronium ion (H_3O^+), which transfers to the cathode, will also stop the fuel cell from working. In general, this is because the oxygen that

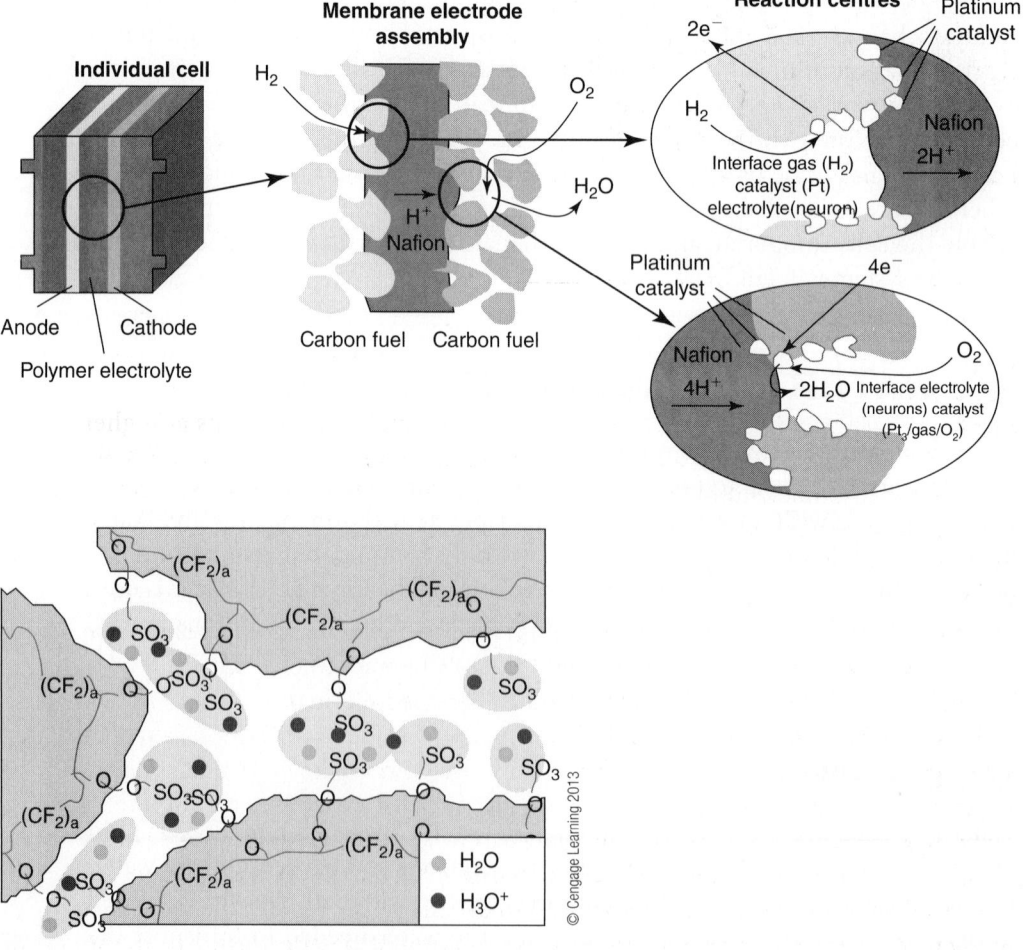

■ **Figure 6-1** Close-up view of water in the PEM membrane.

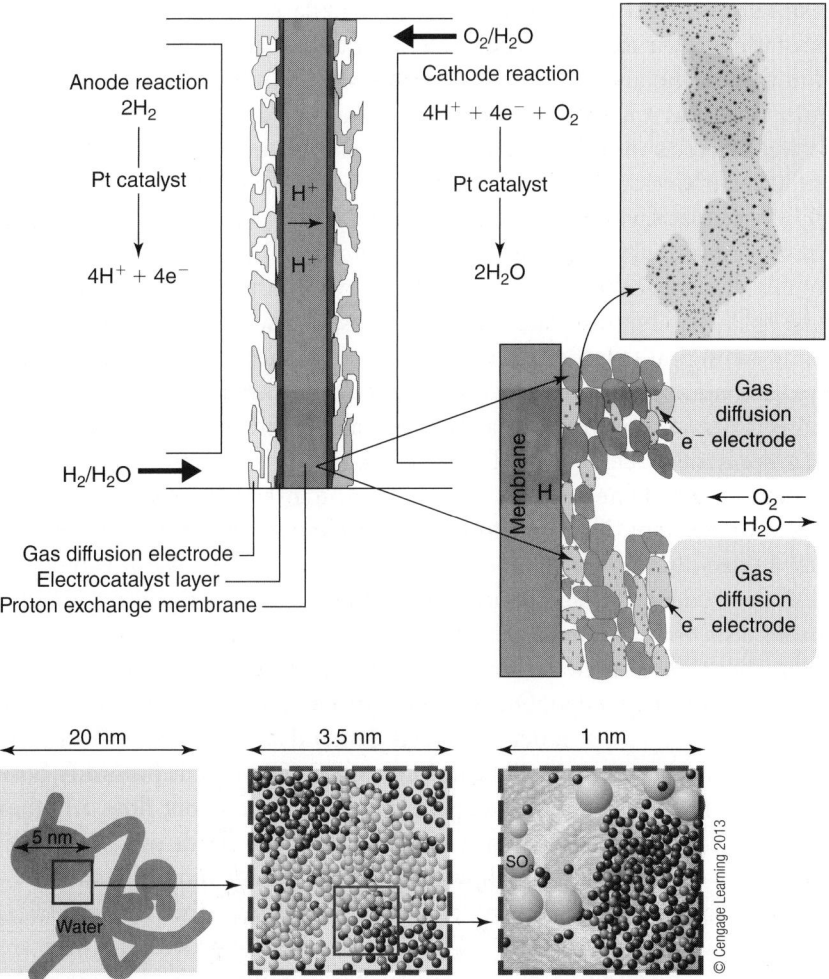

Figure 6-1 (Continued)

combines with the proton (H^+) and the electron (e^-) to make the water molecule comes from air in most SPFC (PEM) fuel cells. This is as opposed to supplied oxygen, which is generally pure oxygen, while there is approximately 21% oxygen in air. This air is usually warm or even hot, since the fuel cell itself runs at higher temperatures, and it will dry out the system unless supplementary water is added beyond that produced at the cathode. If the cell runs at temperatures over 60°C, then supplementary water is usually needed in a SPFC (PEM) fuel cell. In many systems, this water is added as a vapor to both the hydrogen gas at the anode and to the air at the cathode to humidify them. This enables a very fine level of control to the system rather than just dumping liquid water in. In general, input air should be in excess of 80% **humidity** for SPFC (PEM) cells to operate efficiently.

All of these issues are well known in other systems and can be dealt with in fuel cells as well. For instance, the positively charged hydronium ion (H_3O^+) moves to the negatively charged electrode at a certain velocity when the system in running at **steady state**. Since the ion has mass and volume and is charged, other molecules will cause resistance just as there is resistance to you moving through a large crowd of other people. This is called **electro-osmotic drag** and is proportional to the current flow in the system. It can be used to calculate how much water and how many protons are moving in the system. The amount of water diffusing from the cathode back to the anode is also dependent on the thickness of the membrane and the gas pressure within the cell and can be calculated. The amount of water lost to evaporation at the cathode is dependent on temperature, pressure, and the relative humidity of the air entering and can be calculated. The results from all these calculations (and a number of others) can then be compared to determine how much water needs to be added as humidifying water to the two input gas streams so the cell does not dry out or flood.

Additional calculations show what gas pressure (fuel and input air) is required to balance the system, what temperature the system will be at when operating, what current can be expected from a given system configuration, and a number of other things. These will be the system specifications, and will be published so that any deviation will signal trouble. Maintenance is improved if technicians understand where such numbers come from and why they are important.

In smaller cells, careful design can balance a system to eliminate the need for added components such as humidifiers. Working backward from the exit air, which is generally maintained at 100% humidity, the internal parts are designed to guarantee water flow but prevent **flooding**. Anode to cathode current sets the flow of water in one direction while the gas pressure along with gas diffusion layer thickness and electrode thickness balance the flow to prevent **backflooding**. Usually, even in smaller cells, lowering the total cost by eliminating a humidifying system is offset by lower overall efficiencies within the cell, so more cells will be required for any given current loading needed. One of the requirements of design is to balance all the variable costs to give a product that meets demand at a cost that will not only make a profit but that the market will bear. Most SPFC (PEM) cells humidify both input gas streams (fuel and oxidant). In general, a smaller current flow, a higher temperature, a lower pressure, or a higher reaction rate at the anode or cathode will result in the system not creating enough water and thus require humidification.

In larger systems humidification systems are used because the heat, electrical load, and gas pressures of the system require extra water be provided. There are a number of ways to humidify a gas stream, and no single method is currently used. Many systems are simple enough, such as bubbling the gas through water or running it over any material that will wick from an accompanying water source to the gas stream. Wicks have also been used directly in the membrane, with water being drawn into the cell from an outside source. Water can be sprayed directly into the gas stream in a fine spray, a method that works very well on large systems since it will also serve to cool the gas stream and thus limit heat buildup in the system as a whole. Water can be pumped directly into the membrane area as well, but this complicates the pressure balancing system since gas and water are both driving into the same spaces. Water can be taken out of the exit gases and recycled to the input gases using wicks or diffusion membranes that take the water from exiting gas and transfer it back to the input gasses. Water can be stored in the membrane itself if suitable materials such as silica (SiO_2) are included as part of the membrane, since these particles will attach water if there is excess and release it if there is a deficit. These types of materials are called **hygroscopic** and are used in things like toothpaste to keep the material from either drying out or getting too runny.

AFC

Water in AFC systems is a problem just as it is in all fuel cells. In AFC systems, water management depends to a great extent on whether the electrolyte is **recirculated** or not. In recirculating systems, the issue is relatively simple because the gas pressure on the fuel side of the anode tends to move the water into the electrolyte. It builds up there and when the system moves the electrolyte out, it takes the water with it.

This water can be removed through evaporation. Keep in mind that even in systems where the alkali (or acid) percentages are high, they are still partly water anyway, since most acids or bases (alkali) are water based. At least these types of acids are, and we won't get into other types of acids. When water moves into the electrolyte, in essence it is lowering the alkali percentage (becoming a less basic, or more acidic, solution) of the electrolyte. When water is removed from the alkali liquid through evaporation, it means the alkali percentage is being raised (becoming a more basic or less acidic solution).

Water management is not just maintaining the right balance. In an AFC system, water is produced at the anode. Producing more water means more electrons are being produced so more current is being produced. There is a limit to how much water can be produced in each cell because there is a limit to how many electrons can be produced. If the water is not removed, it builds up. If there is too much water in the AFC cell, not only will it block fuel gas access to the catalyst sites, but it will force the reaction producing electrons back the other way, in essence breaking the water back apart. This is one of the fundamental principles of chemistry, in that having too much of what is on one side of the double arrow in an equation will force the reaction back the other way to even things out. As mentioned above, the buildup of gas bubbles is one way electrodes can be destroyed, and this reversal of the anode reaction to produce hydrogen and oxygen gas (**electrolysis**) is how the gas bubbles occur. Figure 6-2 shows an electrolysis cell. Notice how similar it is to a fuel cell.

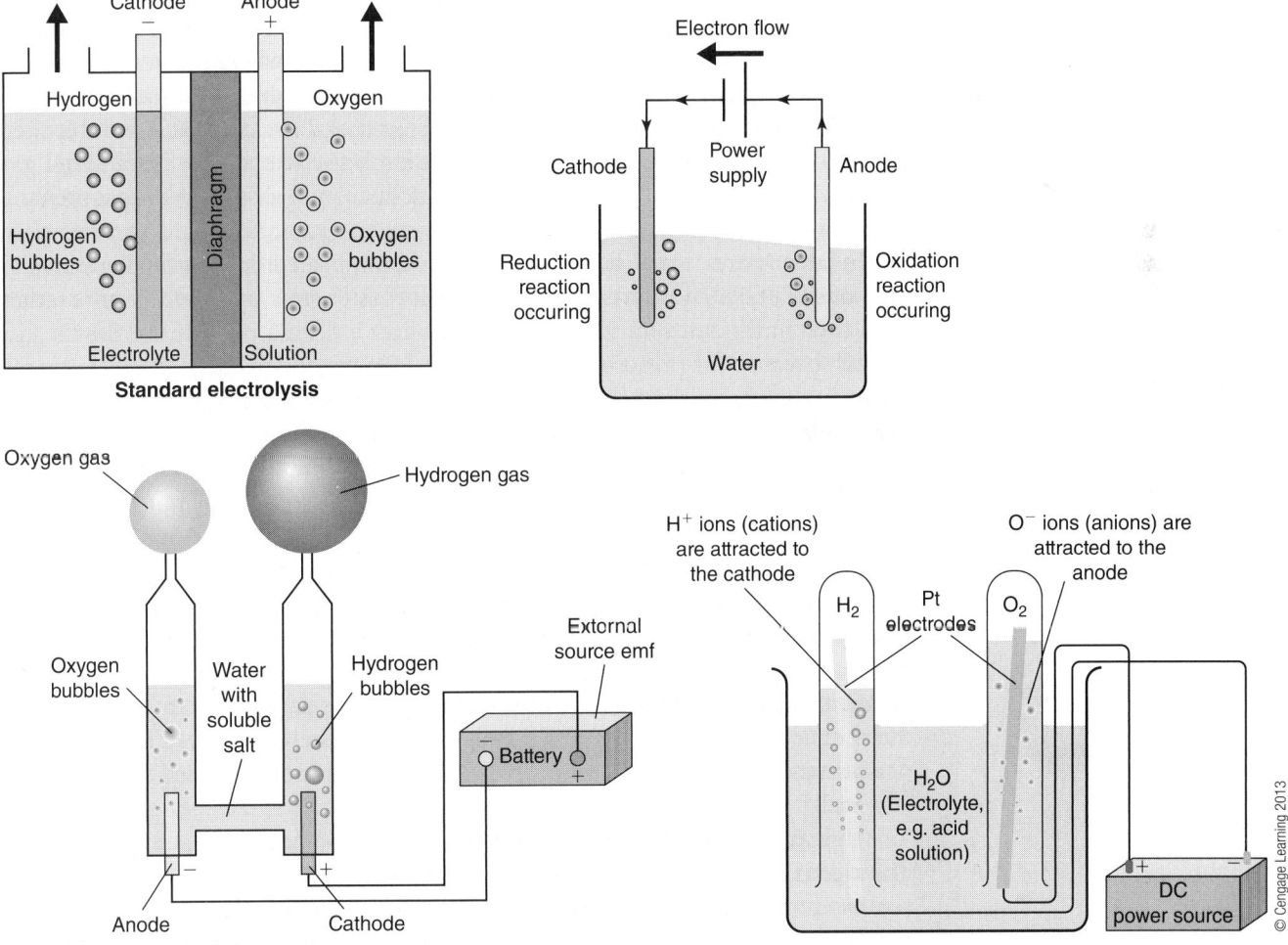

■ **Figure 6-2** Electrolysis cell.

While too much water at the anode can occur relatively easily in AFC systems, too little water at the cathode is also a problem. The chemical reaction at the cathode consumes water to form the hydroxide ion (OH^-) and lack of water there means no hydroxide enters the electrolyte. Since the electrolyte is made up of hydroxide ions (OH^-), potassium ions (K^+), and water (H_2O), drying up the cell also removes hydroxide ions, since they are still being consumed at the anode. Thus, the concentration of two of the three species that make up the electrolyte soon falls, leaving the potassium ions (K^+) getting more and more **concentrated**. At some point, the electrolyte breaks down because the positively charged potassium ions (K^+) are in such large numbers negatively charged ions cannot pass through.

Water management in fuel cells is more than just moving water then; it also requires a good grasp of fuel cell chemistry to handle correctly. That is not to say that fluid flow can be ignored though. The movement of fluids in fuel cells is also complicated. Remember that two pore sizes are required in most fuel cells, a small pore to transport the fuel or oxidant gas to either anode or cathode and a larger pore to transport the water.

If pores are too small, water cannot pass through them. The water **drags** along the side of the pore just like a small stream of water will not run down a rough surface. Since pores have surfaces on all sides, once they are small enough, the drag from the surfaces will overlap and the pore will plug. The plug will force water into other channels and the only way to move it is to increase the pressure behind the plug. Once the force of the pressure exceeds the drag on the water along the pores walls, the plug will start to move. If too much pressure is used, the water moves in past the catalyst sites and into the smaller pore structure where gas normally flows. The water will then plug these given the much smaller size of the gas flow pores. The only way to remove it so gas flow is restored and the catalyst site unblocked is to increase the gas pressure pushing it back the other way (input fuel gas pressure). That starts not so much a tug-of-war but a push-of-war as the gas pressure fights with the water pressure.

In larger pores, water flow is much more complicated than in smaller pores. If the water is under constant pressure, it will flow uniformly in the entire pore rather than flow in the bottom of the pore, like water will in a pipe. If the flow is fast enough because the pressure is high enough, the flow distributes relatively evenly around the walls of the pore and leaves an open space in the middle of the channel. That open channel will allow gas to flow within the moving liquid. As long as the gas pressure is not too high to turn the liquid flow **turbulent**, the liquid can flow in one direction and the gas in the opposite direction. Both liquid and gas can move in either laminar flow or turbulent flow. In fuel cells, maintaining consistent laminar flow is critical to the operation of the cell. Figure 6-3 shows the difference between laminar and turbulent flow.

If the pore sizes, the gas pressures, and flows as well as the liquid pressures and flows are all balanced, then all the catalysts sites will be available for reactions to occur, all the sites will have the three requirements for reaction to occur, and all the chemical reaction species will have paths to go where they need to go. The cell performance will be optimized. If balance is not achieved, the cell will not work very well.

This multiple flow within an AFC cell was important to space applications. The electrolyte was not recirculated in those designs, but excess water could be removed by using excess hydrogen fuel at the anode. This excess gas would not react (all the catalysts sites are taken up with reactions) and could move through the larger pores used to transport liquid, picking up excess water (being humidified)

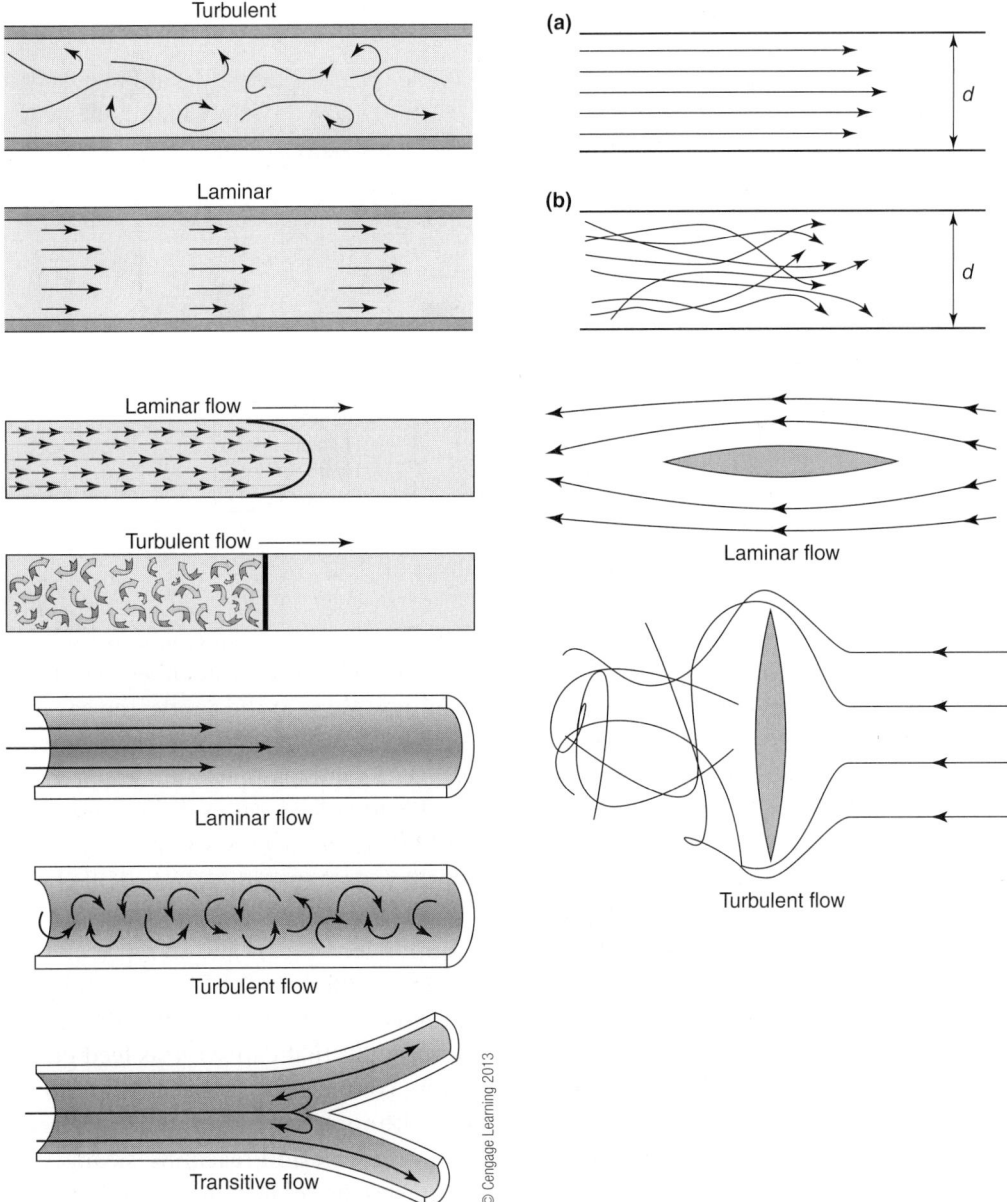

■ **Figure 6-3** Laminar and turbulent flow.

as it moved. The gas flowed back out of the cell and the water in it was evaporated out to supply the crew. Many fuel cells systems, especially smaller ones, still run versions of this type of water management, either adding water in or taking it out with the gasses.

MCFC

Water is produced at the anode according to Equation 8 (a and b). It is not used in the cell itself and must be removed. A number of different methods are used to remove the water based on the size of the overall system, but all involve what is in essence a gas venting. The water is produced as steam and mixes with the unreacted fuel gas (either carbon monoxide or hydrogen) in the pores of the anode itself. This gas is then vented out as part of the pressure control system of the stack.

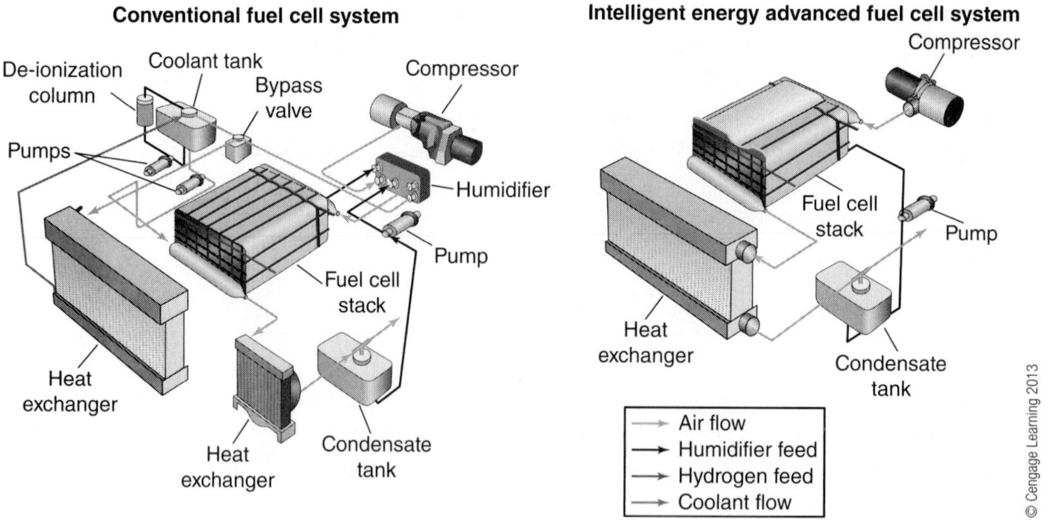

Figure 6-4 Conventional fuel cell liquid loop.

Separate systems may also be in place if an indirect internal reformation system is in place, since other gasses will be produced in the reformation chamber away from the anode chamber. If direct internal reforming is done at the anode, the gas mix removed from the main anode side of the cell will contain additional constituents and may require subsequent treatment, such as a combustion chamber where the unreacted gasses can be used for a combined heat and power cycle or a recycling unit, since the steam can be reused in the reformation reactions that produce the hydrogen or carbon monoxide. Figure 6-4 shows a conventional, simplified fuel cell liquid loop.

Both of the reactions used in reformation involve water and so represent the single most important requirements for water handling in MCFC systems where the main fuels of carbon monoxide or hydrogen do not need to be fed directly to the anode. While there are many hydrocarbon fuels that can serve as feed gasses for reformation, the simplest is **methane**. More complicated gasses such as natural gas or propane are treated in the same way although they can require pretreatments or multiple reaction steps. In the **steam reformation** reactions, steam is recycled through the system at high temperatures to the reformation anode (usually with a nickel catalyst) and the following reaction occurs:

$$CH_4 + H_2O \rightarrow 3H_2 + CO \qquad \text{Eq. 16}$$

The products of this reaction can be fed directly to the anode in a MCFC since both can fuel the cell. If a pure hydrogen feed is needed (as in other systems), then a secondary reaction, called the **water gas shift reaction** (water is shifted into a gas) takes place, shown below. Both of these reactions must have steam and occur at temperatures lower than the operating temperatures of MCFC systems. Both can also occur directly at the anode or in a separate chamber.

$$CO + H_2O \rightarrow H_2 + CO_2 \qquad \text{Eq. 17}$$

PAFC

Water management in the PAFC cell itself consists of moving the water produced at the cathode out of the cell. Water is removed to keep the electrolyte from being diluted and thus reversing the reaction at the cathode as we have discussed several times. As mentioned above, the cathode reaction does not happen as fast as the anode reaction in the PAFC, so when water is removed, the reaction tries to make more to compensate. Because the electrolyte is a liquid, water is not needed to the extent some other cells require. Water must be removed from the cells though, since it is not generally recycled or needed to keep the cell from drying out. It can be used to preheat fuel gas at the anode or oxidant gas at the cathode though since the cells usually operate above the boiling point of water, some preheating is needed.

SOFC

The water in SOFC systems must be removed from the anode or recirculated to **hydrate** the feed gas. Other than balancing the humidity of the fuel, it is not used in the cell itself and so management consists almost entirely of providing a pathway for it out of the individual cells and into either preheat or humidifying subsystems. It is a high temperature steam product and so it also has considerable use in larger systems where cogeneration, steam plants, or heat exchangers can be used to extract the energy and increase the overall efficiency of the system.

THERMAL MANAGEMENT

Managing heat is as important as any other aspect of a fuel cell, and in fact can be one of the main reasons for adopting the technology. As with most engines, fuel cells must operate within relatively narrow ranges and the operating temperature is one of those ranges. Unlike most engines though, fuel cells can be as efficient at lower temperatures as they are at higher temperatures. Managing the heat needed or produced in fuel cells is probably the one thing that produces the most return on investment in fuel cell systems.

SPFC (PEM)

Most SPFC (PEM) cells systems use graphite to transfer heat from inside the cell to where a cooling liquid (usually water) can be used. These plates usually perform several functions, collecting the electrons from the anode or transferring them to the cathode, acting as a sealing unit between cells in a stack, and distributing gas to both electrodes.

SPFC (PEM) systems are considered more efficient than say an internal combustion engine, but that is really not saying much. Even an engine considered very efficient does not do a very good job of getting all the energy in a fuel out. In general, SPFC (PEM) cells (which are considered very efficient) might convert only half the **available energy** to electricity with the other half converted to heat. Electrical generators (or any industrial process that produces heat as a **by-product**) can reclaim excess heat by using **cogeneration** processes where the excess heat or steam is used to drive a turbine (to create more electricity) or **combined heat and power** processes where it can be used to heat a building or water to create steam. Figure 6-5 compares the relative efficiency of conventional heat and power generating schemes to CHP schemes.

Smaller cells (100 watts and up) can be cooled using air. These are cells usually considered for commercial use in personal devices and operating at low

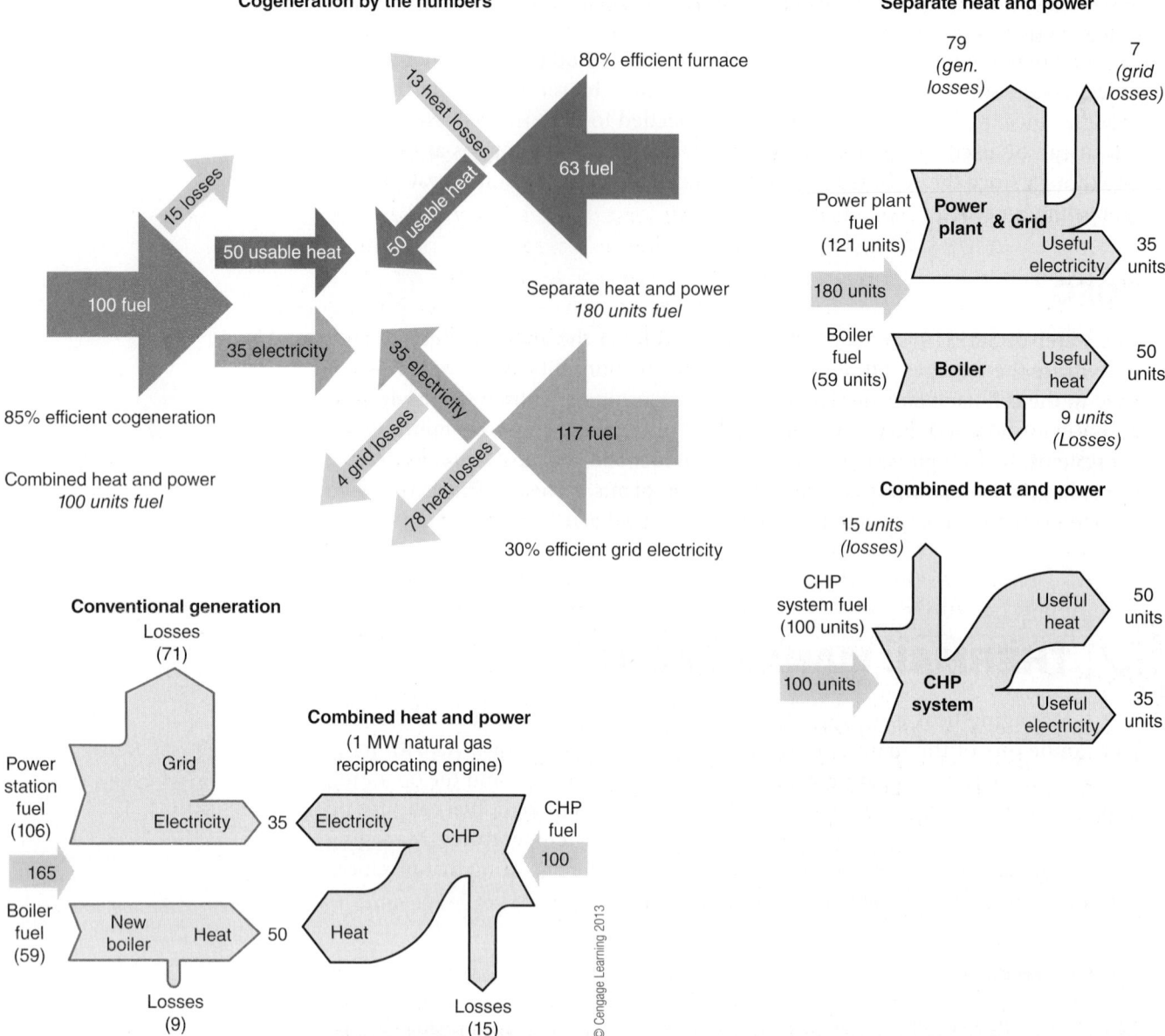

Figure 6-5 CHPO versus conventional heat and power systems.

temperatures so they limit the possibility of burning the owner or accidentally starting a fire if left unattended. The most common methods of providing air cooling capabilities to small cells is to flow air over the whole stack or provide cooling fins for the stack. When those methods are not good enough (above several hundred watts produced), cooling air can be run through channels in the bipolar (interconnect plates). The size or number of channels in the bipolar plate is limited by several design factors though, so that method too has its limits. Even though the bipolar plates are over three-quarters of the volume in most SPFC (PEM) cells, above about 2000 watts (2 **kilowatts**) produced, air cooling struggles with removing the heat generated without making the plates larger than other design criteria call for. Above 4 kilowatts, water cooling is generally required.

In combined heat and power (CHP) or hybrid systems (combining fuel cells with other power generating units like wind or solar), these numbers don't mean much, since the heat generated can be used for other purposes and not just vented to the atmosphere.

AFC

Fuel cells have relatively narrow temperature ranges where they operate most efficiently. This is a critical issue with fuel cells because the cost of the electricity they produce is high enough that even small losses in efficiency make them less attractive. Thus, the primary purpose of thermal management systems in fuel cells is to guarantee operations within a window that optimizes system efficiency and voltage produced. These two are not the same things. Raising the operating temperatures will produce less voltage but increase the efficiency (power produced per unit of fuel) up to a point. Further increases in temperature can improve secondary waste heat operations in any cogeneration system associated with the fuel cells but decrease the cell voltage and cell efficiency; the net result though might be lower fuel cell efficiency but an increase in the overall system efficiency. The acceptable range of temperatures then is not so easily determined and managing thermal operations does not necessarily mean keeping everything cool.

Heat is generated because there is no free lunch. Atoms and molecules interact with other atoms and molecules. The more atoms and molecules there are to interact with, the more expensive the lunch is. If you drag a piece of metal across another piece of metal, both pieces will heat up. You have used energy stored in your body to move the metal. Part of that energy moves the pieces from one place to another. Another part of that energy is turned into heat when the two metals drag against each other and some creates heat in your body. If you take your piece of metal and move it the same length but move it against a liquid, you do not generate nearly as much heat because the molecules in the liquid are not pinned in place like those in a metal; they can move away from the metal so they do not interact with the molecules in the metal. Move that same metal through the air and even less heat is generated because even fewer molecules exist in the air than are in either a liquid or a solid.

In an Alkali Fuel Cell, gas atoms move in a gas, whether the fuel or the oxidant stream; that generates some heat but not much. Ions move in a liquid alkali solution; that generates a little more heat. Electrons move in a wire; that generates much more heat. There is a list of things moving and each movement generates some heat. In general, if a fuel cell operates at 50% efficiency, it generates a watt of electricity and another watt of heat. Fuel cells run at efficiencies of around 60% (4 watts of heat generated for every 6 watts of electricity) if they are not partnered up with some method to recover the energy in the heat. In low temperature fuel cells like AFC systems, it is hard to recover that heat because while there is a lot of it, it is not at a high enough temperature to do anything with other than perhaps warm water up.

Because there is not very much of it though, low temperature fuel cells can manage their heat without much secondary equipment, which lowers the initial capital cost of the unit. Smaller AFC systems are perhaps the easiest systems to manage heat since recirculating the electrolyte is often sufficient to handle any heat that is generated. Since these systems usually run above room temperature, a simple heat exchange or water loop using the hot exhaust gas can just preheat the other gasses entering the cell to help increase overall efficiency a bit. An even cheaper way to handle limited heat is to use cooling fins. Simple **convection** is a cornerstone of low temperature operations for many types of equipment.

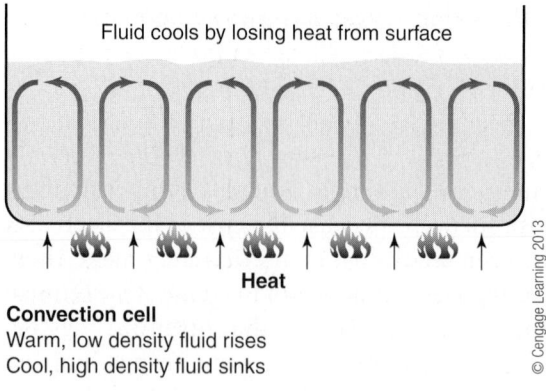

Figure 6-6 Heat convection circuit.

The fins heat the air and the air rises because it is warmer than the surrounding air; the resulting movement draws in cooler air to be heated and the cycle goes on endlessly. In some of these cases, the heat management is then transferred to the room the system is in, requiring an air conditioner to cool so the system does not overheat. Even that can be made part of the system, however. Remember there is excess water that has to be evaporated and that can be run through an evaporative cooler that then is used as an air conditioner in the room the system is in. Good **systems design** uses every part of the system to either reduce overall cost or to increase overall efficiency. Simple convection is shown in Figure 6-6.

MCFC

One of the most important thermal management schemes in MCFC systems occurs if reformation is done within the stacks to provide fuel at the main anode. The breaking up of hydrocarbons to form the two fuel gasses for the fuel cell itself (Eq. 16) takes heat from the surroundings to proceed (is endothermic). The reaction occurring at the cathode (Eq. 8b) produces heat (is exothermic). If direct internal reformation is used, the two parts of the anode (the outer part where reformation occurs and the inner part where the fuel cell reactions occur) see different temperatures. Since the thicknesses of these cells are relatively small, the internal stresses arising from the temperature differences and the cycling that occurs in between them can damage the anode. Equalizing this inherent temperature differential is key to not only a long lasting MCFC but also to an efficient one and much of this work is done by recycling the internal gasses at both the anode and cathode.

Within a cell, the anode and cathode are hotter than the electrolyte. No chemical reactions occur in the electrolyte, so no heat is generated there or consumed there by chemical reactions. The delivered air or fuel gas is usually heated so the two ends can be kept relatively uniform using good thermal management. Individual cells though are very thin and while there is a temperature differential, it is generally not of concern. When the cells are assembled into a stack though, the middle part of the stack becomes a thermal management issue. Hot gas is circulated around the outside of the stack; heating plates might surround the stack. The coolest part lies in the middle. Maintaining uniform heat there is a function of the fuel feed and air feed going into the individual cells. This is a difficult application, in that if the input streams are overheated to keep the interior up to temperature, they may in turn overheat the outer cells, lowering overall efficiency.

The cathode does not have the same issues as the anode because no reformation is occurring there (or near there). Instead, the cathode needs to keep an optimized temperature to minimize resistance. MCFC cathodes operate more efficiently (more voltage produced) as the temperature goes up. Above 650°C, this improvement flattens out and the added cost of more expensive high temperature materials outweighs the cell improvement, so that is considered by most to be the optimal operating temperature. At the cathode, air is mixed with carbon dioxide to form the carbonate ion. The carbon dioxide is recycled from the anode and the air is brought in from outside the cell. The air supplied to the cathode must be heated and a common way of doing this is to take the exhaust from the anode, **combust** what is left of the fuel (no fuel cells use all the fuel gas) to form additional CO_2 and H_2O, and heat. This can mix with the incoming air to heat it up. Some systems separate the unburnt anode fuel and recycle it back to the anode, sending only the anode product CO_2 to the cathode. Some larger cogeneration systems use CO_2 from an outside source such as the effluent gas from a completely different process.

While operating temperature generally determines the overall thermal management schemes, it is the start-up cycle that determines the most critical mission of temperature control. Even small **transient temperature differentials** (50°C or so) during the initial start-up and setting of the seals (some systems even do the final placement of catalysts during this phase) can have a substantial impact on the future operations of the MCFC system as a whole. Mistakes during this phase cannot be fixed, so a fine control of temperature throughout the stack is critical. Even though operating control may be relatively straightforward, additional control abilities must be put in place (at even greater expense) to guarantee critical control during system start up. Since the solid carbonate shrinks in size when it turns to a liquid during start-up (the solid to liquid transition almost always is accompanied by a reduction in the volume occupied except for special cases like water), the thermal management scheme must account for this so the ceramic holding the electrolyte is not cracked or bubbles formed. These start-up cycles can take well over 10 hours in larger systems and temperature control during that time is critical. In addition, if at any time during operations the temperature drops below the melting point, the liquid carbonate will freeze and expand, cracking the ceramic matrix, so provisions must be made to handle upset conditions of this kind as well. Many systems have secondary heating in place to keep the system at some minimum temperature even if operations are interrupted during startup.

MCFC systems are not designed for applications that required any type of cycling but rather for continuous output, and much of the thermal management system is used to guarantee long-term temperature stability of the entire system.

PAFC

As in other low temperature fuel cells, the main thermal management issue is limiting heat buildup. The heat generated during the chemical reactions taking place in the cell must be removed and the internal operating temperature maintained over a relatively narrow band. Both liquid and gas cooling systems are used, with the choice being made based primarily on cost. Air cooling is relatively inexpensive, requiring flow channels and fans for the most part while liquid cooling systems must have piping, manifolds, valves, and control systems. Liquid systems are considerably better at cooling than air systems though, so the reduced cost of air cooling is offset by larger internal flow structures and an upper limit to the size of the cell, since that will depend on the ability to circulate air within the stacks.

Water is the most common cooling liquid, but in most PAFC systems it is in the form of steam rather than a liquid. Most cells operate around 200°C, so liquid water would boil in the cell, expanding to cause troubles within the flow system and causing variable cooling rates in the cell, since areas nearer where the water enters would cool quickly while areas past the point where water boils would cool slowly. Cooling is done with a steam usually about 50°C lower than the operating temperature, which allows for reasonably quick, uniform cooling. Steam is corrosive though, so the same sort of precautions taken with boilers must be employed with the piping and steam production system. Liquids can also be used to cool, but they must have a boiling point beyond the normal or **upset temperatures** seen during operations.

Air cooling is done using fans to force ambient air through the stacks. Usually, the same carbon or graphite plates that collect electrons to transfer them into the wiring system are used to channel the air through a stack. Not every cell needs be cooled for the most part and some systems run up to five cells between cooling channels. As in most fuel cells, smaller units tend to use air cooling while larger units employ liquid or steam cooling.

SOFC

The main issue driving thermal management in high temperature stacks such as those operating in SOFC systems is the stable operating range of the stack itself. Although other factors such as inlet temperatures needed for contaminant removal systems (mainly sulfur) and fuel reformers are important for the overall operations of the fuel cell, the amount of air needed to control the temperature rise across the stacks drives the management system.

Optimizing the operating temperature is not a trivial matter in SOFC systems, since the associated equipment (**balance of plant**) is not well matched when it comes to the various operating temperatures needed. If gas turbines are used in cogeneration, their inlet temperatures should be greater than 850°C. If internal steam reformation is done at the anode, then the stack temperature should be above 700°C. The stack itself can operate across a wide range of temperatures running from approximately 500°C to over 1000°C, but the most efficient operating temperature for the electrolyte is generally above 900°C. The fuel reformation system in use, the output requirements of the particular application, the type of fuel used, the particular (and probably proprietary) electrolyte in the cell, and the requirements of the downstream integrated systems are all taken into account when setting operating temperature of a SOFC system. Even though the electrolyte is the most important variable, it can be better to give up some efficiency at the individual fuel cell level to make a larger gain with a downstream cogeneration system or upstream with improved fuel reformation.

In general, if the SOFC system is not going to be used with gas turbines in a combined heat and power cycle, then operating temperatures are kept as low as possible. In those systems, the lowest temperature is determined by the **ohmic resistance** of the electrode/electrolyte assembly, the **kinetics** of the reactions, and the resulting **cost of power** economics. These stand-alone systems are generally smaller units supplying power outputs of less than 15 kilowatts.

Two primary thermal management schemes are in use then, one for the higher temperature operations of systems integrated into larger combined cogeneration facilities and the other for lower temperature, units primarily meant to supply limited power meant for consumer application. Both systems typically use air as the cooling medium.

Thermal management of the fuel cell stack is relatively complicated in SOFC systems in part because the higher operating temperatures for these systems require input and output temperature streams that are balanced. Many of the SOFC systems use complimentary gas systems to accomplish this, one for the fuel/air inputs and one to recover heat. For instance, most fuel cells do not completely use all of the fuel supplied to the anode. SOFC systems take any hot air in the system, mix it with the remaining fuel coming unreacted off the anode and combust them in a separate chamber. The rapid expansion of the combustion reaction drives the compressor used to pressurize the new fuel gas going into the anode. This not only regulates the pressure of the system but preheats the fuel entering the cells. Higher temperature systems may recirculate the unreacted fuel gas back to the fuel feed stream to assist in the internal reforming of the new fuel entering the systems and provide thermal management of this stream as a preheating mixture. Waste heat from the fuel cell and reformer system is then transferred from this internal loop using heat exchangers. This heats an external air loop which regulates heat throughout the entire system and provides a ready supply of preheated air to the cathode. An external loop can also act to preheat the entire system during start-up to lessen thermal shock, an important consideration when dealing with solid ceramic oxides, since they do not tolerate fast heating or cooling at all well.

For SOFC systems, thermal management is not just about temperature control during operation. In planar systems, for instance, the seals between the individual cells in the stack are made of glass, which softens at operating temperature to seal the stacks. Many systems do not shut down even if fuel is interrupted or problems occur in the balance of plant to interrupt service. In these cases, especially for high temperature systems, thermal management systems act to keep the unit hot enough so the internal seals and solid oxide cells do not suffer damage during a rapid cooling event. The internal gas environment of the cells are also an issue, since the temperature and hydrogen atmosphere can chemically attack the oxide cells if too much fuel gas is unreacted or the temperatures get too high in localized spots.

Thermal management in all higher temperature fuel cells also have to do with materials of fabrication. Metals used in gas lines or shells or interconnects can only withstand so many degrees, and then they begin to slowly flow or their alloying elements diffuse through the system or they crack from the combination of stress and corrosion. The silica used in glass seals will migrate under pressure (up to 15 atmospheres) and temperature (>1000°C) to contaminate the cells. Close attention must be paid in higher temperature cells to how these materials behave when an upset condition occurs and a spot gets too hot or too cold for even a short period of time. Keep in mind that under the operating conditions used in most SOFC systems, any **breach and release** can cause serious damage to both personnel and facilities. The thermal management system must be capable of responding to problems, not just keep the system cool.

GAS SYSTEMS AND PRESSURES

The gas pressure inside a fuel cell is one of the most important variables for optimizing operations. It is also one of the most difficult to control given the nature of the **gas phase**. The various equipment and schemes used to control the internal pressures in a cell, stack, and system can range from nothing more than a small arrangement needing little more than a side mounted fan to miniplants tied to grid connected installations.

The type of gas control system used depends on the size of the unit and the type of fuel cell used as well as installation pricing needs. In this section, important different concepts will be introduced for each type of fuel cell, but the student should realize that all of these concepts are important to all of the other fuel cell types as well.

SPFC (PEM)

SPFC (PEM) systems can operate at various pressures depending mostly on the cost of the system and the electricity required for delivery. Usually, smaller systems of less than 10 to 15 kilowatts (for comparison, it takes a kilowatt or better to run a hair dryer) do not operate under pressure. The larger the system is, the more likely it will be run under some sort of pressure. Increased pressures are used because that increases the voltage produced by the individual cells. The increased voltage in general results from the chemical reaction at the cathode, where increased pressure lowers the energy needed to start and maintain the reaction. Remember that pressure is a form of energy and adding energy into a system always has some effect on what happens there. Many of the gas pressure issues in these systems arise from the need to keep water within the SPFC itself or the cell will stop working.

Pressurizing any fuel cell takes money. The cost may be in the supplied **raw material**, such as when a **compressed gas** is being used or in system equipment such as when a compressor is used. In systems where exhaust gas is available, then compressing gas is not too expensive other than the initial capital cost in equipment. Fuel cells create exhaust gas but at relatively low pressures, so if the system is going to be used to supply its own pressurized gas, it may need to compress the exhaust gas as the first part of the recycle loop. The system then has to supply the electricity used to run the compressor and that takes away some part of what the cell is generating. Smaller fuel cells running compressors using their own generated electricity typically consume around a quarter of what they make, depending on the size of the cell and how much pressure is needed. Compressed air can also be too hot to be used in a low temperature cell, not hot enough to be used in a high temperature cell and contain contaminants not suitable for any fuel cell. Additional heat exchangers might be needed in those cases as well as systems to clean the gas stream.

Pressurizing fuel cells can become a complicated issue in design, engineering and maintenance. There are thresholds involved in almost all machines and fuel cells are no different. You can make dump trucks larger and larger, but at some point, the tires will give out from the load or the amount of torque needed to start moving after the dirt is loaded will break something. In fuel cells temperature is one of those threshold issues. At around 60°C, the SPFC (PEM) cell starts to lose more water from evaporation than is being produced at the cathode. If it operates above around 80°C, some internal pressure must exist or the cell will dry out. In those circumstances, a lower pressure of around 3 **atmospheres** (44 **psi**) is used to keep the water from evaporating too quickly. Increasing the pressure increases the **boiling point** of a liquid because it is harder for the gas molecules to free themselves from the liquid surface if they have to move against a pressure. Again pressure is energy, and the energy of the pressure above the liquid must be overcome for a gas molecule to jump from the liquid into the gas above the liquid (boil).

Students should understand that boiling is not really the bubbling you see in a pan of water when it "boils" since that is water at the bottom of the pan (where the heat is most intense) actually boiling the liquid there. Since the resulting gas is lighter (and hotter) than liquid water, that bubble of water gas rises to the surface. The boiling point is the temperature where there is enough energy (as heat and so boiling point is temperature driven) available (usually in the liquid but not always) to free molecules of water (H_2O) from their neighbors and transition them into a gas, where they have no neighboring molecules. Remember that solids are chemically bound to the neighboring atoms and are pinned in place; liquids are bound to small groups of their neighbors with the groups not bound and able to move freely around each other; a gas is not bound to any neighboring molecules and can go wherever it likes (and does constantly). Increasing the pressure above the liquid means it take more energy (heat as temperature) to pop a liquid molecule up into the gas and that is why water boils at higher temperatures at sea level than in the mountains. The difference can be substantial. On the beaches, you boil spaghetti in water that is at 212°F (100°C), but cooking along the highway on the highest mountain pass in America (just under 15,000 feet of elevation) is done at around 185°F. By the way, it takes a lot longer to cook spaghetti in the mountains than on the beach, so be prepared.

Cell size also limits whether water management requires an internal pressure. If the cells and stacks get too big, then the area over which water can try to escape the surface to transition from a liquid to a gas will grow too large. When that happens, the water being produced at the cathode will not be able to get to the entire cell before it evaporates. Remember that water is being drawn into the gas as it moves through the cell (humidification), and the more gas there is, the more water goes in. The gas is going from anode to cathode so it starts to pick up water molecules as soon as it comes in contact with water. The larger the cell, the harder it is to balance the liquid and gas moving countercurrent within the polymer membrane and so the more likely water is being added to the fuel input gas as well.

It is important to understand that liquids do not need to boil to escape as a gas. Leave a glass of water out at room temperature and it will disappear; pour that same amount of water into a cake pan (increase the area) and the water will evaporate much quicker. Liquids evaporate for much the same reason that an electron will travel from a place where there are too many electrons to a place where there are too few electrons (as long as a path exists): in the liquid are a lot of some molecules, say water, and in the gas above the liquid there are very few of those molecules. In any setup like that, things will move to balance out. Evaporation has to do with the liquid and the gas above it trying to maintain a balance of pressures. In air, there is pressure on the liquid from nitrogen gas and oxygen gas and humidity as well as a bunch of other gasses (**partial pressures**), and since gas molecules are in constant motion, these all move. If a water molecule as humidity moves away, the liquid/gas interface will try and replace it with another one to maintain the balance of temperature, pressure, and volume in the system. The more area there is, the larger the system is and the more molecules are involved for both liquid and gas.

Size also requires pressure for a different reason. If the cells are too big, the water will need pressure to move through the very small channels within the cell and the longer the distance, the more the need for pressure. Large systems used for industrial production of electricity or very large systems used to generate power for the electrical grid will need a substantial amount of fuel and air (or oxygen) pressure just to keep everything moving along. This is an important issue for cells that will cycle on and off, since any vapor will turn back to water and remain in the channels. If the working pressure of the system is not high enough to overcome the force keeping the water on the channel (**surface tension**), then the cell will not be able to move air out and pressure will build up, damaging the cell. Surface tension is very important in fuel cells since water as both a liquid and a gas is so important. One of the defining characteristics of liquids is that they flow (move under shear stress). Inside a liquid, each water molecule interacts with other water molecules and they tend to move in unison when some force acts on them. But the surface water molecules only interact with half water molecules and half other molecules, whether that be a gas above them or a surface below them. The forces between water molecules and other molecules is different than between water molecules and other water molecules and so the surface molecule move in different ways and along different paths than the interior water molecules. This has several effects, ranging from bugs walking on water, since they do not exert enough force downward to break the forces between the surface water molecules, to water beading up on your car instead of flowing because the force between the surface of the car and the water is strong enough to hold that much water in place. Figure 6-7 shows this interaction of forces in liquids.

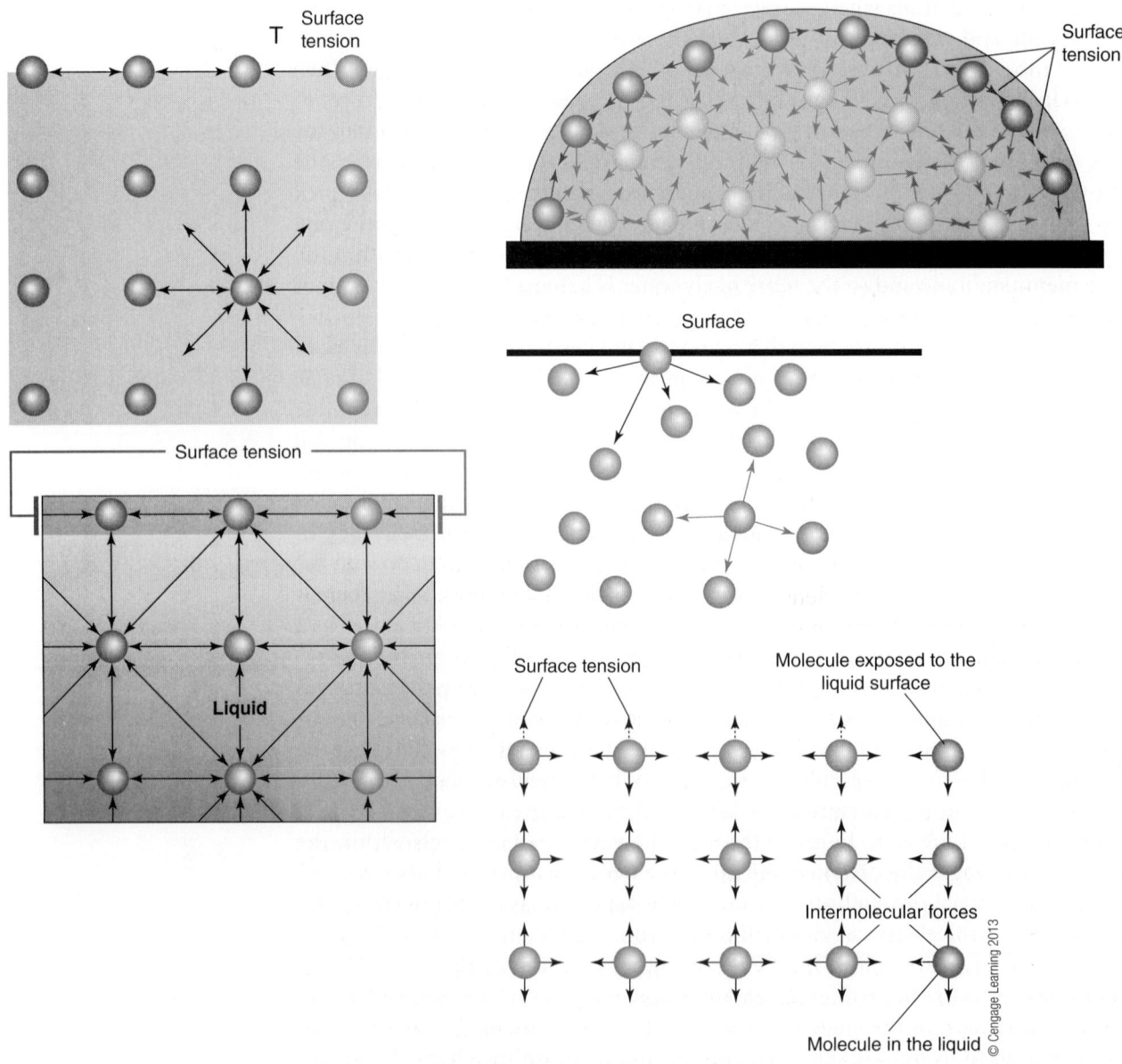

■ **Figure 6-7** Surface tension.

Pressure will also change the gas and water within the cell. If you take a handful of snow, it occupies so much volume. Density is the weight of the snow divided by the volume it occupies. By using your hands to pack the snow, you reduce the volume the weight exists in to get a snowball. The pressure within the cell will do the same thing to gas and water, making the gas and liquid flow characteristics within the cell somewhat variable as pressure changes. Gas is affected the most by this effect since gas can be compressed while liquids really cannot be. This has particular importance when either gas or water is removed from the cell, since moving from the high pressure cell to the lower pressure atmosphere outside the cell will cause expansion. This is why carbonated beverages will foam out of a can or bottle. The carbon dioxide (CO_2) gas inside the can is put there under high pressure, causing the gas to be compressed into essentially very small bubbles. When you shake the can, you add energy into the system, making the gas bubbles push back at the liquid, increasing the overall pressure within the can even more. Opening it up exposes the high pressure inside the can to the low pressure outside the can, and like everything in the universe, the high energy system inside the can will try to lower its energy by expanding into the lower energy system outside the can. Once the bubbles get outside the can, they no longer are being held down by the higher pressure and will expand into large bubbles. That is the foam you see. If the SPFC (PEM) system is not controlled, the same thing will happen to it when gas and liquid are discharged. The higher the operating pressure, the more complicated the associated systems (balance of plant) become.

Keep in mind that the water is not really lost from the cell even if it does transition to a gas. It is still in the cell and can be reclaimed from the exit gas. Many times, the water in the exit gas is used to humidify the incoming fuel and air gases to keep the cell from drying out, eliminating the need to use new water. This also helps to keep the water used in the cell clean, guaranteeing that contamination of electrolyte or electrodes is minimized.

AFC

Like temperature, the gas pressure within the cell will affect both the voltage produced by the cell and the efficiency of fuel conversion to electricity. In AFC systems, pressure is particularly useful used in conjunction with the effectiveness of the cathode reaction when it is occurring in an alkali liquid such as KOH. Increasing the pressure will increase the density of current flowing within the system, which is to say that more electrons (and more ions as well) will be moving in any given volume of material. Increasing the pressure in the system (since it is a form of energy) also helps to reduce the particular voltage requirement (**overvoltage**) needed at the cathode to get the reaction started as well as the energy needed to maintain the reactions. The reasons behind this are beyond the scope of this book, but can be looked at as a crowd at a sales counter. The more people in the crowd pushing to the counter with money in their hands (pressure and opportunity), the more sales there will be as opposed to there being only one or two people at the counter with money or a whole crowd but none with any money.

Like most things in fuel cells, there are trade-offs to pressurizing the AFC. As the pressure inside the cell increases, the cost does too, especially if the electrolyte is recirculated, since the entire liquid system has to be at the same pressure. If you use a lower pressure system to decrease the cost, you generally have to raise the temperature to get an acceptable voltage and efficiency. Raising the temperature means you have to increase the percentage of potassium hydroxide (KOH) in the liquid alkali electrolyte or it will boil off (the cell is above the boiling point so it is not evaporating but actually boiling). If your alkali solution is too basic (as opposed to acidic), it will be solid at room temperature, meaning you have to keep the recirculating system at the higher temperature. Just as mentioned above, you will then have to heat it all back up if you turn the system off and that is difficult. Remember that one of the advantages of an AFC system is that you can turn it on and off more easily than most other fuel cells. A system that first has to heat and then pressurize the cell and associated systems may cause the economic advantages inherent in a quick-start cycling system to be lost. System temperature, pressure, and volume must be balanced depending on the size and application of the fuel cell for it to operate correctly.

For smaller systems that produce several thousand watts (several kilowatts), the pressurization may be free. Smaller systems might use a pressurized tank of pure hydrogen for a fuel to eliminate the need for purifying gas. The pressure for the fuel cell is then provided by the tank, lowering the capital cost considerably, since both pressure and purity requirements are provided by the fuel gas. This in turn raises the cost of the fuel gas though. Even without the tank, pressure can be provided if the fuel and oxidant gas are first heated, since heating causes expansion which raises the pressure.

MCFC

The principle issue with the gas supply in MCFC systems is essentially the same as in other fuel cells. Gas must be supplied uniformly to all the cells in a stack and to each stack in a system, no matter what type of fuel cell is being used. Not only must the amount (**volume**) of delivered gas be uniform across the system, but a constant pressure must be maintained as well to ensure efficient operations (pressure in part determines the ohmic loss across a cell). In general there are two ways to supply gas to fuel cell stacks, **internal manifolding** and **external manifolding**.

In external manifolding, all the stack components are the same size and sealed. A bipolar plate (interconnect) is at one end, then a cell anode, an electrolyte, a cathode, and another bipolar plate (interconnect). The anode is fed vertically or horizontally and the cathodes are fed at 90 degrees, so that if the anodes are fed from the top and bottom, then the cathodes are fed from the two sides. This allows fuel to flow in one direction and air to flow in the opposite. External manifolding is cheaper and allows for a lower pressure gas supply to the cells but has troubles maintaining seals due to the temperature differential between the inside of the stack and the outer manifolding brought on by the two-directional gas flow since fuel gas will not be at the same temperature as the oxidant gas. Figure 6-8 shows a typical external manifolding arrangement.

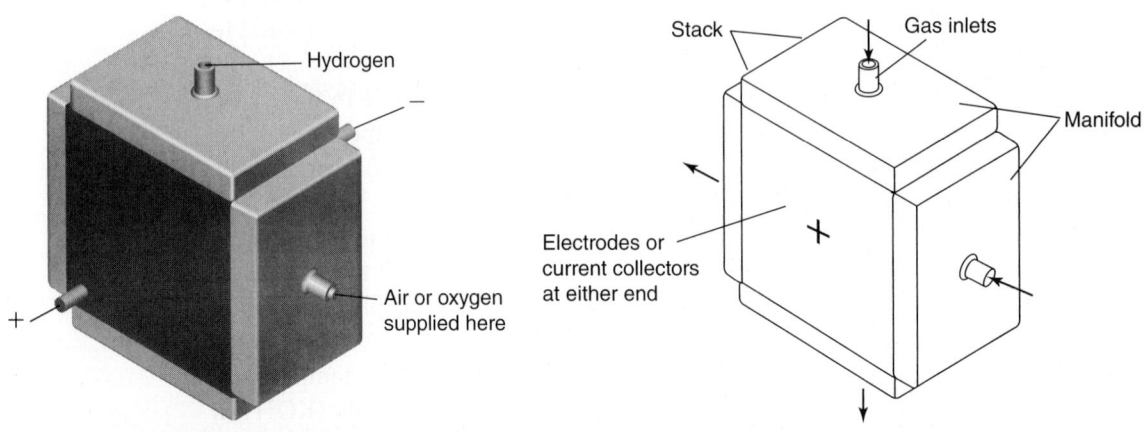

■ **Figure 6-8** External manifolding.

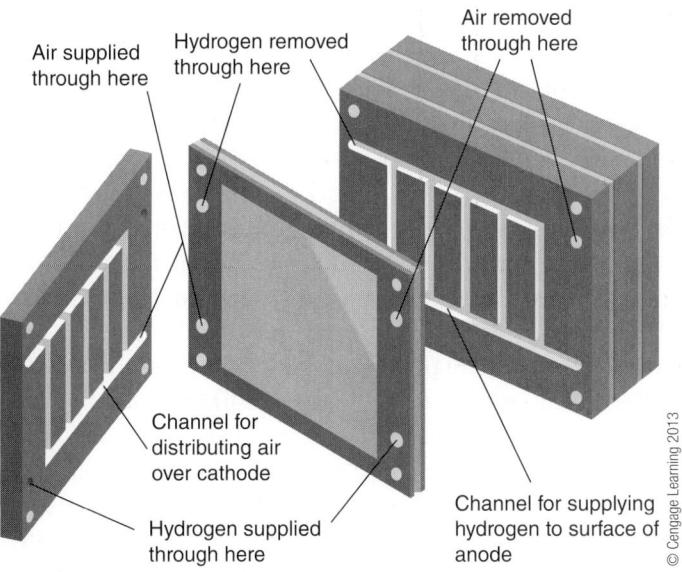

Figure 6-9 Internal manifolding.

In internal manifolding, the individual cells will be smaller than the electrolyte plate and the bipolar plates (interconnects). Holes are drilled through the stack (usually at opposite corners) to create a flow chamber for the fuel and air supplies. The internal flow channels in each bipolar plate (interconnect) are hooked up to one of the internal supply holes in the corners. Since there are two flow channels on either side of the plates, the fuel gas will connect to one set of holes and the air flow will be hooked up to the other. Stacks can pile individual cells one on top of another (vertical configuration) or one beside another (horizontal configuration) but most stack designs use horizontal configurations to limit the weight pressing on the bottom cells in the stack. Figure 6-9 shows a typical internal manifolding arrangement.

Internal manifold is considerably more expensive to produce (the more complex the design, the more expensive the product) but allows for much better flow, temperature, and pressure control for each cell in the stack. The flow channels can be designed in a number of configurations to improve cooling or pressure or flow as well. The channels can be round or square flow paths or even corrugated plates set tip to tip. Gas velocity and pressure can be changed by changing the area of the flow paths (larger areas leading to smaller areas increase pressure and velocity) and paths can be changed as they progress to either increase or decrease the amount of gas leading to a particular place in a cell. Recall that one of the problems MCFC systems face is different temperatures inside the stacks so that manipulating flow channels to provide more cooling or increasing the pressure to increase efficiency at cooler spots can be done using internal manifolding.

The pressure within the cell is very important in MCFC systems as it is in all the liquid based electrolytes. As in all fuel cells, one stream of gas enters at the anode side of the individual cell and a different stream of gas enters the cathode side of the individual cell. If one gas pressure is higher than the other, then the high pressure gas will move into the electrolyte and push toward the opposite electrode (**gas crossover**). This is particularly detrimental to the carbonate cells since the high pressure side can move electrolyte away from the high pressure electrode and toward the low pressure electrode, altering the dynamics of the chemical reactions and the movement of electrons into the wiring system. Differences in pressure can also damage the ceramic matrix containing the electrolyte. Since the seal in MCFC systems is the electrolyte itself, even smaller pressure differentials across the cell run the risk of breaking a seal when the electrolyte flows under the pressure.

Keep in mind that while high internal pressure cells may see no difference between anode and cathode pressures if care is taken to balance them, they will see differences between the internal pressure of the cell and external pressure of the atmosphere. Liquid based electrolytes in particular might have problems because seals not only keep what is inside the cell in but keep what is outside the cell out. Internal cell leaks will affect a system, but leaks to the outside will shut a system down. To limit this problem, some high pressure systems contain the stacks within a pressure chamber, usually with a nitrogen gas (nitrogen wont react with anything inside the cell and is already there in large amounts since the air is mostly nitrogen) maintained at the internal pressure used in the cell. This adds substantial cost to the system though.

High pressure improves performance and efficiency, so systems tend to push the limits of pressure within a stack to maximize electricity production. Keep in mind that balancing high pressures within a stack of cells is more complicated than simply running everything at atmospheric pressure, particularly with external manifolding. Tracking pressures and balancing flows is expensive and in general, smaller systems producing several kilowatts of power will not be attractively priced using the more complex components. Larger systems producing hundreds of kilowatts are more likely to have control systems in place for higher pressure systems. High pressure systems usually run below 5 bar (100 psi) as the capital and material cost for higher pressure systems is too great.

Somewhat beyond the scope of this book but important enough to mention is the contribution of gas pressure to whether the liquid electrolyte will be acidic or basic. Usually, molten carbonates are considered acidic. The amount of gas at any time in a closed container is determined by the pressure of that gas in the container. Higher pressures mean more atoms or molecules of a gas will be present within the individual cell and just like liquids evaporating molecules up into a gas, the gas can pass molecules back into the liquid to balance the system as well. Some gasses, such as carbon dioxide (CO_2) will cause the electrolyte to become more basic, and the more molecules present, the more basic the electrolyte is. Putting it differently, the higher the CO_2 pressure in the cell, the less effective it will be at passing ions from one electrode to another and thus the less efficient the cell will be.

Pressures also increase as temperature increases. One of the laws of chemistry (the ideal gas law) relates temperature, pressure, and volume of a gas. If the volume of the container holding the gas does not change (like in a stack), but the pressure goes up, the temperature goes up. Temperature and the pressure will go up in unison as well as going down in unison. Consider our liquid passing molecules up to a gas sitting above it. If the temperature of the liquid increases, there is more energy available to kick more molecules up and the more gas molecules there are in a sealed (constant volume) container there are, the more pressure there is in the container. That is the definition of pressure, more or less, the number of molecules available to bang into the container walls.

Thus, since MCFC systems operate at very high temperatures, the gas pressure will already be higher than atmospheric pressure. This is further complicated by the fact that as the lower temperature feed or air gas enters the higher temperature cell, it will try to even things out by expanding into the flow paths in the bipolar plate. Moving into a lower temperature part of the stack will cause the gas to decrease in pressure and pull back from the flow channels. Maintaining constant flow and pressure within a stack, no matter what kind of fuel cell, is not easy because there are so many issues to keep track of.

Higher gas pressures will cause the nickel oxide (NiO) cathodes to degrade faster as well. In particular, carbon dioxide (CO_2) and water (H_2O) will react with the nickel oxide and move nickel into the electrolyte, eventually destroying the cathode. Cathode lifetimes can be severely compromised (lasting only one quarter of the time) if excessive pressure builds at the cathode for long lengths of time.

This same problem can occur at the bipolar plates (interconnects) since some metallic component must be present there to provide a path for the cells current. This is a particularly complicated issue at the electrodes, since one electrode operates under a **reducing atmosphere** and the other under an **oxidizing atmosphere**, placing the bipolar plate that is in contact in with both under different chemical conditions.

PAFC

As in many fuel cells, the gas system within the cells and stacks of cells are controlled by what goes between the cells in the stack. Plates that distribute the gas (bringing it in or taking it out), collect current, serve to separate the anode and cathode so gas does not mix or even provide structural strength are one of the most important parts of the stack system fuel cell systems rely on. In PAFC systems, graphite plates are common, with one side providing gas to and removing gas from the anode while the other side does the same at the cathode. These provide an added feature in PAFC systems (but no others) in that they also are used as an acid reservoir. That makes them one of the more expensive of such plates, since they have to resist acid attack as well as not react chemically with either the fuel gas on one side, the air on the other or any of the other gasses produced in the reactions. Figure 6-10 shows the graphite structure, composed of carbon atoms arranged in a stacked, platelike arrangement. Carbon can arrange itself in other types of structures as well, such as diamond.

Graphite works well until the cells are pressurized. Even lower pressures (three atmospheres) will cause a substantial decrease in the operating lifetime of PAFC systems, since the graphite rapidly degrades in a pressurized fuel/air/acid mix. Heat treating (at 2700°C or 4892°F) the graphite increases their durability but also adds cost. Note that temperatures of that magnitude are very high temperature for any industrial process and require special furnaces.

Gas manifolding for PAFC systems is generally external manifolding with the vertical component feeding one electrode while the horizontal component feeds the other. Since the cells operate at low temperatures, fueling the cells does not require internal gas loops from one electrode to the other; the only gas formed is steam at the cathode, which can be handled as a single output. As in most fuel cells, the hydrogen fuel is not entirely used so the anode side also requires an output gas loop to handle the unreacted gas. This represents the simplest gas handling arrangement possible for a fuel cell, where a gas (either fuel or air) enters the cell, contacts a porous electrode, participates in the catalyzed reaction and then is carried back out of the stack as a different gas mix. The temperatures involved require either heating or cooling, especially on the input side, but in many cases, this can be assisted by heat exchangers between input and output gas streams. As the cell temperatures go up, this handling system becomes more complicated and more costly.

Some of the systems will run under pressures of up to 8 atmospheres. As in AFC systems, when the strong acid system is pressurized, there is a safety enclosure around the cells, sometimes as part of the stack itself, which keeps hot acid gas or liquid from escaping. Many of these surroundings are pressurized as well using an inert gas such as nitrogen as a further safety precaution.

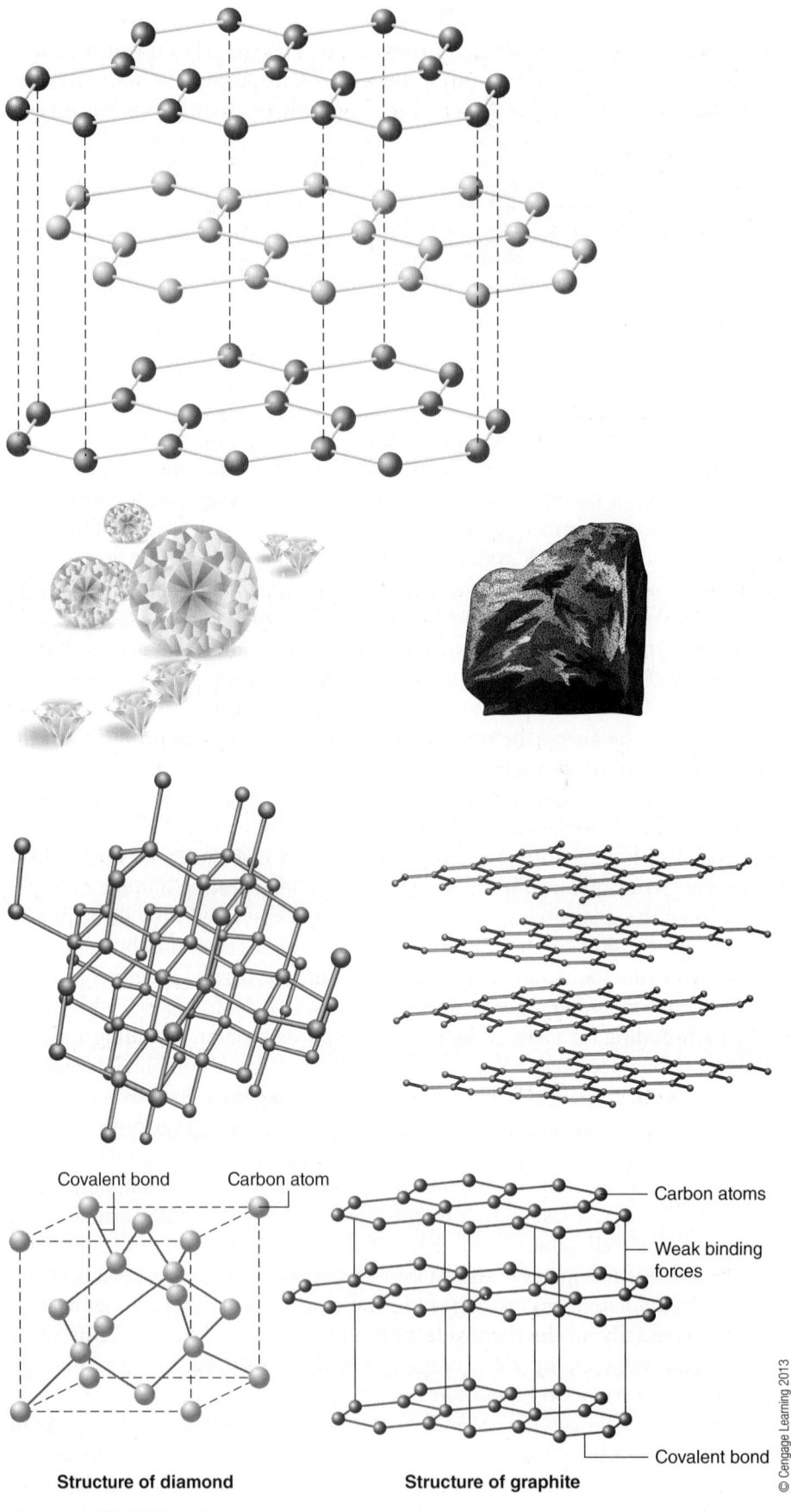

Figure 6-10 The graphite structure.

KEY WORDS

Knowing the terminology used is critical when dealing with fuel cells. Following is a list of the important terms in this chapter, which are also in bold typeface within the chapter. It is recommended that students are required to submit definitions of some of these words as homework assignments in which they look the terms up in other books, articles, or on the Internet.

- atmospheres
- available energy
- backflooding
- balance of plant
- boiling point
- breach and release
- by-product
- cogeneration
- combined heat and power
- combust
- compressed gas
- concentrated
- convection
- cost of power
- dehydration
- drags
- electrolysis
- electro-osmotic drag
- external manifolding
- flooding
- gas crossover
- gas phase
- humidity
- hydrate
- hygroscopic
- internal manifolding
- kilowatts
- kinetics
- methane
- ohmic resistance
- overvoltage
- oxidizing atmosphere
- partial pressures
- pound per square inch (psi)
- raw material
- recirculated
- reducing atmosphere
- saturation
- steady state
- steam reformation
- surface tension
- systems design
- transient temperature differentials
- turbulent
- upset temperatures
- volume
- water gas shift reaction

DISCUSSION QUESTIONS

1. What would happen if a SPFC(PEM) cell became dehydrated?

2. What type of cooling system is used to maintain the right operating temperature in a low temperature, recirculating AFC system stack?

3. Why is the cooling system of an SOFC system more complicated than the cooling system of a SPFC(PEM) system?

4. What gas pressure is used in the electrolyte area of a SOFC?

5. If I put a lid on a pot of water, does the water have to reach a higher temperature to boil or a lower temperature?

6. Which is more expensive to produce, internal manifolding or external manifolding?

CHAPTER 7

FUEL CELL SYSTEMS AND SUBSYSTEMS

objectives

This chapter introduces the student to the systems used by fuel cells that are outside the cell itself. Producing, cleaning and storing the fuel for fuel cells is discussed in this chapter, with examples of several different types of systems shown.

INTRODUCTION

Fuel cells are small, produce limited amounts of electricity, are expensive to produce, might last only half a dozen years and don't really use the right kind of fuel for the infrastructure now in existence. For a century and a half those were compelling arguments against them but that is changing. If one cell does not produce enough electricity then stack several together, tie those stacks together into a box, and then tie those boxes together until you get what you need. If they are too expensive, improve them to lower the cost, sell them not as supplied power but as political and environmental necessities, then once enough have been sold, **economies of scale** will bring that cost in line with other electricity. If the fuel cells themselves have a limited lifetime, then make them **modular** so the cells can be replaced while the associated items remain in service, and then recycle the cells to lower the cost even further. If you cannot get fuel to them then make the fuel they need where the cell is.

These actions are beginning to happen as the industry moves from laboratory equipment to demonstration systems to commercial products. In general, systems consist of several modular subsystems that can be assembled into a range of products. Keep in mind that all of these subsystems are connected and that computer control is required as well.

First, fuel must be provided to fuel cells. Fuel cells rarely have access to a supply of pure hydrogen, and pure is an appropriate term here. Recall that some cell types must have levels below the parts per million (ppm) level or 99.9999% pure. External reformation of fuel gas such as methane or natural gas is generally done on-site to supply the hydrogen needed or internal reformation is done at the cell in higher temperature system. The type of fuel reformation system used is many times dependent on the amount of fuel needed.

Once the fuel is available, some type of fuel **quality control** subsystem is almost always used. Fuel cell systems must purify the feed gas to keep the cells from being destroyed by the contaminants. A number of systems are employed, some as simple as a **palladium membrane** and others involving multiple units such as **filter beds**. These units must be able to maintain a steady flow of gas, under variable conditions in cases where current draw is not uniform, at a steady pressure even while filters plug and feed consistently into the preparation units. Input and output channels must interface as well as control systems and possibly in-line analyzers to track quality.

The input streams of fuel gas and the oxidant stream (air for instance) must be prepared before being piped to the electrodes. Preheating or humidifying the gas streams is most commonly done. Depending on the cell type, the heat may come from a heat exchanger using hot gasses from the reaction zones, or steam might be delivered from the electrode reactions. Internal cell connections leading to the external preparation zone must be in place if external systems are used. Temperature regulation is required for these as are flow rates into and out of the reaction zone so instrumentation and control loops would be active.

The core of the system is the reaction zone, usually at some elevated temperature, containing the **membrane electrode assemblies** and the internal gas supply systems such as the **gas diffusion plates**. In addition, the current is produced and gathered in the core and external connection must be provided to the electrical system. This core must be accessible and modular, in that these are the units that will probably need replacement during the life of the fuel cell system. Reaction and current flow monitoring and control are required in this area.

Power conditioning is usually done at the cell itself. **DC to AC conversion** is the most common but other, more complicated conditioning is many times required depending on the installation. Uninterruptible or back-up power loops that kick in when the main supply fails are one such application where additional control is required. One of the important uses of smaller fuel cells is in transportation applications where mobile emergency or industrial units run the electrical systems when parked to limit the use of gasoline or diesel. The power system in these applications must be capable of suitable voltage output but also control loops that can kick in additional cells or stacks under increased load and cut out completely when the mobile unit starts the main internal combustion engine.

All of these areas require some sort of processing, either signal or computer, to enable stable cell control. A wide variety of control systems are used, ranging from simple PID or set-point controllers to complex processing interfaced with electrical grid systems.

PRODUCING THE FUEL USED IN FUEL CELLS

Fuel is expensive in large part because it has to be made from something else. One of the reasons why fuel cell systems have been around so long but have never been commercially successful is that the fuel they use is not generally available and what is available is considerably more expensive than competing fuels. That is changing but in reality, the cost of hydrogen as a fuel will probably be more than other competing fuels for years to come. This is because most hydrogen comes from those same competing fuels. Natural gas is a good example, since it is used in a number of fuel cell systems. The cost of natural gas to a fuel cell installation is the same to a conventional turbine based installation but the initial capital costs, the longevity, and the maintenance of the turbines are lower than the capital and maintenance costs of fuel cells. Natural gas can also be used directly in an internal combustion reaction, eliminating the need to reform it into hydrogen gas. However, costs change and other issues such as environmental regulations, scaling capabilities, or political pressure can take precedence over simple cost-benefit analysis. Turbine systems, for instance, are not generally available for light industrial uses such as hospitals or buses and by themselves are not as efficient as a fuel cell/turbine **hybrid system**. Fuel cells are becoming increasingly common and that will probably continue in the future and even accelerate because the number of possible applications for them is growing.

The fuels used in fuel cells are generally carbon monoxide (CO) and hydrogen (H_2). These are produced in a limited number of reactions and cleaned and handled in a limited number of ways. Hydrogen (H_2) is the principle fuel used in fuel cells. Carbon monoxide (CO) is also used in high temperature cells as a fuel. Producing CO is easy enough to do, just burn something; H_2 is not so simple. Several different methods exist to produce hydrogen, **reforming fuel gasses** (chemical reactions that split the hydrogen and carbon in fuels apart), **electrolysis** (using electricity to break the hydrogen and oxygen in water part), **gasification** (turning coal or biomass into a fuel gas), **liquid reforming** (turning coal, biomass or crops into a liquid fuel), and **biological production** (using microbes to produce hydrogen). The major problem with all these technologies is the cost of the energy produced when compared to conventional liquid fuels like gasoline or electricity from coal fired or nuclear power plants.

In the past, hydrogen production has been an industrial process where the gas is produced and then used as a feedstock for a **following process** such as producing ammonia. In the United States, over nine million tons of hydrogen gas is produced industrially as a feedstock every year. The ammonia (NH_3) produced by combining the nitrogen (N_2) in air with the industrially produced hydrogen (H_2) gas using a catalyst is similar to the types of reactions occurring in fuel cells. Over 100 million metric tons of ammonia is produced worldwide each year, most of it requiring hydrogen gas as a feedstock. Producing hydrogen then is not difficult, does not require developing new technologies, and major facilities exist in most of the world to do so. The major problem with current hydrogen gas supplies is that they are produced and consumed at major facilities and are not generally available in other locations. Fuel cells are being designed, built, and operated to take advantage of existing supplies of hydrocarbons rather than existing supplies of hydrogen. **Fuel reformation subsystems** become part of the fuel cells systems rather than requiring a massive infrastructure to deliver hydrogen. This approach allows a transition to fuel cell technology within the existing power generating grid and transportation network, minimizing the cost of transitioning to a different technological base.

Two primary sources are currently used to produce hydrogen, methane, and natural gas (natural gas is primarily methane with up to 20% "higher" hydrocarbons). Having mentioned them numerous times, this is a good place to discuss **hydrocarbon based fuels** in a little more depth. Hydrocarbons are generally liquids that have hydrogen (the hydro part) and carbon atoms. They exist in **molecular chains** of various lengths with the carbons atoms tied to other carbon atoms to form the chain. The single atoms of hydrogen are then attached to each of the carbon atoms in the chain.

Figure 7-1 shows the first two hydrocarbons, those with one and two carbons in the chain. Notice that two have similar names (-ane) but one name is slightly different (-ol). **Methane** is the simplest hydrocarbon, consisting of one carbon atom with four hydrogen atoms, and is thus the easiest to take apart, since all you have to deal with are carbon to hydrogen bonds. Carbon atoms usually bond to four other atoms

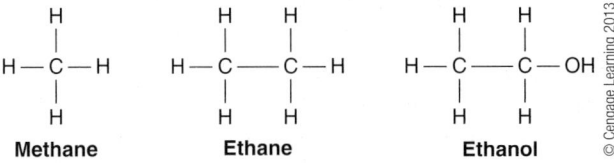

Figure 7-1 The first three hydrocarbons.

because the rules of chemistry dictate that form will be the most **stable**. Add another carbon atom to start the chain building, and you get **ethane**. Notice that the carbons now have only three hydrogen atoms each because one of the four bonds is needed to attach the two carbon atoms of ethane together. Hydrocarbon chains can have a number of other atoms involved as well. One of the most common carbon chains around (and particularly important to most students) is **alcohol**. These are simply hydrocarbons where instead just hydrogen atoms attached to the carbon atom, a molecule of oxygen bonded to hydrogen takes the place of one of the hydrogen atoms. The number of carbons in the chain determines the name, with methane having one carbon atom, ethane having two carbon atoms, **propane** having three, **butane** having four, and so on. The student should find these molecules familiar. The structure of Nafion shown in the last chapter is a hydrocarbon structure. Plastics and polymers are just modified hydrocarbons, and they are one of the main products produced by refineries along with fuel gas like gasoline, diesel, and jet fuel.

Hydrogen is produced when the hydrocarbon molecules (or other molecules that contain hydrogen like **ammonia**) are taken apart in chemical reactions. Many reactions will break hydrocarbons apart, like combustion where oxygen is added to make carbon dioxide (CO_2), water (H_2O), and heat. There are three types of industrial processes currently used to take apart the hydrocarbon molecule to produce a simple hydrogen gas: **steam reforming**, **autothermal reforming**, and **partial oxidation reforming**. Reforming is a general term that refers to breaking the hydrocarbon chain to get hydrogen gas as a product. Steam reforming, shown in Figure 7-2, is used extensively in industry; it is the lowest cost of the three and converts about 98% of the fuel gas if methane is used.

Autothermal and partial oxidation methods are not used as much in industry in part because they are more expensive and more complicated processes. All three of these methods are **endothermic** (requiring supplied heat for the reaction to occur, in general around 800°C) and **catalysts** in the case of partial oxidation (POX). Currently, about half the hydrogen produced in the United States is produced using steam reforming. A comparison between steam reformation and autothermal reformation is given in Figure 7-3.

In the past, the fuel used for fuel cells have been an incidental part of the systems, since many of these systems were used for space, research, or targeted applications where cost was not a primary factor. In recent years, as fuel cell systems have moved

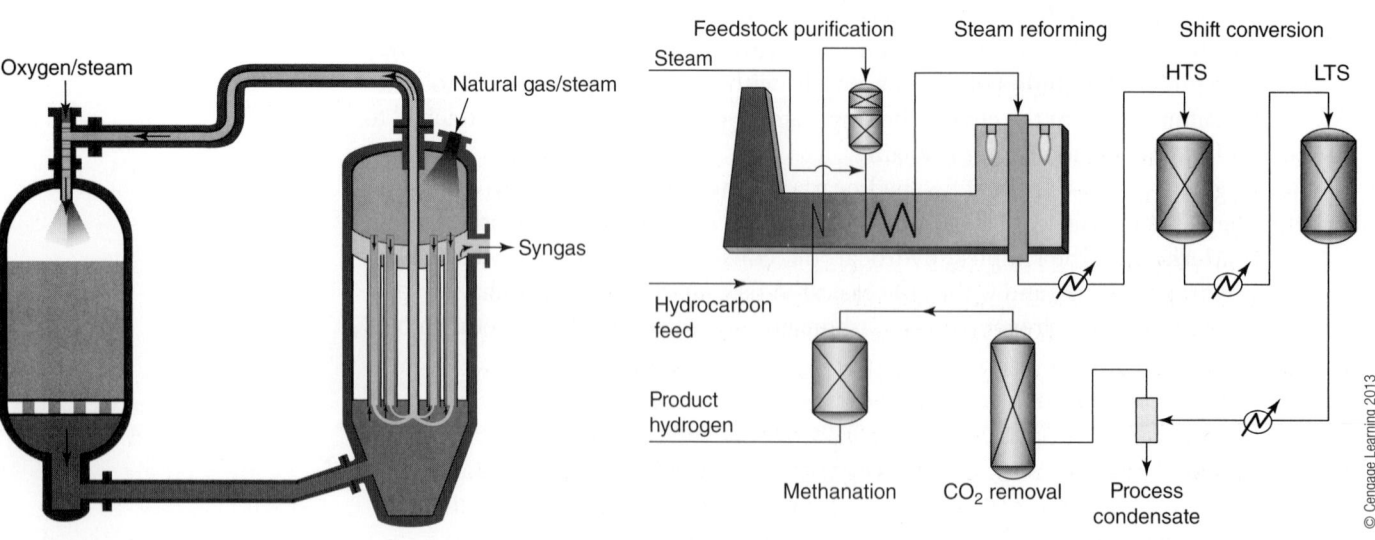

Figure 7-2 Steam reforming.

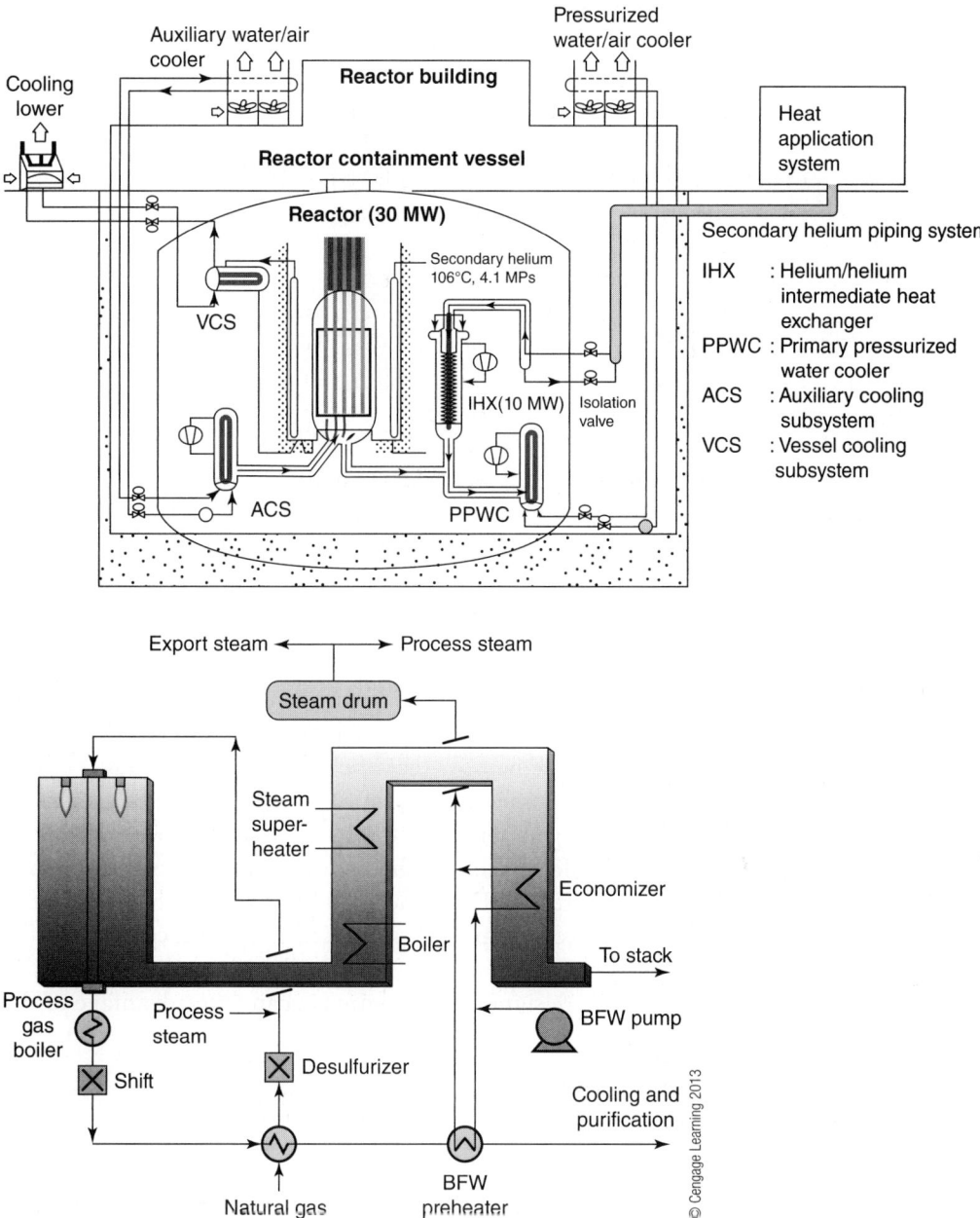

Figure 7-2 (Continued)

toward commercialization on a larger scale and their component costs have dropped, the fuel itself has become more important. Producing and then fueling thousands of fuel cell systems is no simple task; just as fueling cars or homes is a complex problem requiring massive infrastructures. Large industrial steam reformers are not well suited to localized systems and considerable work is being done in developing suitably sized hydrogen production systems that match the size of the installed equipment. In areas where industrial production is available, hydrogen gas distribution systems are being built to supply fuel gas as well. European countries in particular are building the infrastructure needed to supply transportation applications, particularly fleets, using large commercial reformers feeding centrally located distribution stations.

It is important to remember that fuel cells are currently limited by the availability of the fuel they use as much as by their cost. The capital costs of fuel cells are going down due to standardized systems, large production volumes, and improved

Comparison of Fuel Processing Autothermal Reforming - Steam Reforming

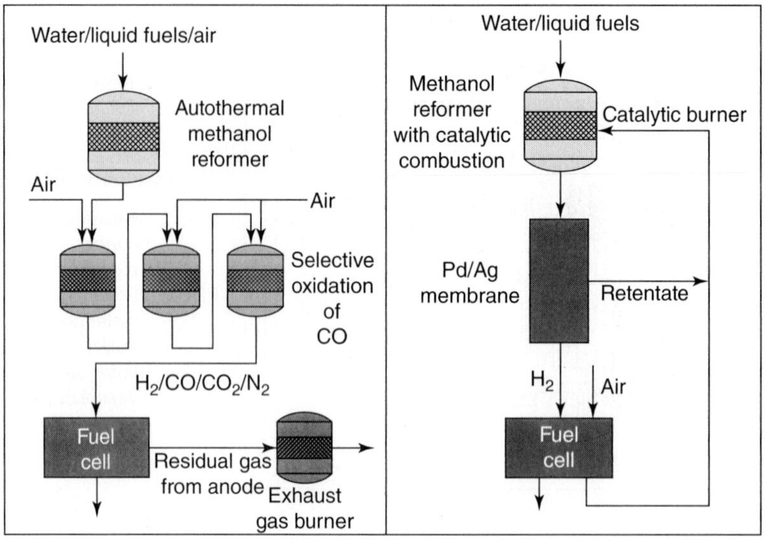

Figure 7-3 Comparison between autothermal and steam reforming.

fabrication methods. At the same time, the cost of all hydrocarbon based fuels is going up. These two factors mean that fuel cells will become more attractive as alternatives to conventional power systems. Remember that one of the main advantages of fuel cells is in how efficient they are. Some estimates put fuel cells at twice the efficiency of current large scale electrical generation methods, since most power plants run at less than 40% efficiency while gasoline or diesel powered cars are not even half that efficient, as shown in Figure 7-4. The student should consider the possibilities available in the fuel end of fuel cells as much as in the cell end of fuel cells.

Since fuel cell systems have to generate fuel in order to generate electricity which in turn generates heat, designers and engineers can take advantage of the

Comparison of Fuel Cell Technologies

Fuel cell type	Common electrolyte	Operating temperature	Typical stack size	Efficiency	Applications	Advantages	Disadvantages
Polymer Electrolyte Membrane (PEM)	Perfluoro sulfonic acid	50–100°C 122–212° typically 80°C	<1 kW– 100 kW	60% transportation 35% stationary	• Backup power • Portable power • Distributed generation • Transporation • Specialty vehicles	• Solid electrolyte reduces corrosion & electrolyte management problems • Low temperature • Quick start-up	• Expensive catalysts • Sensitive to fuel impurities • Low temperature waste heat
Alkaline (AFC)	Aqueous solution of potassium hydroxide soaked in a matrix	90–100°C 194–212°F	10–100 kW	60%	• Military • Space	• Cathode reaction faster in alkaline electrolyte, leads to high performance • Low cost components	• Sensitive to CO_2 in fuel and air • Electrolyte management
Phosphoric Acid (PAFC)	Phosphoric acid soaked in a matrix	150–200°C 302–392°F	400 kW 100 kW module	40%	• Distributed generation	• Higher temperature enables CHP • Increased tolerance to fuel impurities	• Pt catalyst • Long start up time • Low current and power
Molten Carbonate (MCFC)	Solution of lithium, sodium, and/or potassium carbonates, soaked in a matrix	600–700°C 1112–1292°F	300 kW– 3 MW 300 kW module	45–50%	• Electric utility • Distributed generation	• High efficiency • Fuel flexibility • Can use a variety of catalysts • Suitable for CHP	• High temperature corrosion and breakdown of cell components • Long start up time • Low power density
Solid Oxide (SOFC)	Yttria stabilized zirconia	700–1000°C 1202–1832°F	1 kW–2 MW	60%	• Auxiliary power • Electric utility • Distributed generation	• High efficiency • Fuel flexibility • Can use a variety of catalysts • Solid electrolyte • Suitable for CHP & CHHP • Hybrid/GT cycle	• High temperature corrosion and breakdown of cell components • High temperature operation requires long start up time and limits

Figure 7-4 Comparison between engine efficiencies.

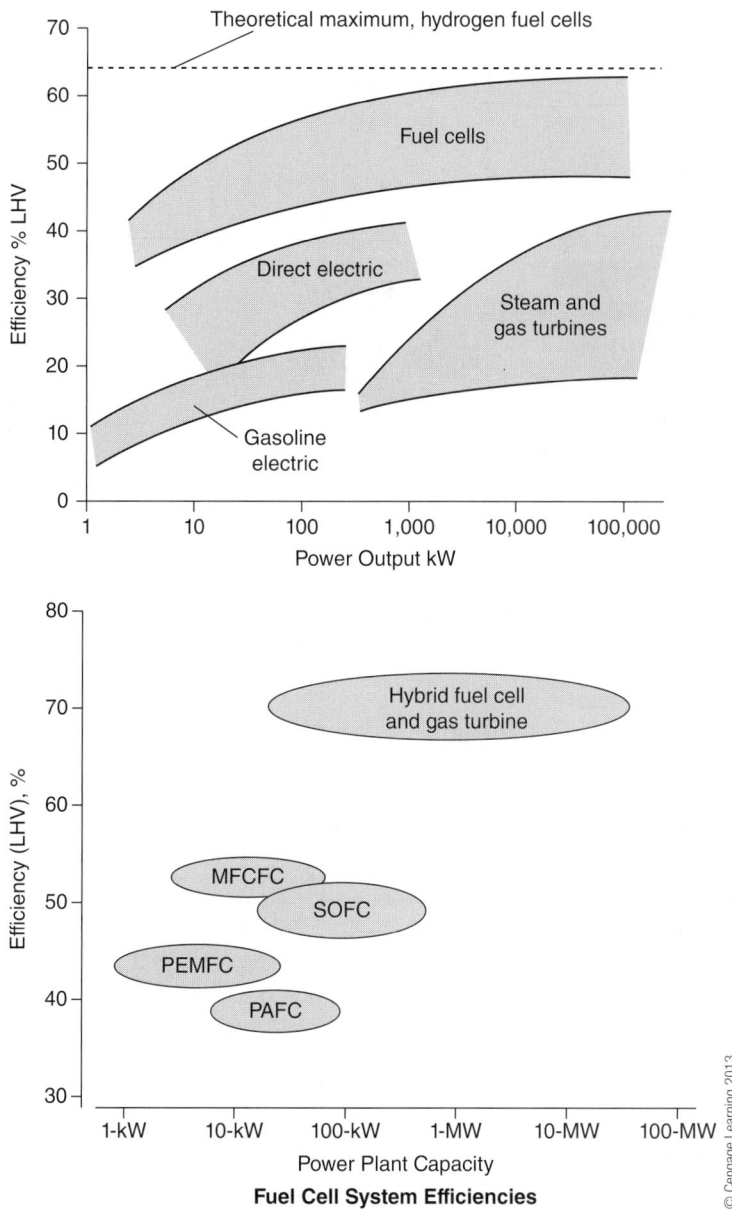

Figure 7-4 (Continued)

entire conversion process rather than just a single part of it. By using the heat generated by the fuel cell reactions to power the fuel reformation reactions, the overall system is more efficient and produces cheaper electricity than it would otherwise. In high temperature systems, there is even enough heat to run secondary turbines. Even low temperature fuel cells can make use of these **combined heat and power** designs, using any excess heat for hot water generation or to supply hot air for local heating in buildings. Figure 7-5 shows a typical combined heat and power cycle.

This approach of producing power in the amount needed where it is used rather than at extremely large power plants near the source of the fuel is a fundamental change in power generation. These **distributed generation** schemes are becoming increasingly common and represent a significant opportunity for support services. Keep in mind that while it lessens the transmission cost of electricity, since current loss over the enormous grid wire system is eliminated, it raises the cost of fuel, since that fuel has to be delivered to the fuel cell location.

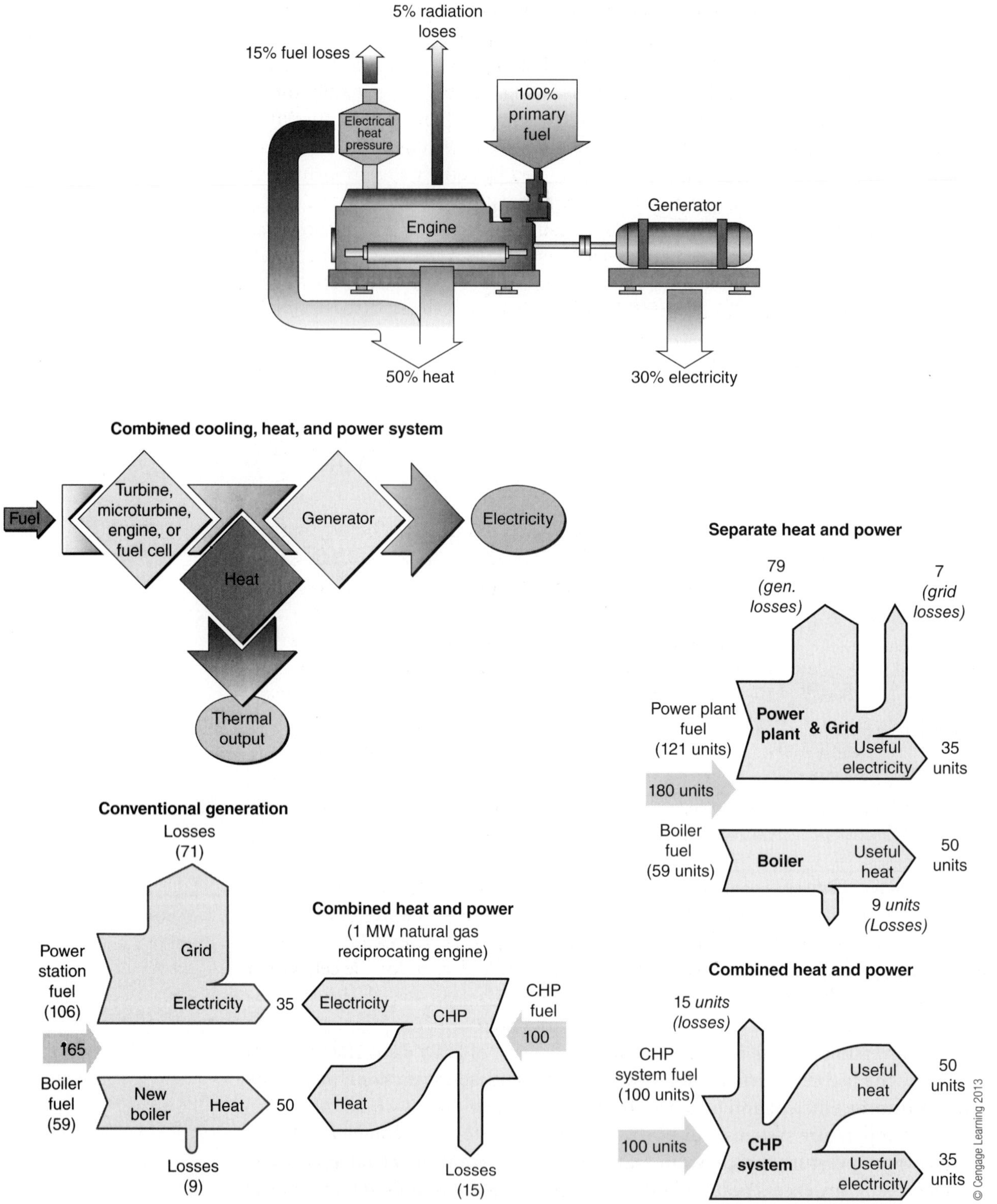

154 FUEL CELL TECHNICIAN'S GUIDE

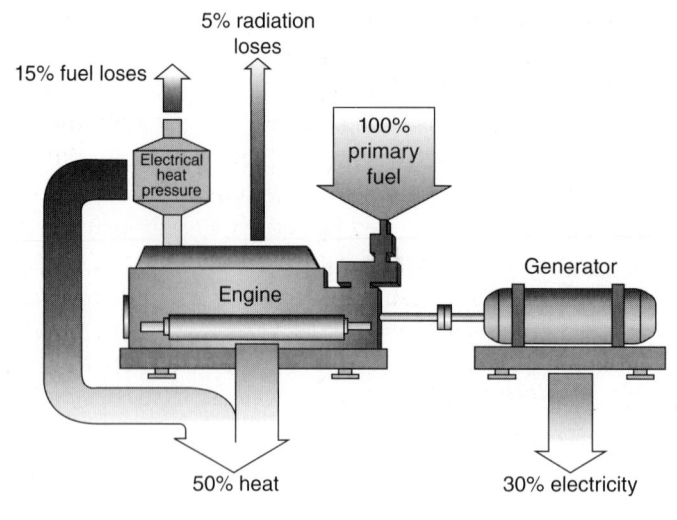

Figure 7-5 Typical combined heat and power cycle.

The main design considerations for on-site fuel production processes for fuel cell installations are different than large generation facilities. Smaller fuel cell site installations must possess high **thermal efficiency** by making use of all the heat generated to supply any heat required and then using any extra heat for complimentary heating systems, preferably at the site using the power. The conversion of the hydrocarbon fuel must result in a fuel gas of pure hydrogen (> 99.9% hydrogen) to maximize the use of fuel and minimize contamination of the cells; again, combustion reactions are good at burning the fuel but poor at reclaiming the energy itself. **Peak load cycling** must be capable of supplying variable power requirements as required current loads go up and down, for instance, when air conditioning is required during the day but not at night. Load cycling requires not only that the fuel cell be responsive, but the system supplying the fuel be equally as responsive. The overall systems must be both modular and **scalable** so different needs can be met at different site installations. These criteria also mean that system weights cannot be so great that they require separate support structures so they can be set on concrete pads or rooftops. Start-up and shutdown cycling of the fuel generating equipment must be appropriate to the application, the site, and the fuel cell system in use. If an AFC system needs long start-up times to bring the installation to temperature, then the fuel generator has to respond differently than if used with a PEM system that comes on and begins working in a matter of minutes at most.

The fuel generation system must supply suitable quality. Large grid systems such as the natural gas supply system supply a gas that varies considerably in quality with the source of the natural gas. That variation in quality is not truly important to most of the customers connected to the grid to heat water for baths or fire a forced-air heating system. In most cases, it is the heat resulting from the combustion reaction that is important rather than the reaction itself and any contaminants are just vented to atmosphere. In fuel cells, the reaction itself is critical and trace contaminants can seriously degrade the operations of the system electrodes where the reactions occur. All the systems that produce hydrogen require additional quality improvement subsystems but not all require the same ones. It is important for the student to realize that different fuel cell types require different fuels and fuel quality, and so what is appropriate for one system may not be appropriate for other systems.

Finally, some fuels are not well suited to conversion from hydrocarbons to hydrogen. In general, the more refined a fuel is, the more likely a candidate it is to be used as a hydrogen source. Fuels like diesel or jet fuel do not reform easily because they contain an excessive amount of contaminants like sulfur. This is a significant issue in systems used in remote locations, since high quality methanol or natural gas may not be available. Even though fuel cells have been around for nearly two centuries, fuel generation systems are much less well developed than the cells they feed. Many successful applications target areas where existing fuels such as natural gas can be reformed to provide the needed hydrogen for the cell. Other applications are harder to design for, not so much because the fuel cell is incapable of meeting the need, but because supplying hydrogen is difficult using current technology. So even though substantial markets may exist, such as in-field military applications or even automobiles, the current inability to supply consistent, high quality fuel may keep fuel cells out of those applications.

Fuel cells then are dependent on their source of hydrogen as well as the application being served. In general, directly supplied hydrogen is applicable in space, in laboratories, and where an existing stream is available, usually in conjunction with industrial generation. Commercial or light industrial fuel cell installations target natural gas, especially since most of these markets are in parts of the world where natural gas pipelines are already in place. Transportation applications are not yet defined well enough to target a specific fuel, in part because the huge variety of uses prohibits any single fuel from meeting all the needs. Compressed hydrogen tanks can supply truck or bus fleets in larger cities but not consumer vehicles in rural areas or trains requiring enormous amounts of the gas as they move across continents. However, auxiliary power units meant to supply electricity to transportation equipment while it is idle, such as refrigeration or communication needs, could use compressed hydrogen in most cases. Military uses require military fuels be used, in particular the JP-8 standard fuel, which is not well suited to reformation. Large industrial power plants supplying the current grid will use either natural gas, coal derived fuels, or biofuels.

The most widely available fuel in the world is currently gasoline, but it is not well suited to reformation. Methanol is the other leading candidate because it reforms relatively easily and returns over 90% of the actual hydrogen in the molecular structure as fuel to be used. Unfortunately, methanol is extremely corrosive to both metals and polymers and readily mixes with water so it cannot be distributed within the current liquid gas distribution network meant for gasoline or diesel. Methanol is also toxic to humans and animals, raising significant health and environmental concerns. Gasoline requires high temperatures or expensive catalysts to reform, fouls the reaction zones by producing carbon as a by-product (coking), produces similar pollutants to the internal combustion engine, and does not offer substantial efficiency increases over present internal combustion vehicles, at least for transportation applications. Gasoline is a reasonable choice for larger installations such as light industrial, commercial, or residential in areas where natural gas is unavailable. Using gasoline means that any additives required for internal combustion engines would have to either be removed or never added to begin with. That could substantially lower the cost of the gasoline, since many regions of the United States mandate specific gasoline blends that generally come from only one or two refineries. Eliminating additives and standardizing blends for fuel cells would lower refining costs. In fact, gasoline itself would not have to be used at all. Refineries could supply some of the precursors to gasoline in the refining process as fuels such as **naptha**.

Systems such as **anaerobic digesters** using living organisms that consume hydrocarbons and produce hydrogen as a gas or **pyrolysis gasifiers** where waste is processed in high temperature reactors to produce methane are also being designed. These can be used with manure, landfills, waste, seaweed, or even cash crops grown specifically as feed to produce fuel where a delivery infrastructure is limited. Figure 7-6 shows an anaerobic digester.

A pyrolysis gasifier is shown in Figure 7-7.

Reforming Reactions

Reforming reactions are the chemical pathways that convert hydrocarbons to hydrogen. In general, several different reactions are needed to get the type and quality of fuel desired so that systems usually have multiple stages and specialized equipment. The general form of the basic reforming reaction is given as:

$$C_nH_m + nH_2O \rightarrow (m/2 + n)H_2 + nCO \qquad \text{Eq. 18}$$

This is how chemistry treats similar reactions. Remember that the term "hydrocarbon" is a general term for any combination of carbon and hydrogen. Because the number of carbon and hydrogen atoms in the large molecule does not change the actual reaction, only the number of molecules involved, chemists use the subscripts seen in Equation 18 rather than list all the possible combinations. So if a hydrocarbon has "n" atoms of carbon in the molecule and "m" atoms of hydrogen, the same number of water molecules ("n") as available carbon atoms gets involved in the reaction. The actual number of hydrogen molecules attached to the carbon is a bit more complicated and the rules involved in determining that number are beyond the scope of this book, but carbon atoms must have four bonds in a molecule. However, carbon can have up to three bonds to another atom (single, double, and triple bonds), so one carbon atom can form a stable molecule

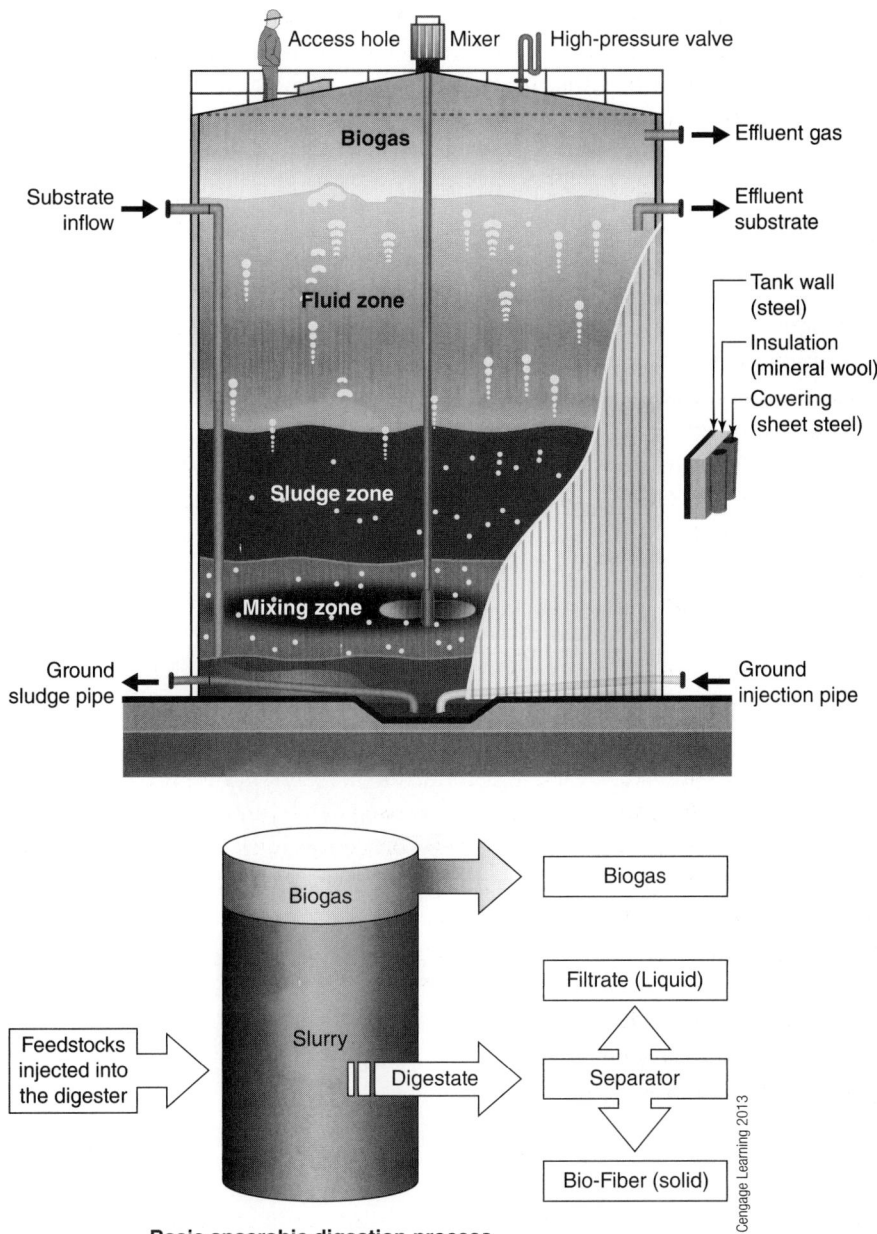

Figure 7-6 Anaerobic digester.

with one other atom but only if there are single, double, or triple bonds distributed to form four total bonds. For instance, acetylene has two carbon atoms and two hydrogen atoms (one triple bond between the two carbons and two single bonds between the carbon and hydrogen), while methane has four hydrogen atoms singly bonded to its one carbon atom. Figure 7-8 shows an acetylene molecule with the three bonds between the carbon atoms.

If you place a specific hydrocarbon fuel in the equation, you specify the "n" and "m" values. Methane as a fuel gives Equation 16, as in the previous chapter. In this case where only methane is involved in producing hydrogen, "n" is 1 and "m" is 4. Thus, "n" (1) goes in front of the water (H_2O) "m/2 + n" (4/2 + 1 = 3) goes in front of the hydrogen (H_2) and "n" (1) goes in front of the carbon monoxide (CO). In this particular reaction then, one molecule of methane reacts with one molecule of water to form three molecules of hydrogen gas and one molecule of carbon monoxide. This shows why methane is such a good fuel, in that not only

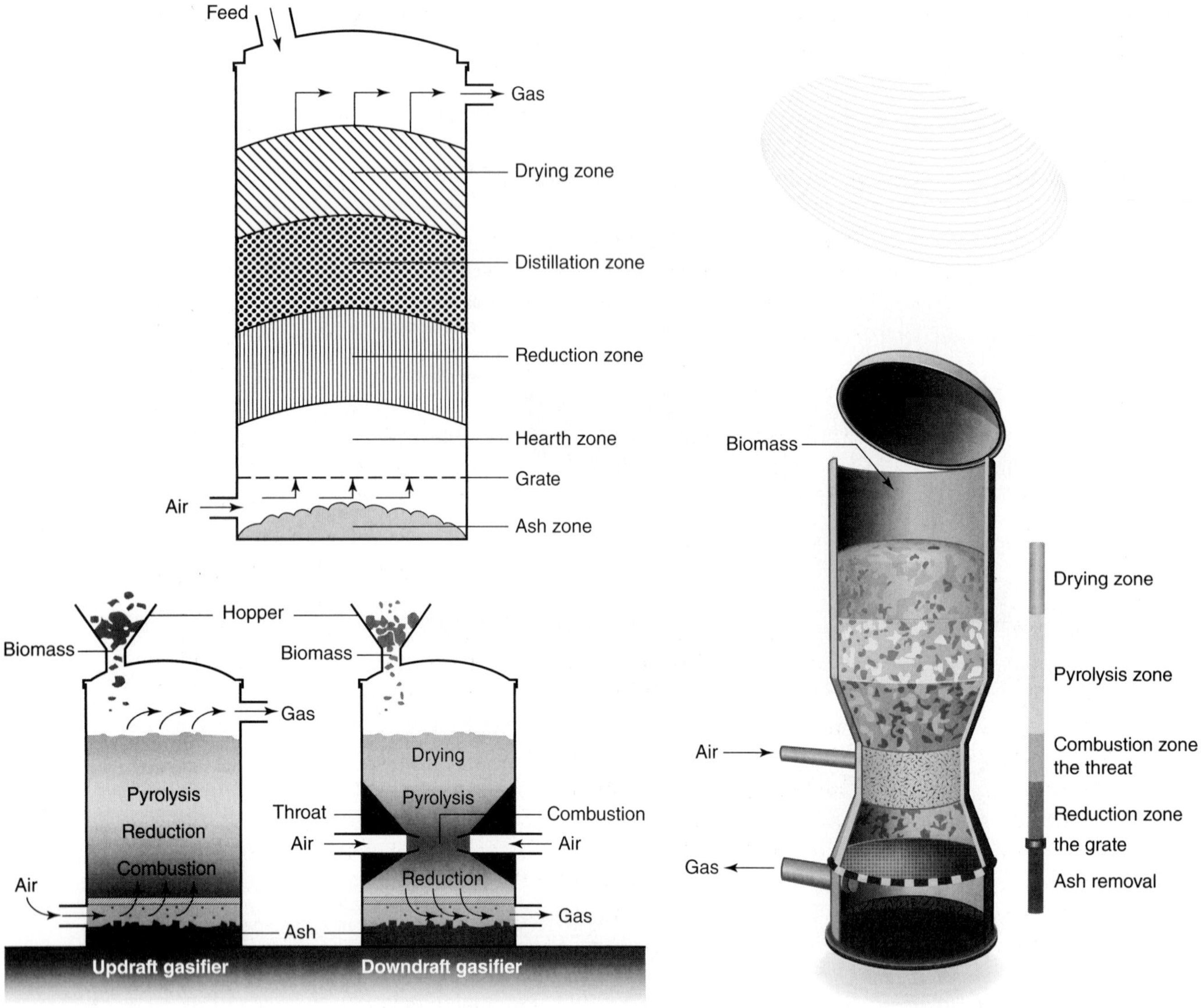

158 FUEL CELL TECHNICIAN'S GUIDE

■ **Figure 7-7** Pyrolysis gasifier.

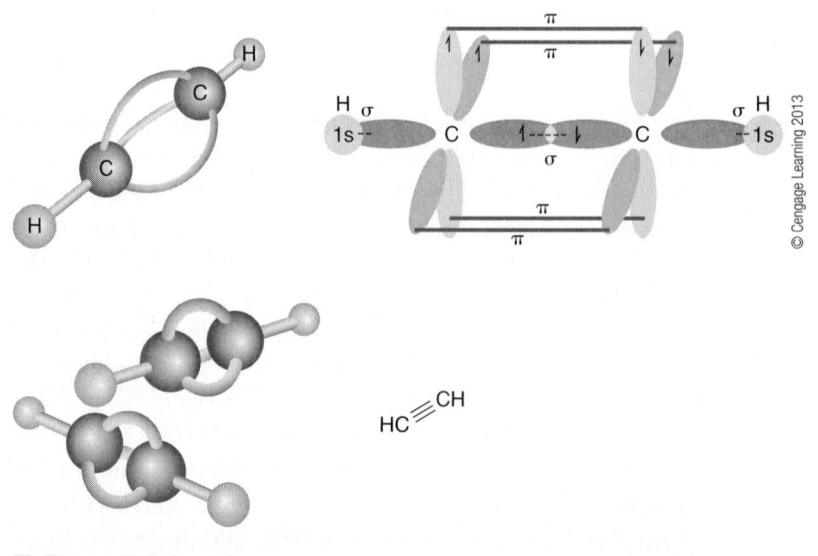

■ **Figure 7-8** Acetylene.

does the hydrocarbon fuel produce hydrogen gas for our fuel cell but in this steam reforming reaction, so does the water.

$$CH_4 + H_2O \rightarrow 3H_2 + CO \quad \text{Eq. 16}$$

When this is done, you are left with carbon monoxide and hydrogen. The hydrogen is used as fuel, but carbon monoxide is a health and safety risk as well as a poison for some fuel cells so something has to be done with it. One of the most important reactions in chemistry is used next, the **water gas shift reaction**. This reaction combines the carbon monoxide with more water to produce carbon dioxide (much less of a hazard) and hydrogen (more fuel). The gas mix is generally about three parts hydrogen to one part carbon dioxide.

$$CO + H_2O \rightarrow H_2 + CO_2 \quad \text{Eq. 17}$$

Steam reforming is not the only method of producing carbon monoxide and hydrogen. Another important industrial process is called **dry reforming**, and involves the use of carbon dioxide (CO_2) instead of water (steam) as shown in Equation 18.

$$CH_4 + CO_2 \rightarrow 2CO + 2H_2 \quad \text{Eq. 18}$$

This reaction may also be used as part of a hybrid system where steam reforming supplies some part of the fuel then any carbon dioxide produced in the fuel cell (in MCFC or SOFC for instance) is used as a supplement. Carbon monoxide can also be removed from these streams using a conventional packed bed filter, but this is usually done only for very low level CO requirements after most (99.999% or so) of the CO is removed. Sulfur removal is usually handled by **packed bed filters** as well. Palladium membranes are used if ultrahigh quality gas is needed or if the physical size and complexity of packed filter beds are not applicable to the cell market, such as with personal devices. Palladium will allow hydrogen to pass but nothing else, so it makes a very effective filtering device. It is very expensive technology, however, and in general has a limited work life.

Many of the reactions where hydrocarbons are used as fuels require heat to be added (are endothermic) and catalysts to be used. In some processes, the two reactions above will occur in the same reaction chamber at more than 500°C in the presence of a nickel catalyst. This single chamber process produces a mixed gas of hydrogen, carbon monoxide, carbon dioxide, unreacted fuel gas and water (steam). Operating temperatures and pressures, the mix of hydrocarbons in the input fuel gas, and the temperature and amount of steam used determines the percentages in the final output stream. The stream is then further processed depending on the type of fuel cell it is feeding.

Methane is a simple chemical (CH_4), but because most fuels are not simple chemicals but rather complex mixes of different types of hydrocarbon chains, reforming them is difficult to do consistently and uniformly. Some fuels like natural gas are blended by the suppliers to account for variation in the supply gas so that these are rarely if ever uniform. In many of these cases, even small amounts of contaminants (oxygen is considered a contaminant in some processes for instance), will affect the reforming reactions, the shift reactions, and the fuel cell reactions. It may be that a multiple stage process has to be used, especially in certain applications where fuel supplies are not steady or where the high temperatures needed for steam reformation is a safety hazard. Remember also that the lower the temperature used in the fuel cell, the less tolerant the cell will be of contaminants, so pretreating (preforming) the incoming fuel gas is critical to low temperature systems.

A typical stationary fuel cell using natural gas may have a preheat unit (300°C), a unit to remove sulfur, a steam reformer (600°C and up but you have to have a high quality steam boiler as well), a cooling circuit leading to both low (200°C) and high temperature (400°C) water gas shift reactors, final temperature adjustment before entering the fuel cell, and off-gas heating/cooling systems. Some systems such as PEM fuel cells may require a carbon monoxide (CO) remover as well. All of these units have to work in unison to supply steady fuel, air, steam, and recirculating streams to and from the cells while operating at several different temperatures.

In one particular case, fuels are first converted to methane or methanol and then converted to hydrogen fuel for the cells. In personal uses such as chargers, houses, or automobiles, having high temperature steam and reactions occurring at several hundred degrees Celsius may not be tolerated, so technologies such as **Direct Methanol Fuel Cells (DMFC)** are being investigated for this market. Methanol is methane with one of the four hydrogen atoms replaced by an OH molecule to form CH_3OH. When methanol is used as a fuel, the reaction is:

$$CH_3OH + H_2O \rightarrow 3H_2 + CO_2 \qquad \text{Eq. 19}$$

This reaction requires heat to be supplied to the system (endothermic) but not too much, so DMFC systems operate around 250°C using relatively cheap catalysts such as copper. DMFC systems are versions of PEM cells, and the lack of carbon monoxide in this reaction set means no CO poisoning can occur.

All of the hydrocarbon reactions have a particular problem associated with them, that being carbon formation. If the steam supply is disrupted or there is not enough steam, hydrocarbons will break apart into carbon and hydrogen above 650°C in a process called pyrolysis. The more carbon atoms in the chains, the more likely this type of formation will occur. Carbon monoxide will also combine with itself so that two CO molecules will form one carbon molecule (C) and one carbon dioxide molecule (CO_2). The lone carbon will then coat catalyst sites, sidewalls, and even gas off-takes, having a significant impact on fuel cell operations. In many systems, excess steam is used to react with any carbon in the gas, reforming hydrogen and carbon monoxide.

Steam reformation is one of the three main methods currently used for hydrogen production. **Partial oxidation reactions** are also large industrial processes, generally broken into two categories, **thermal** and **catalytic**. Catalytic reactions are done at lower temperatures (800°C–900°C) and thus require catalysts; thermal reactions occur at high temperatures (1200°C–1500°C) and are driven by the energy available from heat (do not require the help of catalysts). Partial oxidation (POX) reactions require oxygen to produce the following reaction:

$$2CH_4 + O_2 \rightarrow 2CO + 4H_2 \qquad \text{Eq. 20}$$

Notice that this reaction is similar to the steam reforming reaction just as the reactions in the different types of fuel cells are similar. Industrial processes are generally standardized to a very great extent. They use the same feed stocks, the same reaction vessels, the same materials handling systems, and produce the same products to minimize the cost of the entire process from start to finish. If you learn about electric motors in the mining industry, your knowledge is just as useful in the petroleum business because of standardization. The major chemical reactions used in industry are not that much different than motors in that once you know how one works, you probably have a good idea how others work as well. Also, notice that this equation uses two molecules of methane so the equation is **balanced**.

POX processes tend to be used for **"heavier" hydrocarbons**. Heavier is a relative term and not really related to weight so much as the number of carbons in the hydrocarbon chain. The more carbons in the molecular chain, the heavier the hydrocarbon. When you see light crude mentioned, that means the hydrocarbon chains in the crude oil are short. Thermal POX methods are used to treat the heavier oils while catalytic POX methods are used for intermediate or light oils. Thermal methods also work better for oils that have contaminants such as sulfur. Sulfur does not foul the nonprecious metal catalyst but instead it comes off as a gas (sulfur comes off as H_2S), which can be removed from the gas in subsequent processing steps.

Autothermal reforming is the third major industrial process. This is a combination of steam and partial oxidation, where steam and air are used in the reaction chamber with the hydrocarbon as shown in Eq. 21. This system balances out the steam reaction where heat must be supplied (endothermic) with the POX reaction, which creates heat (exothermic) so no external source of energy is required. The catalyst used in the reaction chamber serves to determine what temperature these systems run at.

$$2CH_4 + H_2O + O_2 \rightarrow 2CO + 5H_2 \quad \text{Eq. 21}$$

Finally, while not currently considered industrial hydrogen production processes, pyrolysis and electrolysis can also be used. In pyrolysis (**cracking**) hydrocarbons are heated with no air or steam present at 600°C–800°C. Without oxygen, neither carbon monoxide nor carbon dioxide can form so the hydrocarbons decompose to form carbon atoms and hydrogen gas molecules. Getting rid of the carbon remaining behind once the hydrogen gas is removed is the principal problem with these types of reactions. Electrolysis is essentially a PEM cell in reverse. Water is oxidized at the cathode according to the following reaction:

$$2H_2O \rightarrow O_2 + 4H^+ + 4e^- \quad \text{Eq. 22}$$

This is an important reaction in that hybrid systems combining renewable resources such as wind or solar are difficult to handle because of their cyclic nature (the sun does not always shine nor the wind blow). Rather than surge electricity into either small or large systems, these can be used to produce hydrogen gas when possible and a fuel cell used to supply steady current from the stored hydrogen.

The final methods that hold promise but are still under development are the biological production methods. There are any number of possibilities but all use some form of organic matter (trees, algae, manure, and even grass clippings) as feed for a process that produces either hydrogen or CO. Fuel gas production using biomass can be done by combustion, hydrogasification, fermentation, bacterial or microbial agents, pyrolysis, hydrogenation, and several other processes beyond the scope of this book.

Many of the designs used in reforming are relatively straightforward. The fuel cell gas that is not converted to electricity (most cells do not fully convert the hydrogen or carbon monoxide fuels) is fed into the bottom of the reaction chambers (usually pressurized), where it is combusted. The main feed gas is fed from the top so it heats as it is forced down. The preheaters warm the second stage located on top of the combustion area. The second stage has the catalysts needed for the reforming reactions mentioned above and that is where the initial reformation is done. As the partially reformed gas flows up, it is gathered at the top and transferred to another reformation chamber where the reactions are taken to completion. This exits to the gas cleaning systems that depend on the type of cell being used. A number of variations exist for this general scheme, including alternating combustion and reforming plates to limit size, centralizing larger pressure chambers to feed multiple cells, and even industrial sized systems meant to supply hydrogen to plants.

While there are a number of associated concepts not covered in depth by this textbook, it is important to understand that dealing with fuel cells is not limited to just fuel cells. The technology requires a basic understanding of numerous other technologies that are included in the fuel cell system as a whole, and fuel production is one of the most important of these groups.

External Reforming of Hydrogen

Reforming one of the various fuels used to provide a fuel cell is done in two general ways, either at the fuel cell area or away from it. External reforming is done away from the main fuel cell area. That may be in a petroleum refinery, a dedicated reactor placed several feet away from the cells and fueling several systems, or even attached to the cell and serving as a dedicated supplier. External reformation has one key advantage in that any undesirable constituent in the supply fuel gas (hydrogen or CO) can be dealt with outside the fuel cell system so degradation of the cells themselves is minimized.

Starting at the large end of the supply chain are the in-place and very mature industries of industrial production, primarily employing stream reformation techniques. These large scale reformers generate up to 30 million cubic meters of hydrogen per day using a pellet-in-tube technology that flows the initial fuel gas down a heated (850°C) and pressurized reactor tube (~12 meters long) made of high temperature alloys. The initial fuel gas contacts the catalyst coated pellets to produce the secondary fuels like hydrogen. These technologies can be scaled down, but only so far (to about 100,000 cubic meters of hydrogen gas produced per day). To put this in perspective, the smallest scale for this type of technology would fuel approximately 4000 cars per day. Scaling down any further requires an almost complete redesign of the system to deal with heat transfer, gas flow, and cost considerations.

The design of large reformers varies as well. There are numerous types, many proprietary, with several common configurations used. At issue is where the input fuel gas stream encounters the reaction zone. In reformers where heat is required for the reactions to occur (endothermic) such as steam reformation, the input fuel is typically fed from the bottom and the reaction zone is somewhere above it. This allows the fuel to heat as it move up and react in the hottest part of the furnace. In reformers where heat is created by the reactions (exothermic) such as partial oxidation reactions, the reactions occur nearer the input areas to allow heat extraction and cooling if needed. Some reformers use single reaction zones or single burners in the top or bottom of the units, while others set several in the reactor to allow for better control within the reformers.

At the small end of the scale are miniature reactors that can be built on a chip to supply fuel to very low wattage fuel cells (a couple of watts supplied or less). These units are meant to supply small amounts of hydrogen to personal or emergency systems of less than a kilowatt but could be manufactured in bulk much like semiconductors and then stacked like fuel cells.

Size is only part of the issue, in that the input fuel is also critical to the design of external reformers. Almost all available fuels are lacking in one way or another. The most common fuel currently used for fuel cell installations is natural gas, coming from massive pipeline systems across much of the Northern Hemisphere. Outside of those areas served by pipeline grids, the most common widely available fuel is gasoline or diesel, and considerable research has gone into external reformers that can accept those as input fuels. Unfortunately, they contain substantial amounts of impurities and many of these will contaminate the reformer and the follow-on fuel cell. Sulfur is the worst of the contaminants seen in internal combustion fuels and removing it from the input fuel is difficult as well as costly. The most abundant fuel is water (H_2O) where electrical currents break the oxygen and hydrogen apart, with the oxygen generally supplied to one side of the fuel cell and the hydrogen to the other. While it seems odd to use electricity to make hydrogen in order to then use hydrogen to make electricity, it makes sense in certain applications. Electricity is not really portable once past the smaller voltages and amperages available in batteries, so mobile applications could use electricity available at a central location to produce hydrogen and then use the hydrogen to make electricity in remote locations where a plug-in is not available. Using nonpeak load electricity, say at night when air conditioning needs shrink, to create hydrogen, which can in turn be used to supply peak hours electricity, can also reduce the loads on the electrical delivery system to minimize the need for new infrastructure. Another useful scheme is to partner water electrolysis systems with electrical suppliers that cannot provide constant current flows, like windmills or solar so that fuel cells can provide power during the night or when no wind is available. Water hydrolysis also has one very important aspect in that the hydrogen produced is very high quality, requiring only limited purification for fuel cell use.

Almost any liquid that contains hydrogen can now be reformed to provide fuel cells. Water, methanol, gasoline, diesel, JP-8 (military fuel), kerosene, butane, propane, and even naptha (with special catalysts) have been used with success. Two primary criteria determine which is used: price and cell location. Larger installations used natural gas if available because it is relatively cheap and it is available in large quantities if it is available at all. Mobile and off-grid systems have yet to settle on a common fuel.

The fuel does not have to be a liquid though. Coal is a primary input fuel for hydrogen production as well. Three general coal gasification systems are used, **moving-bed**, **fluidized-bed**, and **entrained-bed**, with the differences between the three primarily being the operating temperature, the size of the coal used, and how the coal is moved through the process. Moving bed systems move coal through a lower temperature (425° to 650°C or 800° to 1200°F) furnace where air and steam are supplied to the bed as it moves, producing a lower grade, mixed gas of hydrogen, carbon dioxide, carbon monoxide, methane, ethane, and various other hydrocarbons. Fluidized beds operate at higher temperatures (925° to 1040°C or 1700° to 1900°F) with the air blown into the bed so it moves like a fluid through the system to produce higher grade gas (less hydrocarbons). Entrained beds operate at the highest temperatures (>1260°C or 2300°F), in which the coal is kept in the reactor until complete combustion occurs to produce a high grade gas of hydrogen, carbon dioxide, and carbon monoxide. These are all exothermic (producing heat instead of requiring it to run) with the excess heat produced available to the follow-on fuel cell systems. As with most processes, contaminants are still present in the gas stream and need to be removed before the fuel enters the cell. Unlike most systems, coking (the formation of free carbon during reformation which then coats the reaction vessel) is not an issue.

All fuel reforming systems have to deal with another basic issue, that being whether to store the hydrogen or risk supplying it as needed during fuel cell operations. Storing hydrogen is not easy. It is the lightest of all gasses and so requires large vessels capable of withstanding high pressures, the means to pressurize the hydrogen to fill the vessel, and vessels with which hydrogen will not react. Most pressurized gasses are stored in steel vessels, but hydrogen will enter the atomic structure of steel and eventually cause it to turn brittle so that the storage pressures will cause cracking. Developing a suitable hydrogen storage tank is proving to be no easy task. If there is no storage available, then the reformer has to supply the fuel cell in real time while it is operating. Any disruption means the fuel cell will shut down, not something that can be easily done in low temperature systems because thermal cycling is bad for all the parts, and something that cannot actually be done at all in high temperature systems since any rapid, unplanned cooldown could destroy the cell.

As mentioned in previous sections, reformed fuels are not just hydrogen. Steam reformation usually produces a mixed gas as do catalyst driven reactions. Hydrogen can be mixed with carbon monoxide and carbon dioxide, ammonia, and methanol, along with impurities such as sulfur, particulates, and tars, depending on the initial input gas as well as the reformation process used. Any gas other than hydrogen has to be removed from the stream before the fuel cell, although some of the high temperature systems use a hybrid system where the mixed reformate gas is fed to the fuel cell, where secondary reformation also occurs.

The reforming scheme used will also determine to a great extent the nature of the equipment. In general, reformation is done at elevated temperatures unless catalysts are used. Steam reformation is done at lower temperatures (760 to 980°C or 1400 to 1800°F) than POX systems (1300 to 1500°C or 2370 to 2730°F) but requires steam and a source of heat (usually combustion of the same fuel being reformed) since the reaction is endothermic (needing a source of heat). POX systems produce an output gas with less hydrogen per unit of input fuel but do so with more efficiency. In POX reforming, the reaction pressure is not critical but it is in steam reforming. Catalysts require a narrow operating range but produce higher quality outputs. Each of these systems operate under different circumstances, are better with some input fuels than with others, require different fabrication materials, use different input and output methods, and produce different products that require different downstream processing. Understanding fuel cell operations requires a much broader base than just knowing how fuel cells work.

External reforming in particular requires systems that have to match the exiting, reformed gas stream to the downstream processes, either to additional purification or fuel cells. This can require temperature and pressure modification to match the required operational needs as well as flow stabilization so the amount of gas exiting one system matches the amount the next system can process. Manifolding is needed if multiple systems are being supplied, safety systems of different size and needs must be in place given the nature of hydrogen and of the reformer feed gasses, shutdown and even blowout systems are needed as well. Most of these systems are designed and manufactured for whatever fuel cell system that is in place, depending on operating temperatures and pressures as well as the initial feed gas, so there is little standardization in current systems. It is up to operations and maintenance personnel to understand each individual system, and as mentioned several times before, what works in one system may not work in another.

Internal Reforming of Hydrogen

Internal reformation systems use the same methods and reactions as external reformation systems do with steam reformation being the most common internal scheme in use. As with most aspects of fuel cells, there are advantages and disadvantages to internal reformation, with high temperature systems capable of using internal reformation and low temperature systems mainly using external reformation.

Reforming the input fuel as part of the fuel cell reaction set has several advantages over external reforming. It requires less equipment, lowering the initial cost of the system. It makes direct use of the heat generated by the fuel cell, so steam generated at the anode in SOFC or MCFC systems can be used as a heat source while at the same time cooling the fuel cell system by removing the steam, effectively eliminating several subsystems. Internal reformation also tends to equalize the supply of hydrogen going to the cells so that manifolding and gas distribution schemes are not needed.

There are two general methods used, **direct** (DIR) and **indirect** (IIR) **internal reforming**. In direct systems, the reforming reactions occur within the same chamber as the fuel cell anode (hence the better gas distribution characteristics mentioned previously). Because of their operating temperatures and catalysts, some SOFC systems actually use the fuel cell anode to both reform the input fuel gas and react the resulting reformed gasses (carbon monoxide and hydrogen) to produce electricity. In indirect systems, reformation is done in chambers separate from the anode. The reforming chambers are usually adjacent to the electrode chambers to make use of the heat generated by the fuel cell reactions to supply the heat needed. Reformation is done and then the gas is transferred to the anode chamber so that reformers alternate with fuel cells in the stack assembly. The piping for indirect systems is thus more complicated since the input fuel gas has to be piped to the reformer along with the steam needed since most of these systems use steam reformation.

Internal reformation generally requires that the input fuel gas be of high quality, since most of these methods do not allow for downstream cleaning of the reformed fuels before they go into the fuel cell. Fuels like gasoline are less likely to be used in internal reformation systems for that reason.

Cleaning the Fuel Stream

Fuel cells require almost pure hydrogen (99.9999% or better) with low temperature systems less able to handle contaminants than higher temperature systems. In general, contaminants poison the electrodes rather than compromise the electrolyte, although electrolytes can also be degraded in some systems. As mentioned above, purifying hydrogen is difficult to do, especially in smaller systems. One of the primary challenges facing fuel cell technologies is in supplying the fuel gas to the cells systems in sufficient quantity and quality.

There are a number of elements and compounds that have to be removed from the fuel gas before it reaches the fuel cell itself. Principle among these are sulfur and carbon monoxide (unless it is used as a fuel as in SOFC systems). Contaminants such as these usually attack the catalysts that are driving the fuel cell reactions, so that lower temperature cells using precious metal catalysts can tolerate a different fuel mix than higher temperature fuel cells where nickel catalysts are used. There are two ways to handle contaminants; the first is to remove them from the input fuel being reformed to provide hydrogen, while the second removes the contaminants from the reformed fuel gas stream going to the fuel cells. Sulfur, for instance, is present in acceptable amounts when methanol is used as the input fuel but is present in excessive amounts when gasoline is used. The amount of sulfur present in gasoline though is being reduced by government standards, so at some point, that may change. This is one of the hallmarks of fuel cell development, in that standards and targeted performances have yet to be fully developed, so many developers are unwilling to spend money in producing systems that may not comply with the ultimate standards. This is particularly true of hydrogen as a fuel, with some governments driving forward while others wait to see how the market develops.

Dealing with contaminants requires an understanding of what the various reformation systems produce. Reforming natural gas using a steam reformer yields about 75% hydrogen, 5% carbon monoxide, and 10% carbon dioxide, along with trace elements that depend on where the natural gas came from. Methanol reformation will produce a higher quality gas with less than 0.1% carbon monoxide and reduce the carbon dioxide to the trace level (<0.001%). For natural gas then, the exit stream from the reformer is cooled and fed to a water gas shift reactor, where the carbon monoxide combines with more steam to produce carbon dioxide and more hydrogen, according to Equation 17. Several water gas shift reactors may be used to lower the carbon monoxide levels to those seen in methanol reactors (<0.1%). This level of CO is acceptable for PAFC systems so no further treating would be required but if systems with platinum catalysts like PEM's are in use, the CO level has to be lowered by another 100 times (to <0.001%) since the CO will absorb onto the metal surface and block the fuel cell reactions.

Ultralow levels of CO required by PEM cells use other treatment schemes in addition to water gas shift reactors. Selective oxidation reactors add a small amount of air or oxygen (~2%–4%) to the fuel stream exiting the shift reactor and run the mix over another precious metal catalyst where the CO and air react at the catalyst to produce CO_2 (carbon dioxide). This is a very effective system, but unfortunately can also be explosive. **Methanation** is another method, where the CO is recombined with hydrogen to form water and methane (the reverse of Equation 16). This does consume hydrogen, but since the levels of CO are already low from the water gas shift reactor, not much hydrogen is actually consumed so the trade-off can be economically worthwhile. The methane formed has no effect on fuel cells and acts as a carry-through, lowering the efficiency of the cells by a small amount but not doing much else. **Pressure Swing Absorption** is the third method commonly used, where the exiting gas from the water gas shift reactor is fed (under pressure) into a bed of material that absorbs hydrogen such as activated carbon or what are called molecular sieves. Once the absorbent material is fully loaded with hydrogen, the system is closed and the pressure bled off. Since hydrogen is the lightest gas, other heavier gasses will come out of the absorbent first as the pressure on the absorbent is lowered. Once a low enough pressure is reached, the absorbent material will release the now purified hydrogen. Membranes such as palladium and even some polymers can also be used, as thin films of these will pass hydrogen through but nothing else. There are other methods being researched as well, such as bacterial or biological systems as well as electrochemical oxidation of CO.

When these methods are used on natural gas or methanol after water gas shift reactors, they can reduce CO to less than 10 **parts per million** (0.001%), so the gas stream can be used in any type of fuel cell. Some of these methods will also remove other contaminants while some are useful only for CO. Selective oxidation is used primarily for CO removal, for instance, while palladium membranes will remove almost all impurities from the hydrogen stream.

Methanol and natural gas, however, are already relatively high quality gasses. The lower the quality of input gas, the more complicated the clean-up becomes. A low quality gas exiting a reformer (using excess steam and higher temperatures to reduce coke formation) might then go to heat exchangers to adjust the temperature, be fed into a cyclone or other type of particulate removal system, go through a hydrolysis reactor to remove bulk CO, run through scrubbers to remove ammonia and acids, a halogen guard to remove the halide series elements (these interfere with the water gas shift reactor), a hydrodesulfurization system to drop out elemental sulfur, recombination with hydrogen to turn sulfur compounds to H_2S followed by zinc oxide fluid beds ($H_2S + ZnO = ZnS + H_2O$) and perhaps activated carbon to completely remove the sulfur (sulfur is very bad for fuel cells and takes several forms that must be dealt with), then to the water gas shift reactor, followed by catalytic reformers and ion or solvent exchangers to catch whatever is left, and maybe a final palladium membrane for ultrahigh purity gas. Most of these systems operate at different temperatures, so heat exchangers may be set between each stage to either lower the temperature of the gas stream or raise it. The cleaning systems used for fuel cell input gas are generally the most complicated part of the entire setup.

Hydrogen Storage

As mentioned previously, producing a high quality hydrogen stream to provide fuel cells is only part of the problem. In many applications the hydrogen supply system is either too large or too small to supply fuel cells on demand so that some form of storage is needed. This is especially critical in high temperature cells where the systems cannot easily be shut down and started up in the event of a supply disruption. In some smaller cells such as those in automobiles, in-house production of hydrogen may not be appropriate at all, so that storage must be in place for the system to operate. Recall though that because hydrogen is the smallest of all the atoms and is thus the least dense of all the gasses, to get any appreciable mass of it stored, it must be under high pressure or be liquefied. Neither of these methods is all that effective either. Many gasses are liquefied simply by pressurizing them, thus propane or butane can be purchased as liquid gasses and kept at room temperature. Hydrogen must be cooled to 22°K in order to be stored as a liquid and then kept at that temperature to remain a liquid. Pressurized cylinders of hydrogen can store about 71 kilograms per cubic meter, not a large amount when compared to other gasses.

Substantial efforts have been made to come up with alternative ways to store hydrogen but with limited success. In part this is because hydrogen is small, moves very quickly (one of the defining properties of a gas is that gas atoms and molecules are in constant motion and hydrogen molecules are the fastest moving of all the gasses) and has the lowest viscosity (how slow something will flow when it is in contact with a solid surface). Combine all these features and you get a gas that in essence will move through almost every material known because the gas (H_2) can break apart on many surfaces such as metal or polymers. It then forms two individual atoms of hydrogen, and those atoms move through the atomic structure of the material, causing substantial damage within the material (palladium is one material it won't flow through and that is why it is used as a purification membrane). Hydrogen gas itself will not only flow through even the tiniest crack or the smallest valve opening, but it will move through solid matter. It will flow through a hole more than three times faster than air will and if you mix enough of it with air, it will detonate. Dealing with hydrogen gas is as dangerous as dealing with other fuel gasses such as acetylene or natural gas, except that it will leak through equipment or cracks other gasses will not leak through.

Recall that hydrogen is an important industrial gas with millions of pound produced every year. If the hydrogen is not used as it is produced to make another product such as fertilizer, then it is most commonly stored as a liquid at very low temperatures (**cryogenic liquids**). Just as the industrial methods for producing hydrogen have limited use in many fuel cell applications, the industrial method of storing hydrogen as a cryogenic liquid also has limited use in fuel cell applications. The method for liquefying hydrogen is not simple, for instance, it has to be compressed and cooled using liquid nitrogen and further cooled by allowing it to expand (pressurizing and depressurizing gasses will also heat and cool the gas, that is why canned air is cold when it comes out). Additional processes are needed as well, but the important thing to remember here is that all of this takes money. Liquefying hydrogen can take from 25% to 45% of the energy available in the hydrogen. To put that another way, using cryogenic hydrogen for fuel cells means that about half the electricity produced in smaller systems would have to be used to make the liquid fuel. While fuel cells are relatively efficient, losing half your energy from the start just does not pay.

Other than pressurized gas or cryogenic liquids, the only other reasonable means of storing hydrogen is as a chemical compound, although some work has also been done that show carbon nanotubes can store hydrogen as well. Unfortunately, nanotubes are not easily produced and may well have substantial toxic properties so that approach is now just a laboratory curiosity. There are several compounds that can be used as hydrogen storage chemicals. Many metals will react with hydrogen (that is one of the reasons why hydrogen causes problems in metal storage tanks) to form metal hydrides where a metal atom is bound to one or more hydrogen atoms. Titanium, nickel, chromium, and manganese form stable hydrides as do other metals. The reaction forming the metal hydride (M + H_2 = MH_2) usually proceeds rapidly and without needing catalysts or heat to derive it from the left to the right (recall that chemical equations go both ways). To break the molecule back up into a metal and hydrogen gas, heat is applied. This method works but is slow, needing several hours, and it requires very high purity hydrogen, since impurities tend to go the metal surfaces and interfere with the hydride reaction.

Other chemical compounds can store hydrogen as well, such as ammonia, methane, and methanol. Methanol is currently the best suited as a fuel, especially in smaller systems, and may well end up being the primary fuel for personal and small commercial scale systems. It may seem odd to consider those liquids as hydrogen storage systems, but any chemical compound that has hydrogen and can be commercially produced will act to store hydrogen. Since much to the world's fuel based infrastructure runs on liquid fuels, reforming liquids may be a more acceptable alternative than storing hydrogen gas itself.

The other leading candidate for fuel cells is currently **borohydrides** (boron in a chemical compound with hydrogen), in particular sodium borohydride. Alkali metal hydrides (the general term used for these materials) generally form molecules with hydrogen and then release the hydrogen gas when combined with water. Equation 23 shows a general form of these reactions using calcium hydride. Hydrides can be solids or liquids, but are generally used in the liquid form since many of the solids are considered flammable, making their transportation more expensive.

$$CaH_2 + 2H_2O \rightarrow Ca(OH)_2 + 2H_2 \qquad Eq.\ 23$$

Sodium borohydride ($NaBH_4$) systems pump liquids on demand over a suitable catalyst to produce the hydrogen needed for the fuel cells. They produce pure hydrogen, depending on the quality of the hydride, the pumping systems and the reaction chamber used. The systems generally produce very high quality hydrogen, operate at ambient temperature, have relatively simple control requirements (as compared to reforming systems), can include extra water to assist in maintaining the water requirements for the downstream fuel cell and are considered to be reasonable safe. The hydride is expensive to produce though, roughly 100 times more expensive than some competing technologies, and these are essentially single use systems. When the hydride is fully reacted, it cannot be recharged and the leftover chemical portion (the $Ca(OH)_2$ in Equation 18 or the borate in sodium borohydride systems) must be disposed of and new hydrides brought in. Nevertheless, there are applications where cost and disposal is a secondary issue. Remember that fuel cells continued to be investigated and improved even though their costs were thousands of times more expensive than competing technologies like batteries because cost was not an issue in the space program. For systems below a couple of hundred watts, methanol or borohydride systems may well be used. Fuel cells, like their cousins the batteries, may end up being sliced into several different forms for different applications. Low temperature systems requiring high purity hydrogen may find on-demand hydrogen of high purity better supplied to personnel or tactical systems by chemical storage units, while larger installed systems are better supplied by scaled down industrial systems operating as part of a larger installation.

KEY WORDS

Knowing the terminology used is critical when dealing with fuel cells. Following is a list of the important terms in this chapter, which are also in bold typeface within the chapter. It is recommended that students are required to submit definitions of some of these words as homework assignments in which they look the terms up in other books, articles, or on the Internet.

- alcohol
- ammonia
- anaerobic digesters
- autothermal reforming
- balanced
- biological production
- borohydrides
- butane
- catalysts
- catalytic partial oxidation reactions
- combined heat and power
- cracking
- cryogenic liquids
- DC to AC conversion
- direct internal reforming
- Direct Methanol Fuel Cells
- distributed generation
- dry reforming
- economies of scale
- electrolysis
- endothermic
- entrained-bed
- ethane
- external reforming
- filter beds
- fluidized-bed
- following process
- fuel reformation subsystems
- gas diffusion plates
- gasification
- "heavier"
- hydrocarbons
- hybrid system
- hydrocarbon based fuels
- indirect internal reforming
- internal reformation
- liquid reforming
- membrane electrode assemblies
- methanation
- methane
- modular
- molecular chains
- moving-bed
- naptha
- packed bed filters
- palladium membrane
- partial oxidation reaction
- partial oxidation reforming
- parts per million
- peak loading cycling
- power conditioning
- Pressure Swing Absorption
- propane
- pyrolysis gasifiers
- quality control
- reforming fuel gasses
- scalable
- stable
- steam reforming
- thermal efficiency
- thermal partial oxidation reactions
- water gas shift reaction

DISCUSSION QUESTIONS

1. How is power conditioning handled differently in fuel cell systems than in grid supplied systems?

2. Name three different methods to produce hydrogen.

3. Name two important issues that currently limit the widespread adoption of fuel cells.

4. Which of the following energy sources are considered distributed sources: fuel cells, batteries, solar power, and wind power?

5. Why are fuels like diesel or gasoline not considered good candidates for reforming in fuel cells?

6. Why is there more than one step to reforming hydrocarbons?

7. Which of these hydrocarbons would be considered heavy: CH_4, C_2H_6, or $C_{10}H_{22}$?

8. Can fuel cells be shut down in an emergency?

9. Name some of the advantages internal reforming has over external reforming.

10. The chapter says that hydrogen must be cooled to 22°K to be in the liquid form. What is 22°K in degrees Celsius and degrees Fahrenheit?

CHAPTER 8

FUEL CELL SYSTEMS

INTRODUCTION

Fuel cells can be categorized by a few major features that distinguish one from another. The two most important features are the electrolytes used within the cell and the temperatures the cells operate at. Earlier chapters have discussed other differences such as the catalysts employed, the types of fuels that can be used, the level of purity required in the fuel gas, the arrangement of the stacks, water balances, and operating differences. Mention has also been made about the substantial differences in where and how the various fuel cells may be used. This chapter will discuss some additional, more general characteristics of fuel cells as well the types of applications they may be suited for. It is meant to introduce the fuel cell as systems of related concepts as well. The initial discussions about systems, applications, operations, fuel supply, maintenance, installation, and safety begin with this chapter. It is important for the student to begin thinking of fuel cells not as batteries purchased for specific applications or equipment and then discarded when used up, but rather as long term investments capable of upgrading and enhancement over a multiyear operational lifetime where subsystems must be maintained and operated in entirely different ways.

AFC Characteristics and Applications

Alkaline Fuel Cells (AFC) operate at lower temperatures, usually around 100°C, but they can be used over a range from ambient to 250°C. Recall that low temperatures such as these generally mean precious metal catalysts must be used so that the reactions can occur. These systems run at elevated pressures as well, 4×10^5 **Pascals** (58 psi), which is not considered high so that pressures of 30–60 psig will produce what is considered good performance of 3.2 A/cm^2 at 0.600 volts (per cell). Increasing the pressure and operating temperatures (300°F at 200 psi) will give 9 A/cm^2 at 0.900 volts. As with most fuel cells, there are tradeoffs for increased pressures and temperatures so that while voltages and currents increase, efficiency and life expectancy tend to decrease. As mentioned several times, though, the AFC systems were used in space where weight and multiple uses (it can produce water) were more important than cost or life expectancy. System efficiency for AFC is in

objectives

This chapter introduces the student to fuel cells as capital investments meant to be used in specific areas and operated under controlled circumstances. Some of the larger issues, challenges, and application of fuel cell technology are introduced with following chapters discussing these issues in more depth.

the 60% range with improvements of another 20% or so if a Combined Heat and Power (CHP) cycle is employed. Many of the systems last only several thousand hours, unacceptable life cycles in the commercial world but acceptable in the space program.

The units generally are higher cost, especially when compared to other low temperature systems, in part because of the designs being developed over the years to meet the requirements of the space program, as cost-is-not-an-issue development usually mean the designs succeed in that mission. The mature systems in use over several decades have evolved to fill a specific and very narrow mission, that being to provide electricity and water in totally isolated environments. In general, they are designed for systems greater than 10 kW and little effort has been made to go below this because of the complexity of the balance of plant equipment. Larger systems have also been left out of many of the design cycles so that AFC systems are a mature technology only within a narrow range. Many of the issue dealing with AFC systems arise from one simple design criteria: the systems must contain sufficient redundancy that noncatastrophic failure will still provide what is required for life to be sustained. **Catastrophic failure** was originally dealt with not by overdesigning the system but by having spare systems onboard. The design and manufacture of these units became so reliable that extra systems were dispensed with while each system became robust enough to handle minor failure. This type of design called for using more catalyst, eliminating any leakage entirely, running on very pure hydrogen and producing very pure water and power. While space applications required these, and were willing to pay the price, terrestrial applications do not need the absolute guarantees or the purity of output.

It is unlikely that AFC systems will be able to maintain their particular design features if they are to compete in a price driven market. High cost, midrange systems have limited appeal outside of mission-critical applications that rarely appear in consumer arenas. Markets that may be open to AFC systems include such areas as off-the-grid heat and power applications where reliability and performance is equal to or slightly more important than price, such as the farm or ranch where any loss of heat and power can result in loss of life or high value crops. Remote housing where heat and power must be supplied, such as high-end resorts or mansions, is another application that may fit well with the current AFC type of design as would hospitals, airports in more remote locations, military operations, or data centers. Whether AFC systems or some other type of fuel cell eventually serve these types of markets remains to be seen.

SPFC (PEM) Characteristics and Applications

SPFC (PEM) systems are low temperature, relatively low pressure systems that produce a large amount of electricity for the size of the cells used (high current density). Unlike some of the other fuel cell systems, they have only been around for half a century or so because the polymers used as electrolytes were not invented until the middle part of the twentieth century.

SPFC units produce roughly 500 amp/ft^2 at 0.7 volts (per cell) running at 65 psig. As with most cells, that number can be improved by increasing the pressure (up to a point) or using pure oxygen instead of air (higher cost). These cells have the same magnitude of power density as AFC installations (better than other cells) but at a lower cost than AFC systems. SPFC efficiency is somewhat below 50%, with that number going up or down depending on the fuel, pressure, temperature, and other operating parameters. Compare this to gasoline (~20%) or diesel (~40%) engine efficiency (the amount of useable energy extracted from a given amount of fuel to do work with) and you can see the advantages of fuel cells.

Again, add a CHP system design and the efficiency of fuel cells can increase by another 20% or so. Keep in mind though that not all installations can make use of CHP cycles. Automobiles, for instance, can only make limited use of waste heat in southern climates and even in northern winters only some waste heat can be utilized. Larger users such as home, commercial, or campus type installations can better use waste heat to increase efficiency but again only to a limited extent since lower temperature cells do not generate sufficient amounts. Typically, SPFC (PEM) systems are used in applications requiring 250 kilowatts or less, positioning them for small to midrange applications.

SPFC systems work in most orientations and have minimal hazardous materials involved in their fabrication and use. As mentioned before, while each individual fuel cell produces a usable 0.7 volts or so, they can be tied together in much larger stacks so the final current output can be manipulated. Most fuel cell systems tie the cathode of the first cell to the anode of the next using a conductive (bipolar) plate that also serves to bring the fuel gas stream to the second anode and the air or oxygen to the first cathode. This scheme then carries on through the stack. The final design of the cells balances the required output by using stacks, running within a narrow operating temperature and carefully designing the electrical wiring schemes. Designs will also minimize cost by optimizing the placement of the expensive platinum catalyst as well as maximize work life and equalize operations by matching liquid and gas flows within the cell. Because of their simple construction, SPFC cells can be produced relatively Cheaply, can be tied together in stacks very easily, and have good working lives in part due to reduced corrosion potentials from the interior parts. This lessens the cost of production and equally as important standardizes the results so that each product behaves the same way, one of the hallmarks of **mass production**. Recall that systems like molten carbonate cells are not manufactured in that way and in fact do not even work until the final operating temperature is first attained during start-up.

Since SPFC systems can operate at low temperatures and have high **current densities**, equal amounts of power can be obtained using smaller cells and stacks than other systems, which in turn lower costs. This is only marginally important for large installations, but is particularly useful in applications that require fast start-up and shutdown, surge uses where a constant output is not required such as transport or battery charging, and consumer uses where the higher system cost cannot be recovered. Two major uses are standby generators and domestic power. Most generators currently use diesel as a fuel, particularly in areas where natural gas is not readily available and SPFC systems offer a quieter, considerably less polluting alternative. Domestic SPFC systems provide not only electricity but can be used to generate hot water as well, since domestic hot water requirements are in the general operations range of even low temperature fuel cells. CHP systems increase the overall efficiency of fuel cells by making use of the fuel cell operating temperatures to offset other costs. Lower temperature cells like SPFC in general lack the capability to compete in larger scale CHP schemes such as steam generation or process heating because they do not generate enough heat. In smaller application such as cars, homes, or campus installations, there is enough heat generated to make a significant impact on heating water, keeping cars or busses warm when parked so the main internal combustion engine can be turned off, or preheating water through heat exchangers for heating steam used in campus-type applications.

As had been mentioned, what is important is not so much the capital cost of the systems, although that is taken into consideration, but rather the cost of any electricity produced. Targeted cost for electricity is in the range of $35 per kilowatt hour, but that number is almost entirely dependent on competing technology such as coal-fired power plants or nuclear energy. Those technologies get more expensive based not only on basic supply and demand pressures but also cultural and political pressures. Nuclear power, for instance, has seen steady technological progress for decades but public opinion of it swings wildly depending on how many years since the last major nuclear disaster, be it Three Mile Island (1979), Chernobyl (1986), or in Japan at Fukushima Daiichi (2011). As the cost of hydrocarbon-based fuels go up and pressures for ever more environmentally benign technology increase, a point may be reached where cost becomes only one of the main considerations rather than the main consideration. Stating costs for kilowatts produced for fuel cells is a gamble at the current time, but in general, rather than producing electricity for tens of dollars, fuel cells are in the hundred to thousands range. It is not to say fuel cells are thus "unaffordable" but rather an alternative where electricity from cheaper sources is not generally available or where personal choice dictates higher costs as being acceptable for other reasons such as political or cultural. Figure 8-1 shows some average costs of produced power, and as can be seen, they vary widely depending on who is calculating that cost.

Other issues also have a direct impact on adopting fuel cells technology for general use. High temperature systems can tolerate lower quality fuels, lowering overall operating costs considerably; however, the higher the temperature, the greater the potential risk. In consumer applications, where people will be exposed

Energy Costs Comparison	
Resource Type	Average Cost (cents per kWh)
Hydroelectric	2–5
Nuclear	3–4
Coal	4–5
Natural gas	4–5
Wind	4–10
Geothermal	5–8
Biomass	8–12
Hydrogen fuel cell	10–15
Solar	15–32
Sources: American Wind Energy Association, Wind Blog, Stanford School of Earth Sciences	

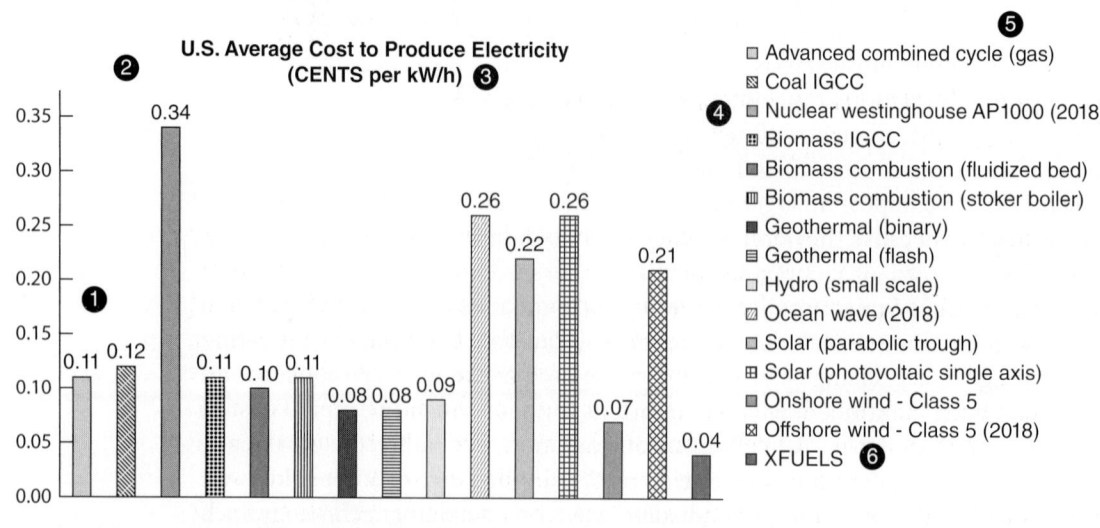

Figure 8-1 The cost of power.

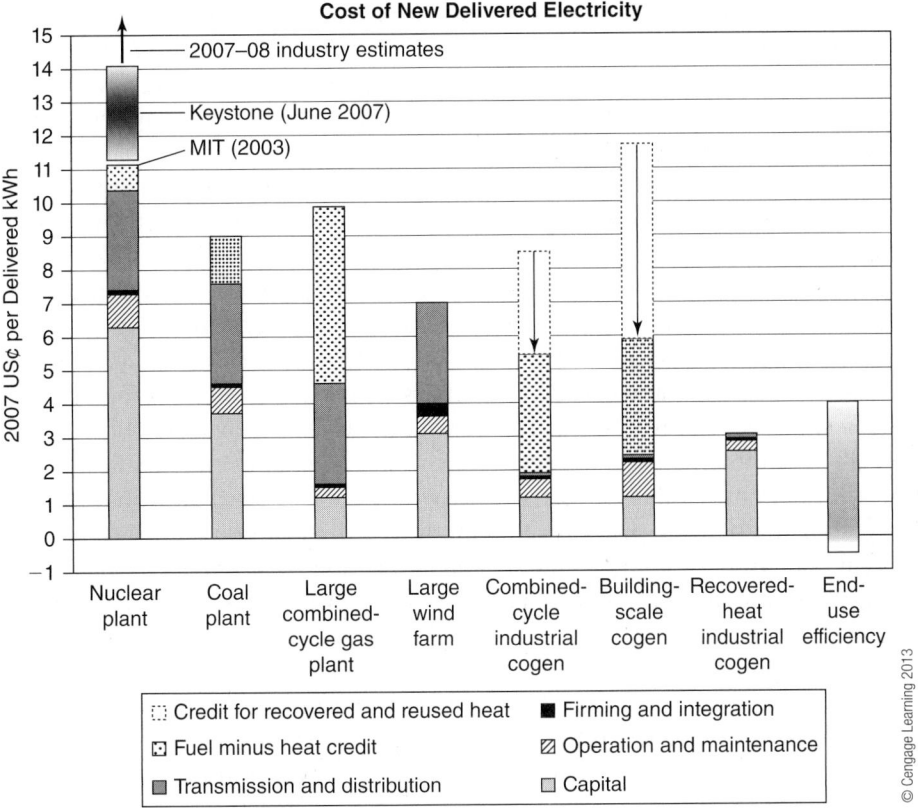

Figure 8-1 (Continued)

to all the potential problems on a regular basis, companies may well chose to avoid being exposed to the risks inherent in new technology being adopted until they know the costs associated with accidents. These types of **risk assessments** done on major technology shifts can take decades to work out in the marketplace.

All of these factors, as well as several others, should make the student aware of the potentials involved in fuel cells. Competent personnel trained and knowledgeable about a wide range of fuel cells will be better positioned in the industry than those trained to a specific type or brand. Understanding the basics involved in fuel cells as a whole will position the technician and engineer within a wider industry that may change as fast as the computer industry changed in the previous century.

Molten Carbonate Fuel Cells (MCFC) Characteristics and Applications

One of the most important characteristics of the MCFC is the material that holds the electrolyte in place. The carbonate is molten (thus the name) so it will act like a liquid and like all liquids, it must be held in place. The electrolyte is molten because the fuel cell operates at high temperature so whatever holds the carbonate liquid must also be stable at that temperature (1202°F or 650°C) and cannot react with the carbonate ion or the ions being conducted. Since the electrolyte must be in contact with the electrodes as well, the container for the molten carbonate has to allow transport across the cell. Contrast that with a solid electrolyte like that in SOFC systems where the anode/electrolyte/cathode cell structure is one solid piece. This has important implications for service in particular. While fuel cells that are monolithic solids have definite advantages, so do cells that have molten electrolytes. Failure in a monolithic structure is catastrophic, meaning there is no repair and replacement is the only option. AFC systems, where the electrolyte can be pumped through the system, allow for in-operation monitoring because the electrolyte solution itself can be monitored. The molten electrolyte systems allow for service options that solid electrolyte systems do not.

Most MCFC systems use a porous ceramic matrix to contain the molten carbonate much like a sponge holds water. The important characteristic then is the ceramic itself, since ceramics are generally brittle and do not tolerate temperature or pressure cycling well. MCFC electrolyte ceramics will crack if the temperature rises or falls too fast or if the pressure at one electrode is much greater than at the other electrode.

Another critical characteristic of MCFC systems has to do with the **start-up cycle**. As mentioned previously, fabrication of the cells themselves are only finished when they are turned on and brought to their operating temperature. Each system then will be slightly different than others, since atmospheric conditions such as temperature, barometric pressure, and humidity will be different during start-up. It is also difficult to cycle these systems on and off and that generally means that these systems are turned on and stay on during their life cycle. This makes them less useful for smaller, consumer-based applications such as cars where cycling is a critical component of use. Applications range from light industrial use to utility applications with outputs in the hundreds or thousands of kilowatts with both natural gas and coal being common fuels.

The differences in electrolytes force differences in fuel cells just as the similarities in chemical reactions force similar forms. In MCFC systems, the amount of molten carbonate is critical to the operation of the cell. A PEM cell relies on Nafion and water to regulate proton (H^+) flow and if the water is not regulated, the cell can both flood and dry up. An AFC relies on water and the hydroxide ion (OH^-) so the same type of problems can arise in an AFC as in a PEM cell. In a MCFC, the molten carbonate is at hundreds of degrees and the water being made at the anode if hydrogen is a fuel rather than carbon monoxide will be in a gas form (steam) rather than in the liquid form. The way MCFC systems handle water is different than the way other fuel cell systems handle water. Water (or steam) handling is one of the critical aspects of system operations, and it is important to recognize that fuel cells systems are similar but different enough that technicians, operators, and engineers must be well versed in the underlying causes of both the similarities and the differences.

Start-up for MCFC systems is critical since it sets the system seals and determines overall electrolyte distribution. Unlike other fuel cell systems, failing to exercise a fine degree of control during start-up can be disastrous to MCFC systems. The same is true of a shutdown cycle. Proper operations and maintenance protocols are extremely important to fuel cells systems, and while computer controls can limit bad choices, they cannot entirely eliminate unwise and even unsafe behavior. Using MCFC systems as an example, the liquid electrolyte is well over 1000°F and if during start-up or shutdown the ceramic holding the molten material cracks, hot liquid may leak.

MCFC systems operate at higher efficiencies over a long lifetime, in part because they are high temperature systems and because they run constantly, eliminating many of the problems associated with cycling. It is critical that the student understand the difference between any system that runs continuously, thus requiring a pre-planned and complicated shutdown plan to service, and a system that is specifically designed for on-call service where shutdown and start-up are routine. Many of the differences between low temperature and high temperature fuel cell systems have to do with design features imposed by **continuous** versus **on-call operations**.

Because MCFC systems are continuously operated, it is generally not economically advantageous to make them small. These systems generate 250 kW at the least and can supply over 2 MW (2000 kW) with efficiency between 50% and 60% (CHP systems can raise those to as high as 85% efficiency). It should be noted that the higher the operating temperatures used, the more effective CHP systems are so that MCFC and SOFC systems operating at high temperatures have the best efficiencies if suitable combined systems are in place. Again, it should be noted by the student that efficiency, operating lifetime, and general availability of generating capacity are considered critical to fuel cell installations. Improving any one of those three categories has the most impact on whether fuel cells are a worthwhile investment because they have the most impact on the overall cost per kilowatt produced over the life of the installation (and thus the investment).

Phosphoric Acid Fuel Cell (PAFC) Characteristics and Applications

PAFC systems are low temperature systems using hydrogen as a fuel. As mentioned before, that means more expensive catalysts are generally used on the electrodes. The main difference then between PAFC and other fuel cell technologies is the electrolyte used. Phosphoric acid is a liquid like molten carbonates are liquids, but they are liquids at lower temperature and so the cells can operate at lower temperatures than MCFC systems do. Like PEM cells, they operate around 200°C and like alkaline cells, they must use hydrogen as a fuel.

PAFC systems are generally sold for stationary applications such as a campus setting where one installation can be used to supply several buildings. Although they operate at lower temperatures, they still generate heat and so can be used in CHP systems, supplying either heat or hot water for internal use in buildings. Hundreds of PAFC systems have been installed worldwide and they are in fact a very successful technology in the range of 200 kW supplied, with larger systems of up to 5000 kW being produced. They are somewhat limited in their applications because of the low temperature operations and thus restrictions on their CHP output, but in general the cells operate at efficiencies of around 40% with CHP efficiencies rising as high as 85%–90% (0.31 watts per square centimeter produced per cell).

One of the most important aspects of PAFC systems is their current operating life of more than 60,000 hours in service. Failure many times results from acid loss as opposed to material failure, as is the case in some other fuel cells, particularly the higher temperature ones. Corrosion and electrode contamination from the input hydrogen fuel are the two other main causes of failures in PAFC systems. Load cycling causes some trouble in PAFC systems but not nearly to the degree that it does in higher temperature systems where expansion and contraction due to heat cycling can rapidly destroy the cells. Most fuel cell systems avoid cycling, no matter what temperature they operate at, since the flow of electrons (electricity) and ions through the workings generate internal heat and stress detrimental to the life expectancy over the long term. Starting and stopping this flow causes problems no matter what temperature of operation. Low temperature systems respond to temperature cycling differently than higher temperature systems, since the greater swings affect the materials in the system to a much greater extent and tend to actually break things. PAFC systems currently in use operate at greater than 95% online service.

In general, PAFC reliability, lack of emissions, ease of operations, low maintenance requirements and maturity of the technology position them well for premium applications where uninterruptible power is required without any accompanying noise or fumes such as hospitals or areas with high population density such as apartment blocks. As with most low temperature fuel cell systems, the initial cost is high because of the precious metal catalysts but this is offset to some degree by the long lifetimes in service. **Cost per kilowatt** produced is generally in the thousands of dollars rather than the targeted goal of $35 per kilowatt produced, but again, the student should keep in mind that these numbers are politically driven to a great extent and can change suddenly.

PAFC systems are **mature designs** currently operating in thousands of locations worldwide due in part to their association with the space program. Reliable units capable of generating power quietly and without pollution for several years with only routine maintenance have great potential even if the electricity produced is expensive. This is especially true in areas that do not have grid supplied electricity but may have a steady source of fuel that can be reformed to hydrogen. That is not to say the technology is fully developed. Even mature technologies are routinely refined to produce better, cheaper products. One need only consider the telephone industry to see how even very mature products can undergo rapid and even transforming change after years of stability. Because of their standing in the United States space program, PAFC systems have been extensively analyzed over the years. The current list of improvements generated by the government can serve as a good look at what needs to be done to make fuel cell systems in general more competitive.

Fuel cell manufacturers must develop electrodes that last longer and cost less. In lower temperature systems like PAFC, just the platinum catalysts alone can represent up to 20% of the total system cost. When producing the electrodes themselves, where positioning them in the cells and guaranteeing uniform gas supply to them also adds cost, the electrodes can represent the single largest cost component for a low temperature fuel cell. It is no wonder then that lowering overall costs must focus on those components.

Fuel cells currently use polymers such as Teflon and Nafion that are well enough suited to the applications but not truly designed for fuel cells. Polymers that handle electron and ion flow better, interact more efficiently with other cell components, and cost less are needed. The student should keep in mind that the development of improved materials is critical to the success of every product in existence, be it higher grade wheat, lighter but stronger polymers for cars, or fuel cell components.

The operating temperatures of fuel cells are too high for many applications, especially that of higher temperature systems. Lowering these operating temperatures while maintaining or even improving the overall efficiency is critical to the overall success of fuel cells. The main reason for this is in the cost of special materials (especially metals) that can endure those temperatures for tens of thousands of hours without failing. Mild steel, for instance, can only be used up to about 700°C (1300°F) before it loses many of its mechanical properties and fails in service.

The manufacturing designs, methods, and systems used to produce fuel cells are currently different for the different types of fuel cells as well as for the different manufacturers. Developing uniform methods and **quality control systems** is one of the best ways to lower costs, ensure quality, and increase in-service life spans. Since designs and methods of manufacture are not yet well established, the waste generated during manufacturing and operations is relatively high compared to other industries. This is due in large part to the improved methods and designs coming out at a steady rate, as manufacturers must retool their systems regularly. More efficient methods and more stable designs will eliminate much of the waste currently being produced, a key advantage of using uniform manufacturing methods. As the installed base of fuel cell systems grows larger and more experience in producing cells for the mass market is gained, as opposed to producing for higher end applications like space, most of the issues currently seen as problematic for the technology will disappear.

Solid Oxide Fuel Cell (SOFC) Characteristics and Applications

SOFC systems are used for large systems of several hundred kilowatts with the individual cells themselves producing approximately 1.8 watts per square centimeter for **linear systems** and 0.25 watts per square centimeter for **tubular systems**. They can reach efficiencies of up to 65% even without associated cogeneration systems and approach 90% efficiency in some CHP systems. They normally operate in excess of 800°C but as mentioned above, one of the main thrusts of fuel cell research is to lower operating temperatures, with 600°C (1112°F) often mentioned as a goal for SOFC.

They have several characteristics in common with other high temperature systems. They do not need precious metal catalysts and instead use much lower cost metal, often using **nickel oxides** on the anode and **lanthanum-strontium-manganese alloys** at the cathode. They are particularly tolerant of contaminants in the fuel gas and can use both hydrogen and carbon monoxide as a fuel. They have long start-up and shutdown cycles, taking as long as an hour to reach operating temperatures for some larger systems.

SOFC systems are currently used mainly for stationary systems supplying approximately a megawatt (going up to 30 MW in some systems) in installations where CHP cycles can make the most use of the excess temperature they generate. Unlike most fuel cell systems, particularly low temperature systems, they can supply relatively cheap electricity, with some sources quoting below $0.10 per kilowatt. Unfortunately the capital costs of these systems is relatively high and even with stack life expectancy of over 40,000 hours in some designs, the overall cost after rolling in initial system and installation costs can still be in the thousands of dollars per supplied kilowatt.

It should be noted again that these systems are competing with generating plants such as coal fired turbines or hydroelectric installations that have operating lives of decades. The cost of such installations far exceeds the cost of a fuel cell system, but the amount of electricity generated and the decades in operation tend to spread the initial **capital cost** out. Fuel cells systems are much smaller and have yet to generate enough installations that economy of scale issues kick in and costs drop dramatically. Whenever the cost of delivered electricity is compared, the student must be aware that most of these comparisons are not nearly as straightforward as comparing the cost of one car to another. It is important to remember that the cost of grid electricity is cheap in developed countries only because the generating units and the grid systems are already in place. In areas where electrical lines do not exist or are not economically viable (isolated areas with low populations), fuel cell systems already compete with other generating systems such as wind, solar, or diesel. This is particularly true in areas where multiple systems can be employed, with solar or wind generating hydrogen from electrolysis, which is then stored as an ultrahigh purity gas so the fuel cell can generate electricity on demand.

While larger (several hundred kilowatts) combined systems partnered with CHP or cogeneration are currently the main thrust of SOFC systems, there is substantial interest in smaller systems because SOFC can tolerate impurities better than most other fuel cells. Charging systems roughly the size of a large suitcase are in use by the military for group application since these operate quietly and can be constructed with minimal heat signatures. Systems of about the same size are being used as auxiliary power units in transportation applications where internal combustion engines can be turned off and the SOFC system used to supply electricity and heat (or air conditioning). These systems are particularly useful for buses that regularly have to idle at stops or utility and construction applications where equipment spends most of the time idling while work is being done outside the vehicle.

While not discussed before, the student should be aware that fuel cells produce various voltages for different applications just a batteries do. Many of the current larger stacks output 14 volts with other running up to as many as 42 volts. As with most portable (or near portable) systems, designers and manufacturers can tailor outputs to meet installation requirements. This is a substantially different arrangement than one finds in grid-supplied electricity, where some variant of 110 volts (in the United States at least) such as 220 (in Europe), 440, or 660 is generally supplied to a building and anything inside has to conform to that voltage or use transformers to change it. The student should be aware that a wide range of voltages are supplied across the world (frequencies are generally 50 or 60 hertz). Fuel cells do not need to conform to these standards but certainly can be manufactured to meet them.

It should also be mentioned here that at one point combined systems had two possible configurations. Smaller systems used heat and power cycles, where the power was supplied by the cell and the excess heat was used to heat air or water which was then used in place of heat or hot water in a general installation. Larger installations could generate enough heat to run turbines (gas or steam) so that these combined systems generated electricity (cogeneration) using both fuel cell and turbine as well as using any remaining heat in the same manner as smaller CHP systems do. In recent years, turbines have become smaller and more efficient as new materials and improved computer control came on the scene. Microturbines are now available that may lower the wattage limit where turbines are economically practical to only a couple of hundred supplied watts. These hybrid fuel cell/turbine/heat systems have efficiencies well over 90% and may be the advance that turns fuel cells from a possible alternative to a direct replacement in commercial, industrial, and even grid applications.

Hybrid fuel cell/turbine systems can also run in direct modes or indirect modes giving them substantial versatility. In direct modes, the input fuel gas is reformed and fed to the fuel cell where it is partially combusted at high temperatures. The cell exhaust and any fuel remaining mixes with any excess oxygen from the fuel cell and is fed into the turbine, which in turn uses the heat energy and velocity of the stream to turn the blades of a conventional electrical generator. The turbine can also be used to supply compressed or preheated air to the fuel cell in a heat exchange. These types of systems are somewhat complicated but in all probability will make up a substantial portion of large installations, especially since the recirculating nature of the fuel's gasses and heat tend toward very low emission systems, one of the major goals of power generation. Indirect systems use the fuel cell exhaust to compress air and that compressed air drives a turbine so it does not need to be made of materials that can tolerate the high temperatures in some fuel cells. There are a number of combined cycle arrangements that can be used in fuel cell installations. In addition to the direct and indirect systems mentioned above, there are topping and bottoming cycles where the excess heat from the fuel cell is sent to waste heat boilers to generate steam which in turn runs a conventional steam turbine. Students who intend to work in fuel cell installations may soon find themselves involved in complex systems involving several different generation schemes. Figure 8-2 shows a typical simplified configuration for a hybrid fuel cell/turbine system.

Combined systems also can use multiple fuel cells. In some designs, SOFC cells are used to treat the gas supplied to SPFC (PEM) cells. This is particularly useful where clean supplies of hydrogen might be hard to get for the SPFC (PEM) cells. In these systems, only part of the input gas (reformed but not to very high quality) is used by the SOFC. The SOFC uses some of the hydrogen but most of the CO with the resulting exhaust gas then fed through a water gas shift reactor and a secondary CO scrubber and heat exchanger before it goes to the SPFC (PEM) cells. The heat from the SPFC (PEM) system is then used in a combined cycle as well. Combined systems exist using SOFC technology that have little or nothing to do with power generation as well, such as the **oxygen sensors** used in cars as part of the combustion control system to improve internal combustion efficiencies and lower emissions. It would be an interesting exercise for the student to investigate how a SOFC turned into an automobile sensor as a way of seeing how technology is never

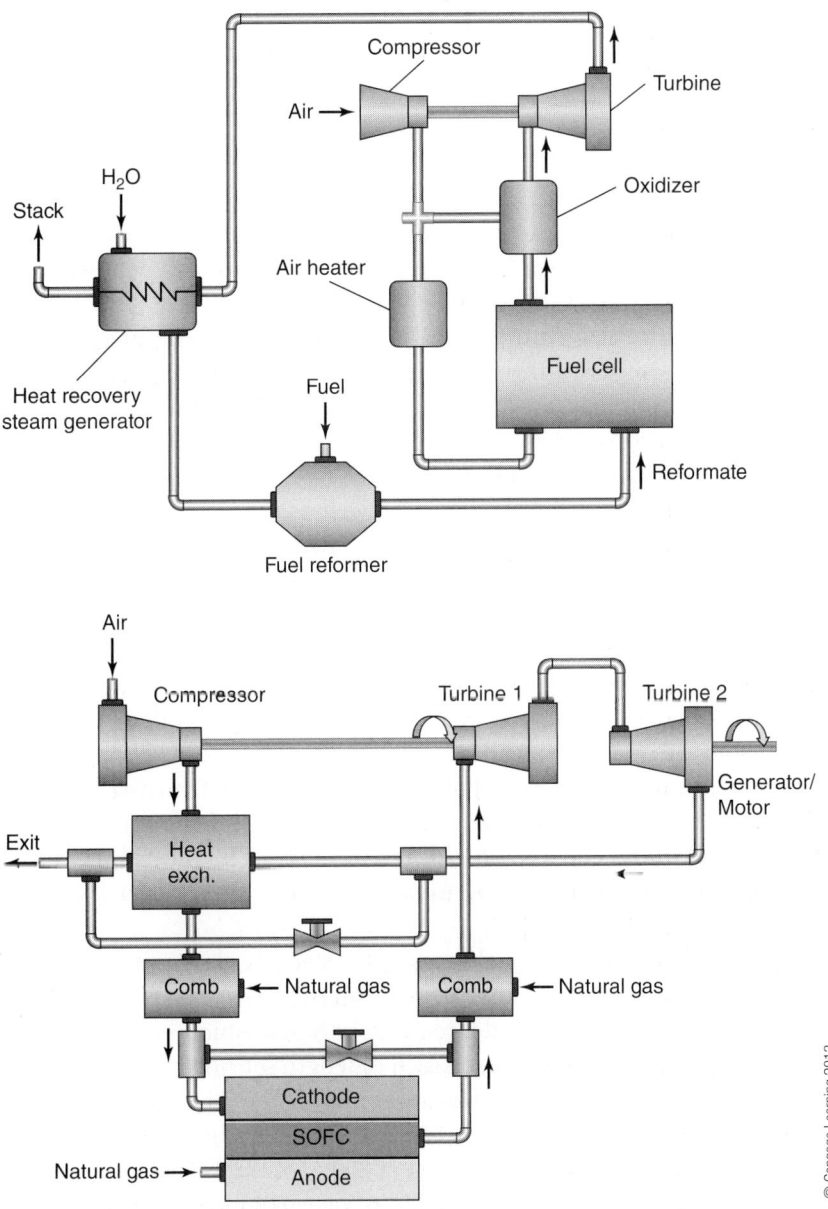

■ **Figure 8-2** Hybrid fuel cell/turbine system.

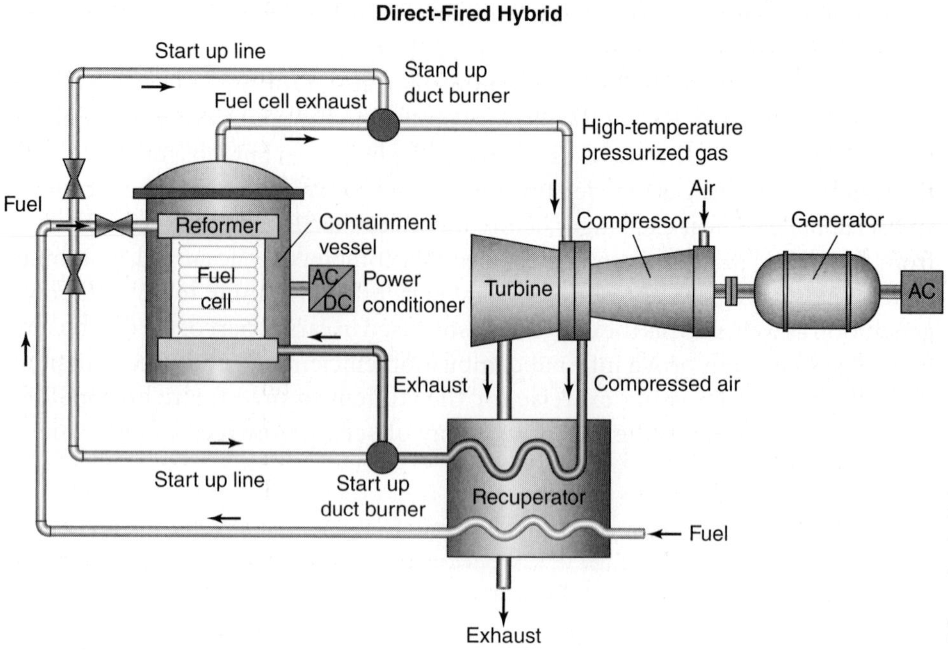

Figure 8-2 (Continued)

static and may well become something completely different so quickly that those who rely on older technology for their livelihood are fast left behind in the unemployment line.

As mentioned in the beginning of this chapter, one of the purposes of this particular discussion is to introduce the students to some of the larger issues, challenges, and application of fuel cell technology. Along those lines, a discussion of how you go from a fuel cell that can be held in one hand to an installation of stacks producing grid level amounts of electricity is in order. This is particularly important in SOFC systems (as well as other high temperature systems), because single cells have definite **size limitations** due to **thermal expansion** considerations, sealing problems, and consistent gas delivery among other things. The **coefficient of thermal expansion** (CTE) is particularly important to high temperature fuel cells so a more in depth explanation of what it is will be important for anyone working on system where temperatures change during service. CTE is the change in size a material (solid, liquid, or gas) sees as temperatures go up. Unlike liquids or gasses, which tend to change their shape when heated, solids keep their shapes but grow in size (mostly, although some things do shrink when heated among them shrink tubing for electrical connections). There can be considerable differences in CTE, and we will use a common problem in SOFC systems as an example to illustrate why size is limited and seals are difficult to get right.

Recall that individual cells are sealed in an SOFC stack by glass, in part because glass will flow at operating temperature to seal microcracks. The CTE for hard glass is 3.3×10^{-6} inches per (inch degree Fahrenheit), which means that for every degree the temperature goes up, each inch of length (or height or width) will contribute 3.3×10^{-6} inches to the overall increase in size. The CTE for 316 stainless which might be used as framework within a cell is 8.9×10^{-6} inches per (inch degree Fahrenheit). Thus, for a 6 inch SOFC operating at 800°C (1472°F), the length of the glass seal will expand by ($3.3 \times 10^{-6} \times 1472 \times 6$) or 0.029 inches while stainless steel will grow by 0.078 inches ($8.9 \times 10^{-6} \times 1472 \times 6$). The glass seal will expand in different directions by different amounts because the length is not the same as the width or thickness on seals.

While this can get very complicated in large systems, it comes down to a simple fact illustrated by the two numbers above. The stainless steel frame is set in place and grows toward the seal to a great extent while the seal pushes back against both the ceramic SOFC and the stainless steel. The frame expands 78 thousandths as the heat comes up while the glass is trying to expand its 29 thousandths. The two push against each other and since glass is not as strong as steel (nor is the ceramic fuel cell by the way which of course is also growing in size) either the glass or the cell will break. Keep in mind that there are usually 25 cells in a stack and that each one of those is expanding as well and in all three dimensions. The design of high temperature systems is extremely complex because of CTE considerations and those who work with such systems must understand just how complex the situation can become. In general, growth is limited to thousandths of an inch and that is why fuel cells (especially high temperature ones) are not 10 feet in size; they would blow apart when brought to temperature. Figure 8-3 shows some of the various sizes of individual fuel cells.

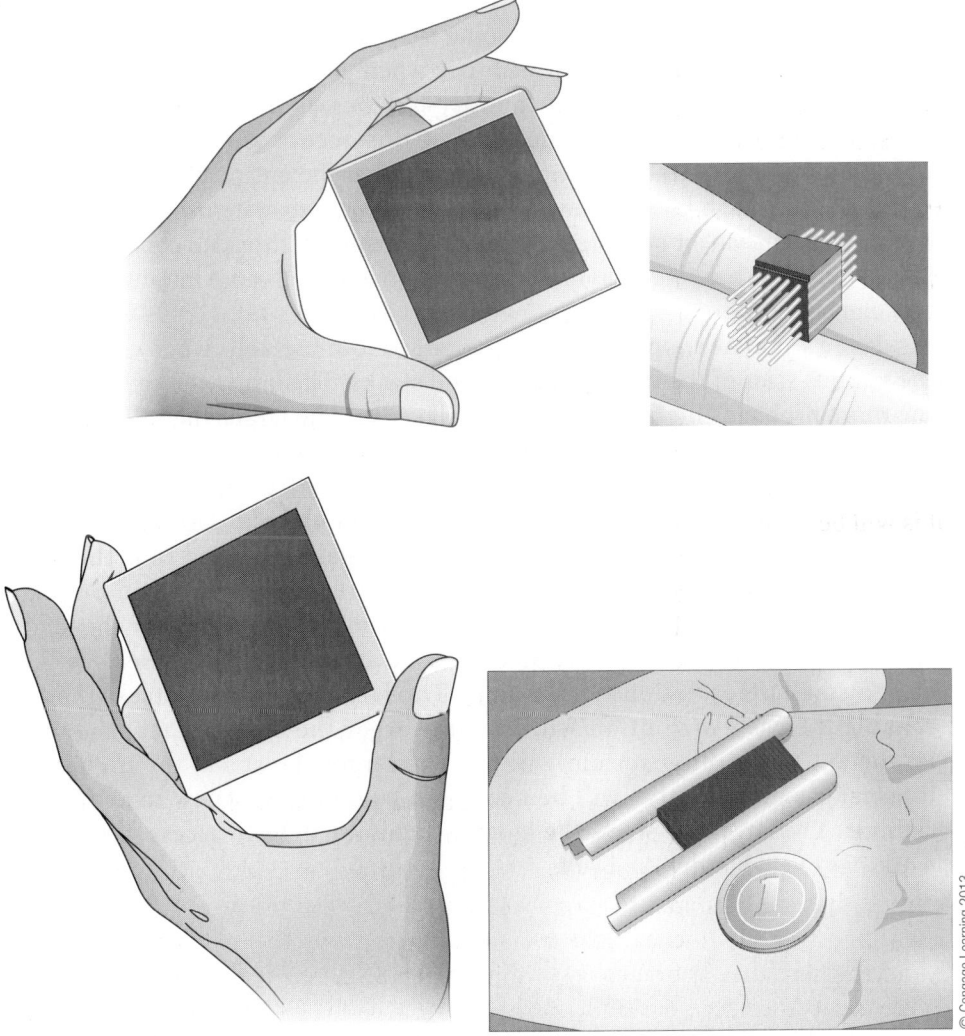

■ **Figure 8-3** Individual fuel cells.

Keep in mind that the fuel gas has to be delivered through the interconnects (bipolar plates) so that there must be different materials in the stack with different CTEs. The numbers quoted above should be enough to make the student see why interconnects are generally not made of 316 stainless steel but more often out of a material (chrome alloys for instance) whose CTE is closer to the glass and to the actual solid oxide fuel cell sandwich of anode/electrolyte/cathode. This issue keeps going by the way. If you stack fuel cell stacks on top of each other, each stack will expand by a considerable amount so that some way to keep the stacks apart must be used. The cells in the center of the stack will probably be a bit warmer than the ones at the edge as well, so nothing in the system is growing (or shrinking if the cell is turned off and cools) at the same rate or by the same amount.

Another issue of concern in this matter is the gas supply. Gas expands as it warms, just as solids do. Fuel gas must be supplied in consistent amounts across the face of the fuel cell anode in part because the cell will experience areas of different temperature and growth causing cracks to develop if the gas supply is spotty. Depending on the design, a stack of 25 six inch by six inch solid oxide cells might use a couple of liters of fuel per minute. That gas has to be manifolded while at operating temperature (1472°F), with the correct portion moved to each anode and then moved into the interconnect for final distribution. As the gas flows through the piping system, it might expand when heated, increasing the gas pressure when it does that, then expand again when going from smaller piping to larger manifold and then contract and increase its velocity when going from larger manifold to smaller internal piping and then contract and increase its velocity again when it enters the interconnect channels. All of this occurs as minor temperature changes take place while operations go on so that the entire system is growing and shrinking much like a bellows. Working on fuel cell systems require that the operator, technician, or engineer realize that tightening a threaded connection by a bit too much or changing a gas line configuration may well cause internal problems that damage cells.

There are some simple rules to remember about fuel cells when considering the various sizes of the individual cells and stacks. The larger the cell (or stack) the more problems high temperatures will cause. The larger the cell, the more pressure changes there will be internally, causing fatigue cycling that may severely shorten the life span of a cell. The larger the cell, the more material there is inside it, increasing weight and thus stress within the cell. The larger the cell, the more likely it was produced as a custom installation, and thus the more likely it will have nonstandard features that are not documented or spare parts that are not easily obtained in a timely manner. The larger the stack, the less mechanically stable the stack will be and the worse the seals will be due to small deformations in the cells. The larger the cell or stack, the less reliable the gas distribution will be. The larger the cell or stack, the more important the cooling system will be. The larger the stack, the higher the internal gas pressure that will be needed to guarantee gas flow to anode and cathode. The larger the stack, the more difficult it will be to access the internal workings of each individual cell.

KEY WORDS

Knowing the terminology used is critical when dealing with fuel cells. Following is a list of the important terms in this chapter, which are also in bold typeface within the chapter. It is recommended that students be required to submit definitions of some of these words as homework assignments in which they look the terms up in other books, articles, or on the Internet.

capital cost	lanthanum-strontium-	pascals
catastrophic failure	manganese alloys	quality control systems
coefficient of thermal	linear systems	risk assessments
expansion	mass production	size limitations
continuous operations	mature designs	start-up cycle
cost per kilowatt	mature technologies	thermal expansion
current densities	nickel oxides	tubular systems
hybrid fuel cell/turbine	on-call operations	
systems	oxygen sensors	

DISCUSSION QUESTIONS

1. What are two of the most important features that distinguish one type of fuel cells from another?

2. If the operating pressures and temperatures of a fuel cell are increased, what happens?

3. Why can't fuel cells used in automobiles add Combined Heat and Power cycles to make the cells more efficient?

4. Name two reasons the ceramic matrix holding a MCFC electrolyte might crack in service.

5. Why are MCFC systems not considered mass produced items?

6. In lower temperature systems like PAFC, how much of the total system cost pays just for the platinum catalyst?

7. Why do fuel cells produce variable voltages?

8. Discuss some of the problems that might arise as fuel cell stacks get larger.

CHAPTER 9

STATIONARY FUEL CELL APPLICATIONS

objectives

This chapter introduces the student to stationary fuel cell systems as turnkey operations set up to provide power to one place. Small to large systems are discussed and comparisons to grid supplied electricity provided. Both the difficulties fuel cells face in providing power and the methods used to overcome these difficulties are introduced in this chapter. Some of the problems associated with electrical systems in general are introduced in order to allow the student to understand the complexity of power generation and distribution.

INTRODUCTION

Stationary applications provide power from a fixed site having the necessary input (fuel, oxidizer . . .) and output (power, heat . . .) channels as well as the equipment required to fully operate and maintain the systems. Sizes can run from very small to very large depending on the particular application being served. In general there are four areas where stationary installations are used—residential, commercial, industrial, and grid. One might just as well term them small, medium, large, and jumbo or less than 100 kW, 100 to 1000 kW, 1000 to 10,000 kW, and above 10,000 kW or even low cost, medium cost, higher cost, and expensive. In addition, a separate category should probably include **military applications**, since many times these require designs and implementations that meet what can be very narrow specifications. They cannot even be categorized in terms of voltage output, since that is up the manufacturer and they can output anywhere up to 150 volts DC.

It must be noted and should be well understood that it is difficult to categorize fuel cells since they are still developing as commercial products. One can think of them as computers in the 1970s, where "big iron" was trying to become personal computing and no one was sure where smaller systems would end up or even if they would survive other than as a niche market. Fuel cell companies come and go on a regular basis, much like early personal computer companies did and even very large companies like Siemans-Westinghouse exit the market. Nonetheless, fuel cells are being produced in large numbers (Ballard produced its one-millionth membrane-electrode assembly in 2010, not long before S-W exited the market with little fanfare). Many of the numerous start-ups arriving on the market are positioning themselves to serve stationary applications, either the smaller consumer type markets or the larger building/light industrial markets. Very little money is currently being made by these companies, but that is not unusual as new markets emerge.

There are thousands of stationary systems now in place across the world. Most use natural gas as a fuel supply, with a number of others placed where biogas supplies are available, such as landfills. Stationary fuel cells offer three distinct advantages over grid-supplied power or conventional generator–supplied power.

Virtually all of the electricity generated by the system is available for use. This is not the case with grid-supplied power, since the **loss on transmission** from **remote generating sites** to the **point-of-use site** is substantial. There is no number defining loss on transmission that covers all or even most systems because each experiences different losses based on the length of the transmission lines, their size, the material they are made of, the quality of the installations, and maintenance, as well as a number of other things. Losses in smaller systems of high quality and good repair, say in small European countries, might only be on the order of 2%, while larger systems of high quality and good repair, say in Canada or the United States, may be on the order of 6%. Some systems in areas where maintenance is haphazard, equipment of poor quality, and unauthorized tap-ins to the system common may see losses well over 50%. While it may seem that these losses are not particularly bad, especially in more developed countries, keep in mind that the United States uses roughly 3.8 trillion **kilowatt-hours** (a kilowatt hour is 1000 watts of electricity delivered for one hour) of electricity per year. A 6% loss corresponds to 2200×10^{11} watt-hours of electricity produced but lost before it can be used. To put this in perspective, the coal-fired plants in Texas generate about 150,000 gigawatt-hours (1500×10^{11} watt hours) of electricity so essentially all the power generated in Texas is lost every year to the national grid. Texas, by the way, generates more power than any other state. Figure 9-1 shows the power grid in the Unites States. It is enormous and extremely complicated and since it is tied together, a disruption anywhere is a disruption everywhere.

There are other losses that occur in large scale generation as well. It takes power to move large turbines and push steam through pipes and pump water to boilers. All of those things use energy that is not available to the consumer. Combustion reactions are not 100% efficient and many parts of the fuel are not combusted at all. Some estimates put the best efficiency possible in large scale generation of power (from generator to consumer) at less than 70% when everything is working perfectly. Fuel cells bypass many of those considerations but it is important to realize that efficiency numbers can be very misleading unless the circumstances under which they are generated are very well understood by the reader.

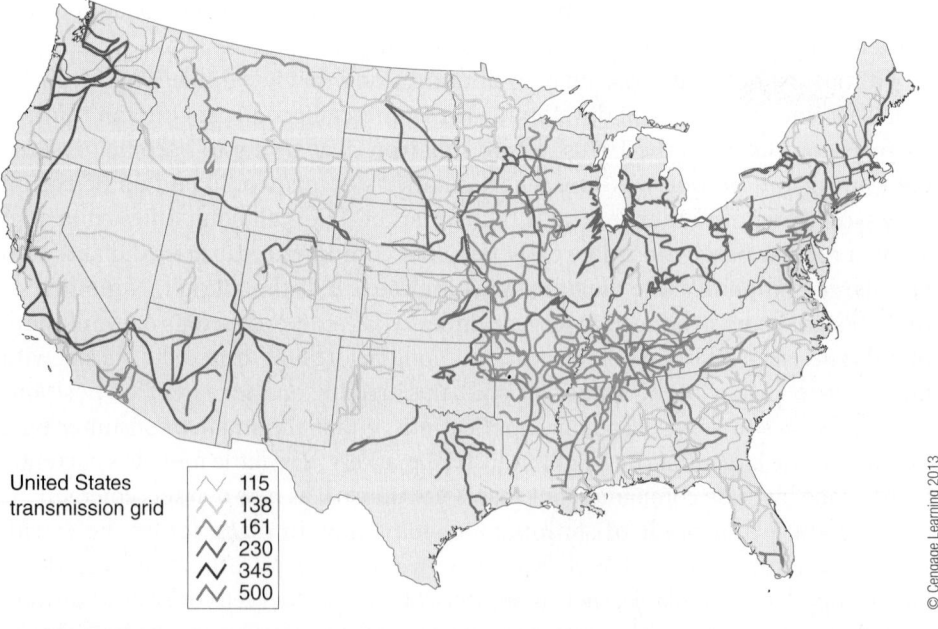

Figure 9-1 The electrical grid in the Unites States.

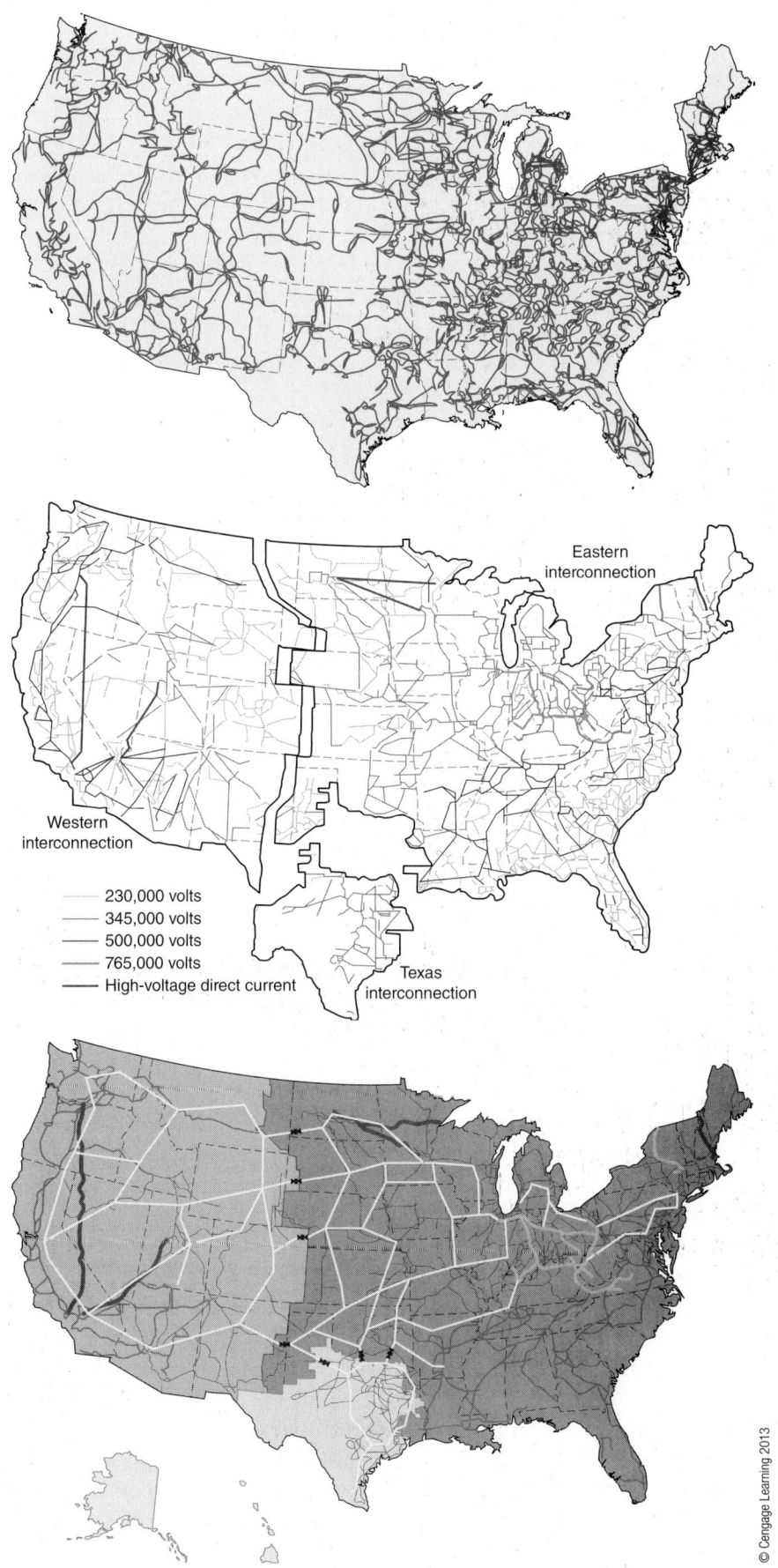

Figure 9-1 (Continued)

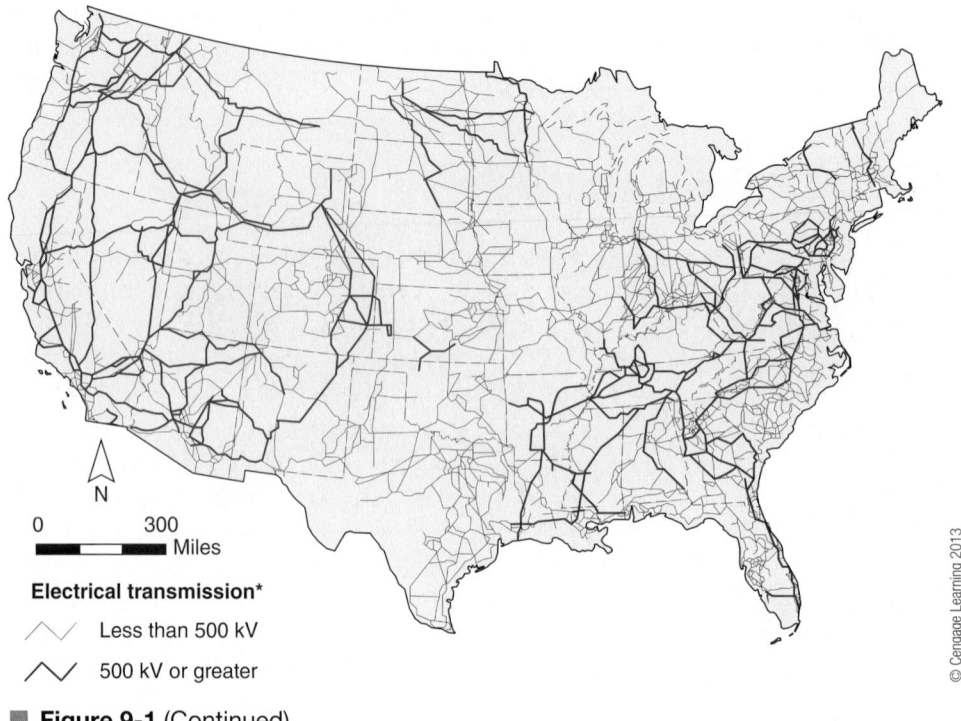

Figure 9-1 (Continued)

The electricity produced by fuel cells is also of very high quality, depending on the associated electronics used to actually produce the current and voltage combinations. While it may seem that your plug in the wall is consistent and reliable, that is only partly true. For many uses, such as lighting or cooking stoves, variation in supplied voltages is not of any great concern, but for more demanding applications like computer systems, medical equipment, and many industrial machinery applications, even minor variation in electrical supply can be disastrous. Some estimates show tens of billions of dollars lost just to computer systems every year to due to voltage fluctuations. Conditioning grid-supplied power to keep fluctuations in line is a very big business. It would be an interesting exercise to calculate the true cost of grid-supplied electricity that includes the cost of secondary markets to condition the power as well as transmission lines and towers, regulating agencies, and engineering staffs as well as the cost of maintaining them to provide a fair comparison to the cost of fuel cell–supplied electricity. There is little if any fluctuation in fuel cell–supplied power, in part because it is consumed where it is produced and because it tends to serve a single customer, minimizing spikes in supply and demand over the system.

Fuel cells are very reliable, with some of the high temperature systems showing reported online availability over 99.999% of the time. This is a key issue for banks, hospitals, computer data centers, and most industries, since unplanned blackouts can destroy equipment and data as well as threaten lives. Blackouts are not uncommon, even in high quality grid systems. The United States experiences a major blackout (over one million sites affected) roughly once every 4 months, with that number dependent to some extent on how many air conditioners are running. In many less developed countries, it is sometimes less a matter of blackouts and more a matter of how many hours you might actually get electricity on any given day. Guaranteeing a steady supply is worth more than just the cost of electricity in many cases, and those who install or maintain such systems tend to be in high demand since they are the guarantors of the supply.

These applications and issues have been mentioned in the previous chapters, but this discussion will go into more depth and detail about design differences, siting issues, and the different ways the systems might be consolidated as their sizes change.

SMALL SYSTEMS

In general, these are the types of systems that can be installed quickly in virtually any location to supply power for specific applications. They are commonly a single piece of equipment containing all the required systems. They should be distinguished from **mobile applications** where a unit can be brought in, used, and then taken elsewhere. Small **stationary systems** go into houses as permanent installation that might be roughly the size of a refrigerator. Provided a fuel source is available, they are set in place, connected to natural gas or another fuel line, and then used more as an appliance. Japan in particular has been in the forefront of such systems with several thousand installed by the end of the first decade of the twenty-first century. One important issue should be noted here: fuel cells supply power at essentially a constant rate once they are turned on. Most residential systems do not use power steadily but rather in bursts, when a heater or a stove turns on. This is an important issue when considering fuel cells for many of the stationary applications to be discussed.

There are of course more to fuel cells than just fuel cells. Even smaller stationary systems at the least will have air and water filters, control units, a reformer unless higher temperature units are set up to do internal reformation, some type of fuel cleaning unit (to take sulfur out in particular), heat exchangers, tie-ins to the residential heating system, a cooling water system, power conditioning that includes AC to DC converters, and a burner to combust any fuel gas not used in the fuel.

Residential systems supply about up to several kilowatts (usually not more than 20 kW or so) depending on the size of the house. They are Combined Heat and Power systems so they provide heat as well, making them ideal for residential applications where hot and cold water as well as household heating and cooling are the main nonelectrical utility requirements. Keep in mind that there is a limit to the size where CHP systems are economical on both the low side and the high side. If the system generates much less than 1 kW, then the heat generated is not enough to do much with and so reusing it does not return the original investment of the CHP equipment. If the system goes over 15 kW or so, the cost of the associated systems may be too much for a residential market to accept as well as generating too much heat for a residence to deal with.

These systems are well suited to providing utility service in areas where existing infrastructure does not exist, usually in undeveloped areas. The student should be aware that this is a typical situation in much of the world where development has yet to reach rural areas. However, it is also common in more developed countries where the urban/wildlife interface is moving steadily out from existing cities or where specific regulations prohibit major development. In many of these cases, the number of people living (or allowed to live) in the area is not large enough to pay for the very high cost of extending the main infrastructure network to the remote location. It is also an issue in many of the rural areas in North America, where populations are in steady decline and the existing infrastructure networks can no longer be maintained as the customer base (and thus the amount of available money) shrinks. Entrepreneurial opportunities for **off-the-grid** alternatives are already available in much of the United States, serving both large retreat homes for the wealthy and smaller primary homes for those who live outside the smaller communities throughout the northern and western areas of the country. Independence is also a strong selling point with systems like this, where users might be willing to pay a premium for not being subject to electrical utilities.

Another promising market for residential applications is the **back-up power** generator. This is a relatively large market in developed countries where the number of electrical units in a household is substantial, ranging from computers to alarm clocks. Many higher end housing developments now include an on-site generator for each individual house, running on different fuels such as diesel, propane, or natural gas. These systems are meant to protect the significant electronic investment in the house in the case of surges or blackouts. Since this market is driven less by installed cost and more by protection from potential loss, a quiet, nonpolluting alternative is attractive. These systems generally run around 1 kW but can be larger depending on the electronics systems in the house. The upscale nature of these systems represents a significant opportunity for maintenance and repair services.

Residences are one promising market, but one that is more compelling is telecommunications. The growing reliance on instant and constant communication requires power be available not only for primary systems in major areas but also in much more remote areas where grid-supplied power is not available. Cell towers, for instance, are required to be essentially **line-of-sight** so they are commonly set in remote locations, especially along the interstate system. Supplying such units from the grid is not possible since the cost of extending grid lines is prohibitive and in many cases there are not even grids to extend. There are virtually no options available to supply reliable power in such situations other than batteries. The limited life of battery power requires frequent charging using solar or wind power, which are not considered constant sources and so may fail in their charging duties if the right circumstances arise. Telecommunication sites get larger and more complex with each additional technology and require more power. Batteries have a limited ability to store current, less than 1 watt-hour per kilogram of battery weight depending on the type of battery, so at some point, the weight of the batteries becomes unmanageable. Voice, data, image, and streaming video are requiring ever more power to properly manage, so that the power supplied to major sites is moving beyond the 5 kW region, which would equate to a couple of tons of batteries. Hybrid systems employing solar, wind, and fuel cells may well become the standard for remote sites in particular and the subsequent servicing of these sites a major business in itself.

Remote sites can range from those having access off a major interstate to those needing a helicopter to service, but they are only one of several possible applications for smaller systems. Since fuel cells are quiet and clean, operate with little emissions other than CO_2, and output very high quality power, telecommunication sites in residential areas may become a substantial market as well. This is particularly true of the lower temperature cells such as a SPFC (PEM) and PAFC systems since the limited heat generated might be used merely to heat an access shed. Larger installations such as those on large city buildings or as part of regional rural sites could use CHP systems as part of the heat or air conditioning requirements in the surrounding buildings, since in these cases the high quality of supplied power would be the attraction and the heat a mitigating factor in cost. Telecommunication installations are a rapidly growing segment of the fuel cell market in areas where grid supplies are unreliable, such as in India where hundreds are already in place. These systems are being set up in more developed countries such as in northern Europe as well, driven by political pressures as much as reliability issues.

It should be noted that political pressure and environmental litigation may well force the eventual adoption of fuel cells for many applications. Extending electrical infrastructure systems, for instance, has become an environmental issue in many areas of both the developed and developing world. Fuel cells generate less CO_2 than any other comparable technology and virtually no **NO_x** (nitrogen complexes from any combustion using air as the oxidizing agents, best known for making smog brown) or **SO_x** (sulfur complexes from any combustion where sulfur is present in the initial fuel, best known for creating acid rain) emissions at all. It may happen that environmental standards in developed countries mandate fuel cell installations as what are called Best-Available-Technology choices in some areas. The start of this drive may even now be going on in the southwestern United Sates, where the haze from coal-fired power plants is obscuring the **viewshed** across many of the unique features of the landscape, including the Grand Canyon. Limitations on operations (or even closures) are being imposed on upwind generators using coal fired power plants to limit these hazes. In cases such as this, cost becomes much less of an issue since installation of nonpolluting generating capabilities would be mandated. In urban areas, noise pollution may accomplish the same thing in the case of back-up power generation for urban installations.

Small stationary installations where cost is less important than reliability and quality of supplied current could be the application that creates a broader fuel cell industry since there are few alternatives available for these types of installations. Hospitals, airports, data centers, telecommunication hubs, critical government services, and even utilities themselves require improved quality levels of supplied electricity as computer and communication technology improves. Fuel cells can provide them at a cost that is currently higher than battery/generator systems but that may change in the near future. To adequately enter most markets, cost is the critical issue, but early entry can also be made in applications where durability, quality, reliability, and integration of services (solar and wind for instance) are important. Smaller installations serving niche markets are gaining in the market faster than any other.

MEDIUM SYSTEMS

The list of potential sites for stationary fuel cell systems is in fact quite large, including:

- any need for electricity in an area not served by a power grid
- hospitals
- airports
- telecommunication towers, both remote and in urban areas
- greenhouses
- residential applications where quality of current and reliability are important such as areas where telecommuting is done
- data centers
- telecommunication hubs
- critical government services such as police and firefighters
- military installations
- peak balancing installations for power grids that could send the heat to cogeneration uses
- remote locations that may be temporary such as mining, logging, or scientific research

Most of these applications involve medium scale requirements and if fuel cells systems are to become a viable technology for supplying energy, they must be able to serve those markets in particular. For the majority of installations, smaller consumer applications like residential installations compete primarily on cost against existing systems that have been in use for decades. Consumer markets are notoriously hard to enter and then to compete successfully in, especially for new technologies. Students can review the history of the fax machine to gain an understanding of this type of market. Larger installations such as peak balancing (coming online to provide electricity when very high demand occurs) compete primarily on cost as well, since they are used sporadically. At least in developed countries, systems like these are also considered regulated utilities and so must meet publicly debated recoverable cost requirements that at the moment are beyond fuel cell ranges. Currently, it is the midlevel applications where fuel cell systems can offer the most to potential customers.

Medium installations can be considered as those that are not limited in footprint and that require site preparation and regular on-site maintenance. They generate several hundred kilowatts of power depending on the size of the building being served and can run into the hundreds of thousands of dollars per installation. The larger units are often coupled with CHP systems, cogeneration turbines to increase power generation and boost efficiency (lower cost), on-site reformation (usually of natural gas) systems, and fairly complex computer control systems that interface with both the building and any existing grid connections. Many times they are coupled with large Uninterruptible Power Source units, redundant utility grid connections, secondary backup generating capabilities using diesel, propane, or natural gas and even flywheel generating capacity to guarantee constant supply.

They are considered primary utility sources, continuously supplying electricity (up to several hundred kilowatts), heat, air conditioning, and even hot water to the location. As such, they must be extremely reliable, operating both for extended times between scheduled maintenance and for what is essentially the life of the building overall. Unlike smaller systems, they are maintained much like boiler or electrical systems, generally with on-site staffs but also with contracted services. Minor maintenance and **scheduled shutdowns** for major repair and replacements are done as part of a comprehensive **operations plan**. In general, the fuel cells themselves operate for tens of thousands of hours (40,000 is considered a benchmark) and then have to be replaced as part of a shutdown routine, in the same manner that boiler tubes are replaced at regular intervals. Much like large industrial boilers, the primary maintenance issues are material degradation due to high operating temperatures and contaminants in the input streams. Other system components such as reformer catalyst units and piping may be matched to the same **replacement schedule** as well. As with other major mechanical and electrical systems, these units require trained and highly skilled personnel to both operate and maintain where smaller systems are meant to run more as appliances with limited lifetimes, little or no maintenance, and no operational support.

The high temperature fuel cells are well suited to medium applications since the CHP systems are very efficient and provide more than one utility stream. Architects and engineers can centralize utility services around these types of systems, lowering both installation and operations cost. While Combined Heat and Power is the term used in this book, it may just as well be Combined Cooling, Heating, and Power (CCHP) and in fact that term is found in some of the literature. These systems offer both heating and cooling cycles and thus can supply substantial centralized capabilities to a building or campus, making them an attractive option when combined with the high quality of supplied power, system reliability, quiet operations, and lessened emissions. Coolers use what are called **absorption chillers**, where a liquid refrigerant is evaporated in low pressure (this cools the area it is evaporating in, say where air lines are located which then feed cooler air into an AC unit), the cooled gas is absorbed into a second liquid (this removes the refrigerant gas and allows more of it to be evaporated, which cools things down even more) and then the second liquid/absorbed refrigerant is reheated (this is where fuel cell heat comes in) to evaporate the refrigerant and send it back to the cycle. The second liquid just has to have a higher boiling point than the refrigerant liquid. You can also use pressure instead of temperature in these systems as well. Figure 9-2 shows a typical diagram of a simple absorption chiller.

It is useful to examine some of the places where medium systems are considered such as apartment buildings, office buildings, strip malls, and hotels. These

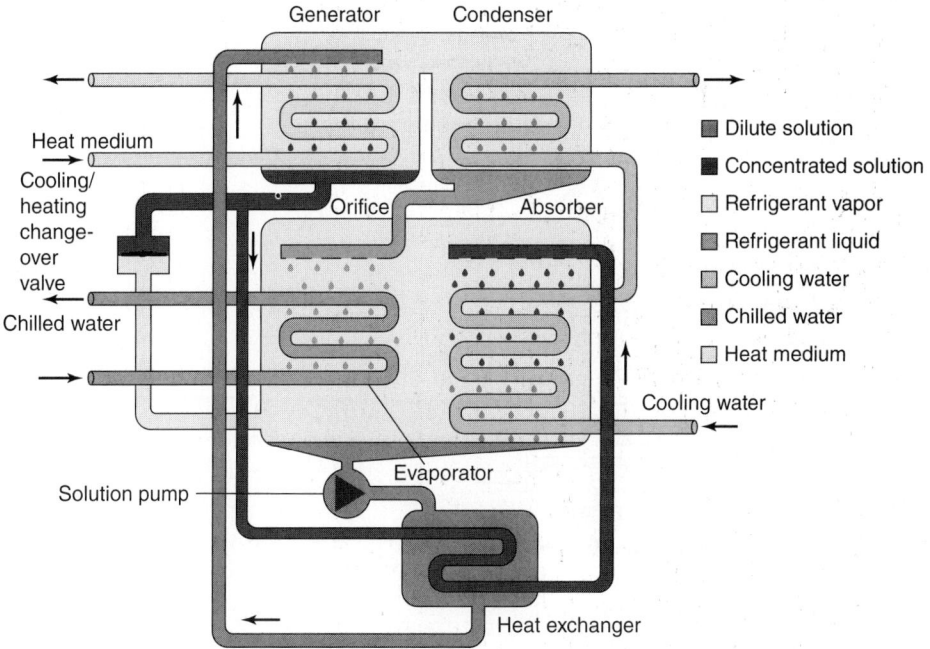

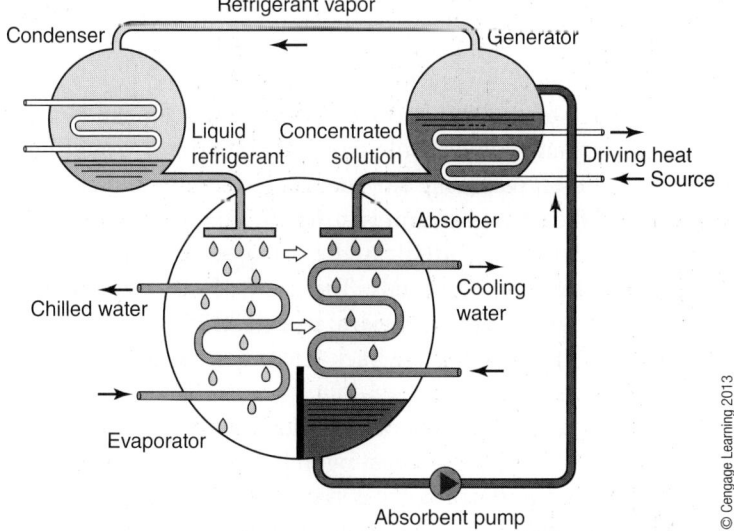

■ **Figure 9-2** A simple absorption chiller.

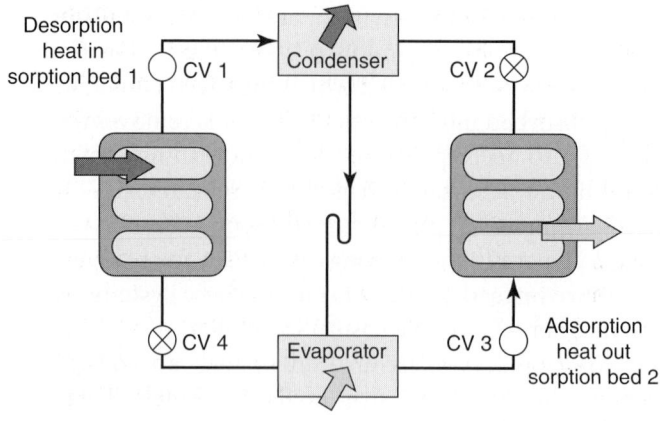

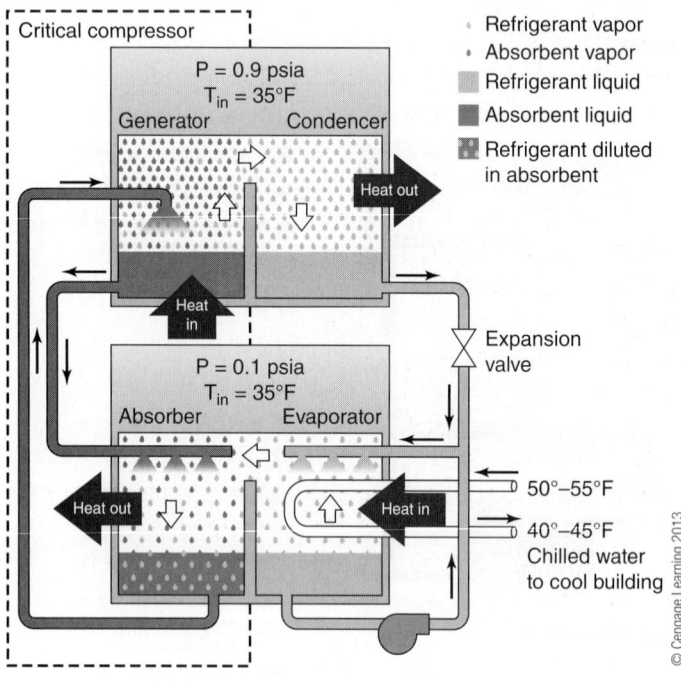

Figure 9-2 (Continued)

are commonly mentioned in literature, advertisement, and government publications as being large potential markets for fuel cells, and all can be considered as medium scale since they are generally single installations providing utility services to a single, albeit large, unit. The fuel cell system would supply electricity but many times not the entire amount required. They may be used for high quality systems such as computers, dedicated electronics such as security systems, or key equipment such as in hospitals. This type of installation would require systems separated from lighting or general use electricity but that would be true of any of these types of system where back-up or conditioned power serves only part of the site.

Medium systems would supply hot water, taking the place of water heaters in smaller installations or providing preheating in larger cases such as hospitals or hotels where substantial amounts of hot water are used. Even if a fuel cell system was not directly replacing a boiler or some other hot water system, using it to preheat water can substantially lower costs. Preheating might minimize boiler maintenance or operations cost or allow for a smaller system so that initial capital or replacement costs are lowered. Boiler installations are costly, in part because there is no real alternative to them, so providing preheated, high quality water (remember some systems produce what is essentially pure water) can substantially lower boiler costs. The fuel cell installation would also work with the heating and air conditioning system. Hot water and HVAC requirements are two of the highest costs in building operations. Medium scale fuel cell installations may not directly replace these units but because they can supply energy in the form of hot (very hot in the case of high temperature systems) air, water, and steam, they can be used to offset installation, operations, and maintenance costs so that the overall cost of operating a building is either simplified or lowered over the life of the installation. Maintenance costs of fuel cells themselves are not excessive, with some installed PAFC systems reporting maintenance costs of roughly 1 cent per kilowatt hour of electricity produced and actual cell/stack replacement occurring only every 7 years.

Fuel cells will not replace mainline utilities in medium scale installation except for cases where utilities do not exist. In developed countries, they will become part of the utility system itself, working in conjunction with conventional power, hot water, heat, and air conditioning systems already in widespread use. It may be that smaller buildings eventually can use fuel cells almost exclusively to supply most if not all of the utility needs but larger buildings do not rely on single systems for their needs and in many cases, **building codes** require multiple and redundant systems. For instance, conditioned or emergency power systems might be required for lighting or security systems, and tie-in ports so water can be supplied to fire suppression systems from outside the buildings are common.

Data centers need to be mentioned as specific applications that fuel cells are uniquely positioned to serve. These are large buildings where computers and data storage is centralized to provide any number of different services. Large corporations maintain billing centers or ordering centers, Internet service companies provide access, supercomputing sites do massive computing jobs, computer animation is done for movie or television studios, the government stores information, stock trades are tracked, and banks maintain accounts in centers like this. There are hundreds of thousands if not millions of such sites across the world and they are critical to the way things are done now. These centers need two things: very high quality electrical supplies and reliable cooling. Computer systems cannot tolerate variation in their electricity nor can they run at temperatures much over what humans consider hot. Fuel cells are one of the few technologies that can reliably supply electricity that does not vary in voltage over extended periods of times. Grid-supplied electricity is not well behaved, batteries fall off over time in somewhat undependable ways, and generators are mechanical systems with smaller units being generally unable to supply constant current. Other systems use power conditioning equipment that is expensive and itself somewhat variable where fuel cells by their nature can supply steady electrical current and voltage. Fuel cell systems can operate at what amounts to less than a minute of interrupted service per year as opposed to roughly an hour per year on average for grid-supplied electricity in the United States. In fact, power outages have become so commonplace in the United States that the Internet offers tracking services for them in virtually every state.

Keep in mind that a single failure in a data center can affect substantial sums of money. If orders cannot be processed or electrical components are destroyed, the costs can be huge to a corporation. If data is lost, there are almost always copies of it in other locations, but accessing it takes time and other issues can arise. If a credit card company suffers even minutes of downtime in a large processing center, many thousand of transactions are affected. Some larger centers handle several million transactions per day and estimates for larger credit card centers are in the millions of dollars lost per hour if transactions cannot be done. Back-up power systems using internal combustion engines cannot come online in the type of time computers deal with, where billions and trillions of operations can be done in a second. Battery systems can supply almost instant backup but must be very large to handle the sort of power computers require. Guaranteeing a steady, reliable source of electricity to data centers is not so much a question of cost per kilowatt but of potential loss to the business. This makes the higher cost of fuel cell–supplied electricity more tolerable in these instances.

In addition, most data centers use the heat generated by computers for other uses in the buildings. Even smaller computer systems can generate enough heat for several floors of a building and very large installations can heat entire campuses. The temperature of this heat is typically not high enough to be used in cogeneration but can be used in conjunction with fuel cells, especially high temperature systems. It can preheat fuel and air supplies, be combined with the higher temperature waste heat streams of the fuel cell to provide both heat and cooling, and be used in turbines to increase efficiency. Data centers may well be the single application where fuel cells are actually the best technology available.

Given the importance of environmental concerns when it comes to power generation, it is appropriate to discuss the matter to some extent and medium sized systems offer a good introduction to the matter. Maintenance, operations, and engineering personnel will have to work not only with the primary fuel cell system but also with associated monitoring equipment. Since fuel cell technologies are being sold in part because of their environmental sensitivity, personnel should understand at least what is being talked about and how much of an improvement might be realized in this area. Studies done in Europe, Japan, and the United States in particular generally consider a list of pollutants. This includes NO_x emissions (nitrogen compounds produced when nitrogen and oxygen combine at high temperatures); since air is primarily nitrogen and oxygen, essentially any combustion (carbon and air reacting to form carbon dioxide and heat) reaction will produce NO_x emissions as a side reaction. It also includes SO_x emissions (any sulfur compound produced when sulfur and oxygen combine, usually at high temperatures as well); sulfur is a common contaminant in hydrocarbon fuels while oxygen comes as part of the air being used in a combustion reaction. It also includes carbon monoxide and carbon dioxide, which are the results of combusting a carbon (fuel) source with oxygen. There are nonmethane hydrocarbons emissions (carbon and hydrogen molecules that have more than one carbon in a chain), which come from uncombusted fuel. There are methane (one carbon atom and four hydrogen atoms in a molecule) emissions, generally caused by fuel that is not fully combusted and perhaps the major **greenhouse gas**. There are also **particulate emissions** (very fine particles), which are health hazards since they can enter your lungs rather than be caught by your natural air filtering system, commonly called the nose. Some studies show that for NO_x, SO_x, and CO, using PAFC fuel cells rather than coal-fired plants to generate electricity can reduce emission by 10 times or more with even better results from SOFC systems. Nonmethane hydrocarbons as well as methane emissions are reduced by nearly that amount as well. PAFC systems show some slight reduction in particulates but SOFC systems eliminate them entirely. Keep in mind that much of the pollution coming from fuel cells is not so much from the fuel cell itself, but from the associated equipment such as reformers.

Those who work with fuel cells systems will be involved with the fuel cells themselves, but these will generally require little maintenance but rather are simply replaced on a regular basis. It is the **balance-of-plant** systems that will be more important. Any time pollution monitoring equipment is part of a system, it automatically becomes a critical component. No matter what capacity the student serves in, understanding how important such monitoring is becomes a critical component of his or her job as well.

LARGE SYSTEMS

Large systems are those that supply substantial amounts of electricity to either the grid or to a large complex such as a campus, skyscraper, or factory. They require a major infrastructure to support them, ranging from electrical substations and high pressure and flow rate feed gas lines to operations and maintenance staffs. They generate thousands of kilowatts (megawatts) of electricity. Many supply primary power to a site and these are many times high temperature cells operating constantly. Others serve to balance the power grid itself, coming on to supply power when demand peaks or a particularly large draw occurs and these are generally lower temperature cells such as SPFC (PEM) that can be brought up and then taken back down in a reasonable time and without damaging the fuel cell system with temperature cycling. While this book cannot fully describe electrical grid systems and how they are operated and maintained, an introductory discussion can be given, especially since so much is made of the reliability and quality of fuel cell–supplied electricity.

Distributed generation moves primary generation capacity away from central facilities and to locations nearer the user. Distributed generation generally supplies less power per installation than the larger centralized facilities because there is less loss-on-transmission. Fuel cells and distributed generation are many times mentioned in the same breath and many consider this good for a number of reasons. There are a number of lists detailing the advantages of distributed generation, many times given as part of the justification for adopting fuel cells. These include such things replacing existing coal-fired plants, expanding the energy supply business into **alternative energy sources**, increasing the use of natural gas since it is considered much less polluting, subsidized installation and operations, and even better environmental considerations. Some even go so far as to point out that encouraging distributed generation will lower the reliability requirements of the major grid, make utility companies more profitable since it equalizes flow within the natural gas distribution system and limits massive infrastructure investment. From a technical standpoint, distributed fuel cell systems are mentioned as installations that can supply both **spinning** and **nonspinning reserves**, **frequency regulation**, grid power quality support, buffering supplies, and **reactive power reserves**. What are all these things? Why are they important to electrical supply systems? Can fuel cells actually do these things?

As discussed in several chapters (and discussed further in a couple to follow), electricity can be supplied either chemically (fuel cells or batteries) or mechanically. The great bulk of electricity supplied across the world today is from mechanical systems (the plug in the wall). Mechanical systems generate power by rotating one metal while holding another metal stationary (remember Faraday from the first chapter). In this simplified version, you control amperage (the number of electrons moving) by controlling the amount of metal being used, with smaller amounts producing smaller amperages. Voltage is controlled using resistances (transformers). The other critical control variable is the frequency of the supplied power, determined by how fast you spin the metal parts.

How much power is needed at any time is neither consistent nor predictable. If too many people turn too many pieces of electronic equipment on too fast, the system cannot supply the needed amount and crashes (blackouts). To avoid this, utilities forecast demand in a number of ways. In simpler terms, summer means more air conditioners and thus more electricity consumed, and the hotter the summer, the more electricity is needed. If industrial plants close, then demand drops; if a new skyscraper is completed, then demand goes up. This type of analysis only goes so far, so utilities maintain plants meant to supply electricity if demand suddenly peaks. There are short term responders and longer term responders. Long term responders come on within 10 minutes while short term responders come on in less than 10 minutes. Short term response is accomplished using spinning reserves, where generators **synchronized** to the 60 **hertz** (Hz) system are already running (spinning at that rate) but not producing electricity. The longer term responders are nonspinning reserves and may be running out of synch or even not running at all but capable of coming online in a short period of time.

Some fuel cell systems can respond fast enough to act as reserves and some cannot. Low temperature cells have that capacity as long as there is a fuel reserve available, since reformers are generally not capable of producing hydrogen that quickly. High temperature cells would have to be at operating temperature to respond that quickly or a stack would have to be kept in reserve, heated but without fuel supplied, probably a cheaper alternative than spinning a generator without getting anything for it. Alternatively, stack heat could be used in CHP cycling rather than in cogeneration, with peak needs redirecting that heat into a turbine if needed for nonspinning reserves.

North American power is supplied at 60 Hz (**cycles per second**) so that the generators in essence spin at the same rate. If every user on the grid consumed exactly the same amount of electricity over the same amount of time (kilowatt-hours) then the business would be pretty straightforward. Instead of that, users consume electricity in a huge number of different ways and amounts and over any number of different time periods; sometimes, they don't even consume any at all. To put it another way, demand (load) does not equal supply (generating capacity). When demand does not equal supply, the frequency of power within the system fluctuates. The reasons behind this are beyond the scope of this book, but it is sufficient to think of it as electrons being pulled out of the system, doing some type of work (heating a pan or a building), and then being returned to the grid but with less energy than they originally had. This can be compared to cars coming on and off a busy highway, where pulling a car out and slowing it down before getting back on will disrupt the entire flow of all the other cars on the road. Figure 9-3 shows a graphical view of this, where if you look at it from far enough back, the system is stable but the closer you get, the more variation you see in the system. The view from far back can be compared to the entire grid where the close-up view may be a town or neighborhood.

When any disruption to the average frequency does happen, the grid must be balanced and that is what is meant by "frequency regulation." The student should understand that this operation applies to every generating plant, whether large or small, the smaller grid it is on, the regional grid it is connected to and the main continental grid as well. In general, this balancing act across multiple systems is done by having several major generators reserving part of their capacity (maybe a percent or two) for frequency regulation. This is done by setting aside some of the physical generators so their rotational speed can be changed, which forces the supply to meet the demand. This is a major cost factor for utilities, since manipulating generators greatly increases the wear on the equipment and throttling up

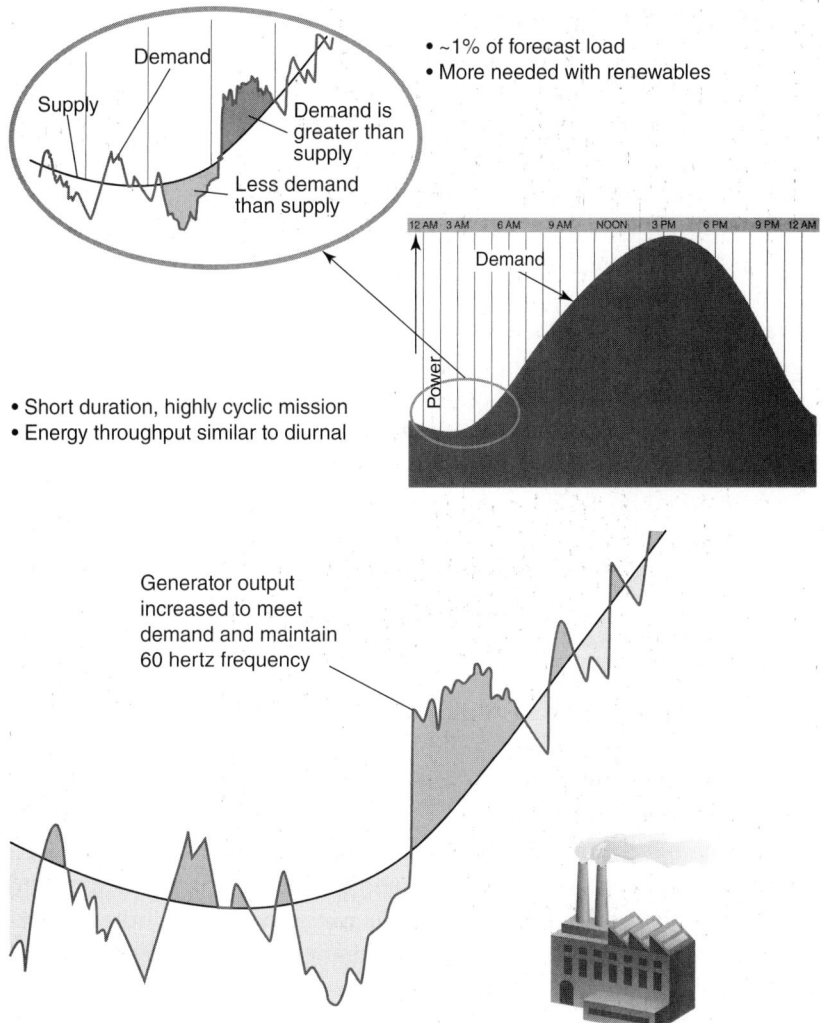

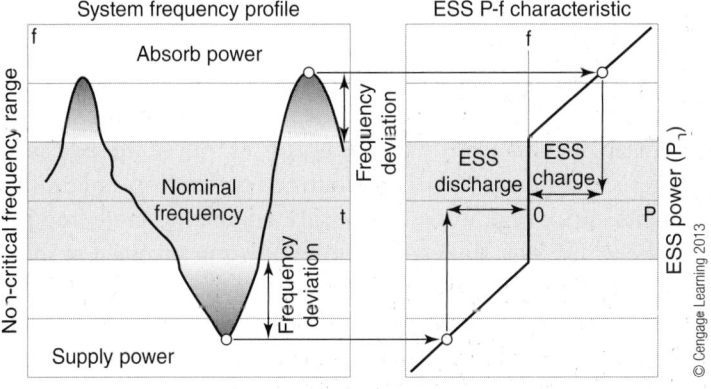

■ **Figure 9-3** Load versus generating capacity.

(or down) steam generating capacity is an upset condition that will increase emissions. You will hear **flywheels** mentioned in connection with the grid because they are one of the few means of storing mechanical energy in large amounts that can be released fast enough to respond to grid needs.

As mentioned though, fuel cells provide very high quality power in part because they do not used mechanical means to produce amps, volts, and frequency.

We will discuss the systems used to condition power in a later chapter, but this is a good place to introduce the student to the difference between electronic methods and mechanical methods. Fuel cells do not spin shafts but use electronic devices to produce the amount and quality of power needed. Where mechanical devices require time to respond and add additional wear to the system, electronic controls are instant and relatively neutral as far as wear goes. If a computer recognizes deviation in grid frequency, fuel cells can respond instantly to balance it. The faster the response is, the less of a swing the grid will see. Many of these are transient loads that come online and then go back offline just as fast. Ramping up a coal generator or using stored steam to increase the rotational speed of a generator means that when the load is removed, the system is now unbalanced in the opposite direction. The faster the current can be brought on and then taken back off, the less disruption the overall grid will see.

Reactive power is quite a bit more complicated, but in essence what happens when AC power is stored through capacitance or inductance and then returned to the grid on the one-quarter of an AC cycle that allows it, is the system is disrupted. In general, the grid deals with this by combining coils and capacitors to "trap" the current and keep it out of the grid, thus, a large induction motor will have a corresponding capacitor installed. Fuel cells are less able to deal with this issue than some others, but because they are inserted locally in many cases as distributed installations and thus control access to the downstream installation, they can be designed to include this feature. This would keep an entire installation or even a large site from returning reactive power to the main grid.

Finally, approximately half of the electricity in the United States is currently generated using coal, but this fuel is becoming increasingly politicized as eastern coal mining cuts mountain tops off and western generating plants obscure even the Grand Canyon on occasion. Fuel cell systems can provide distributed generating capacities such as those discussed above where electricity is produced and used at the same location, but they can and do also generate grid-level amounts of energy (megawatt ranges). Fuel cell systems on the grid generally act to provide peak power loads in order to manage the grid and provide secondary power units of smaller size when upgrading is needed. But either cost or regulatory concerns make such upgrades difficult to obtain, minimize pollution in sensitive areas, or provide a source of high quality power to a particular site. Another application where fuel cells can serve where other systems cannot is in what are many times called applications of opportunity. These are cases where some type of fuel is generated but cannot currently be used for any useful purposes. These include such things as flare gas in industrial plants as well as methane generation in landfills or sewage treatment plants. Since high temperature fuel cells can tolerate impurities better than low temperature cells, these are the types commonly used in these applications, as the fuels generated are usually mixed with other gases. Landfills, for instance, are the single largest source of methane (one of the worst of the greenhouse gases) released by humans into the atmosphere and fuel cells can use a significant amount of it. This is another application where cost may not be the defining issue but rather environmental and political pressure forcing the type of change that only fuel cells can take advantage of.

These types of systems are large in footprint, in feed supplies needed, in cost, and in power generated. Keep in mind that the market for fuel cells is not a small, niche market. Estimates of the total market just in the United States range up from 3 billion dollars per year and that is growing steadily.

KEY WORDS

Knowing the terminology used is critical when dealing with fuel cells. Following is a list of the important terms in this chapter, which are also in bold typeface within the chapter. It is recommended that students be required to submit definitions of some of these words as homework assignments in which they look the terms up in other books, articles, or on the Internet.

absorption chillers	kilowatt-hours	remote generating
alternative energy	line-of-sight	sites
sources	loss on transmission	remote sites
back-up power	military applications	replacement schedule
balance-of-plant	mobile applications	scheduled shutdowns
building codes	nonspinning reserves	SO_x
cycles per second	NO_x	spinning reserves
data centers	off-the-grid	stationary
flywheels	operations plan	applications
frequency regulation	particulate emissions	stationary systems
greenhouse gas	point-of-use site	synchronized
hertz	reactive power reserves	viewshed

DISCUSSION QUESTIONS

1. What is the primary feed fuel currently used in stationary fuel cell installations?

2. Coal-fired power plants in England generate approximately 14 gigawatt-hours of electricity each year and the loss on transmission for the English national grid is 2.3%. How many watt-hours of electricity are lost each year form the coal fires' generating capacity?

3. What associated units might be included in a smaller stationary system used for consumer applications and set up for plug-and-play use?

4. In a small, stationary consumer system meant to service a single house, would it be better to use a CHP subsystem or a cogeneration subsystem with the unit?

5. For a unit to be considered as a primary utility source, what needs must it meet?

6. What are two primary requirements of data centers?

7. The computers in data centers generate large amounts of excess heat. How might this heat be used if fuel cells are installed in a data center?

8. Why is maintaining a synchronized spinning reserve a very expensive proposition for a utility?

CHAPTER 10

MOBILE FUEL CELL APPLICATIONS

objectives

This chapter introduces the student to mobile systems as both technology packages and expressions of culture, discussing the various needs required of portable applications and how fuel cells can meet those needs. The size and types of fuel cells required for mobile applications, the fuels and systems that might be used, the variations in operating conditions and issues regarding pollution and pollution control are introduced and discussed.

INTRODUCTION

Perhaps the most demanding fuel cell application is in supplying power in a **mobile system**. Trucks, cars, boat, ships, trains, buses, forklifts, and even motorcycles all contain design, fabrication, and maintenance elements unlike any other. Transportation itself is seen as one of the main achievements of a civilization and a culture and even now sets both civilizations and cultures one against another. Anything that represents change in the way a society transports its freight and its people already contains issues unlike any other technological change or advance.

For fuel cells, the matter is further complicated because stationary systems and mobile systems are different in design and implementation. It may be that the two end up employing two different sets of principles in design, engineering, fabrication, and maintenance, mirroring the current state of engines. In stationary applications such as coal-fired power plants, the engines that produce electricity run continuously for years while the engines used to produce power for automobiles may run for hours at a time, but rarely for more than that. Diesel engines exist in between these two extremes, running for days or even weeks at a time, but all mobile applications must be able to startup and shut down from cold (sometimes very cold) conditions thousands of times where stationary applications are not designed for those conditions. Mobile engine designs typically aim for several thousand hours of life, while larger stationary engines, due to the constant operations, may target 10 times that. Mobile applications must supply power very rapidly, whereas stationary applications have the luxury of coming to full load over extended periods of times. Mobile applications must respond to a wide range of load requirements from idling at low power to fully engaged, many times in short order, while stationary applications generally come to an optimized level of operations and then stay there if at all possible. Mobile applications generate large amounts of heat that must be dealt with in varying circumstances while stationary systems generate as much if not more heat but can set systems in place to deal with them that do not need to respond to rapid or cyclic change. Mobile systems have to carry their own fuel with them while stationary systems do not. This is a very important issue with mobile systems, since it requires that the engine in use has to be able to both support the fuel load and carry it to the next available location where fuel is available.

In general, it can be said that low temperature systems are better suited for mobile applications while high temperature systems are better suited for stationary applications. The main reasons for this are the time it takes for the systems to come up and then shut down as well as the efficiency of the system. High temperature systems take too long to start up and shutting them down is not good for the fuel cell stacks in particular. High temperature systems are in general more efficient than low temperature systems. For larger stationary systems, efficiency is more important since they tend to be used over a much longer time. Low temperature systems can endure the fast start-up and shutdowns required in mobile applications and can operate within the possible **cooling parameters** of mobile applications (it is hard to cool an exhaust gas from 1000°C while idling at a stoplight and people are walking through the exhaust stream). It should be noted that low temperature fuel cells are considered well suited for mobile applications not only because of those reasons but also in part because their competition, the internal combustion engine, has a very low efficiency.

As has been mentioned, there are several general types of mobile applications to be considered. The fuel cell's system can supply **primary power** to drive a platform, be that an unmanned aerial vehicle, automobile, forklift, delivery truck, or train. The system can supply **auxiliary power** (usually in the form of electricity) for a platform where the primary power for movement comes from a different engine. The system can supply power (usually in the form of electricity) used for personal applications, ranging from charging batteries for teenagers' cell phones to providing soldiers with the power needed to communicate on the battlefield.

The very broad range of possible applications means the wide range of options currently available will have to be duplicated by fuel cells. It should be stated that fuel cell technology in its current state is not capable of supporting many of the portable applications currently existing. Supplying a fuel cell system that can be used for a flashlight at a cost comparable to current battery units is not possible, even if viewed over the course of a service life such as that seen in police or fire applications (hundreds if not thousands of batteries over years of flashlight life). However it is possible to supply a forklift at a **competitive price** that is to be used inside a building, especially when the issues of carbon monoxide emissions inside the building are considered.

One other major issue for mobile applications is in the **costs of fabrication and disposal**. While fuel cells are said to be low (or even zero) emission technologies, that is not strictly true when they are considered in total from mining the raw materials used through fabrication and finally to disposal. This is especially true when considering mobile applications, given the short life span of many of those systems; the shorter the life span, the more products need to be consumed and thus the more overall impact from fabrication and disposal. Compared to conventional ICE systems, the simplicity of fuel cells gives them a distinct advantage when considered from beginning to end of a raw material, fabrication, operations, maintenance, and disposal cycle.

This chapter will bring up many of the various needs in portable applications and discuss how fuel cells can meet those needs.

SIZE

Any engine meant to drive a **moving platform** has to be **compact** but still **robust** and yet provide sufficient power to drive the platform over whatever terrain it is meant to travel in. Personal mobile engines for various uses that can be transported, set up quickly, used, and moved again if need be are even smaller in footprint than those used on moving platforms. Units that are meant to be carried and thus independent of a moving platform will also have a wider range of power output. Moving platforms transport people or goods, and as such perform within a narrow range, while devices meant to be transported and set up can run from those carried in a pocket to those carried by a semitruck. It might seem that **transportable engines** would require a less robust design than engines powering a moving platform, but as anyone who has repaired consumer items can attest to, that is not necessarily true.

There are several types of power plants to consider when discussing automotive type applications. Some systems will drive the vehicle by supplying direct power to an electric motor, which in turn runs a drive train. Some will charge batteries, which then drive the electric motors running the car. Some will provide electricity for uses within the car while a different engine drives the car. Each of these will offer different levels of power and thus be different sizes.

For portable or transportable applications, there is no real limit on size, either too small or too large. Personal chargers using Direct Methanol (PEM) Fuel Cells may fit in the palm of your hand and be used primarily to charge batteries or run small consumer electronic devices. Site specific power units may be so large they require heavy equipment to move, much like generators, but still be considered portable, since they will be set up and taken down on a regular basis.

No matter how large or small the portable application makes the fuel cell system, there will be some commonality involved, as there are in all engines. There must be some form of energy storage, usually fuel in the case of fuel cells, but this can be collaborative storage as well, with fuel storing some of the energy and batteries storing another form of energy. There must be the engine (fuel cell). There must be a way to transmit the energy created by the engine to whatever will use that energy. There must be temperature control. There must be access to the internal workings. In some cases, the ultimate size of the system is determined more by the auxiliary requirements than the engine itself while in other cases, there won't be much else but a fuel cell with everything else supplied from outside the system.

The smallest fuel cells are those designed to go on computer boards and meant to power microchips. These would be supplied with gas, probably hydrogen, operate at the lowest possible temperatures to minimize the heat generated in the computer and supply electricity directly to a chip. They would be supplied with fuel from outside the computer board, be cooled by the computer itself, and need only power conditioning. Repairing them would not be done, but rather they would either be disposed of with the board or replaced in kind. Figure 10-1 shows some smaller fuel cells.

Slightly larger might be those systems meant to supply power to microelectronics such as hearing aids or monitoring devices (some of these are versions of fuel cells themselves) used for environmental control or health sampling, such as those shown in Figure 10-2. These would replace batteries and probably

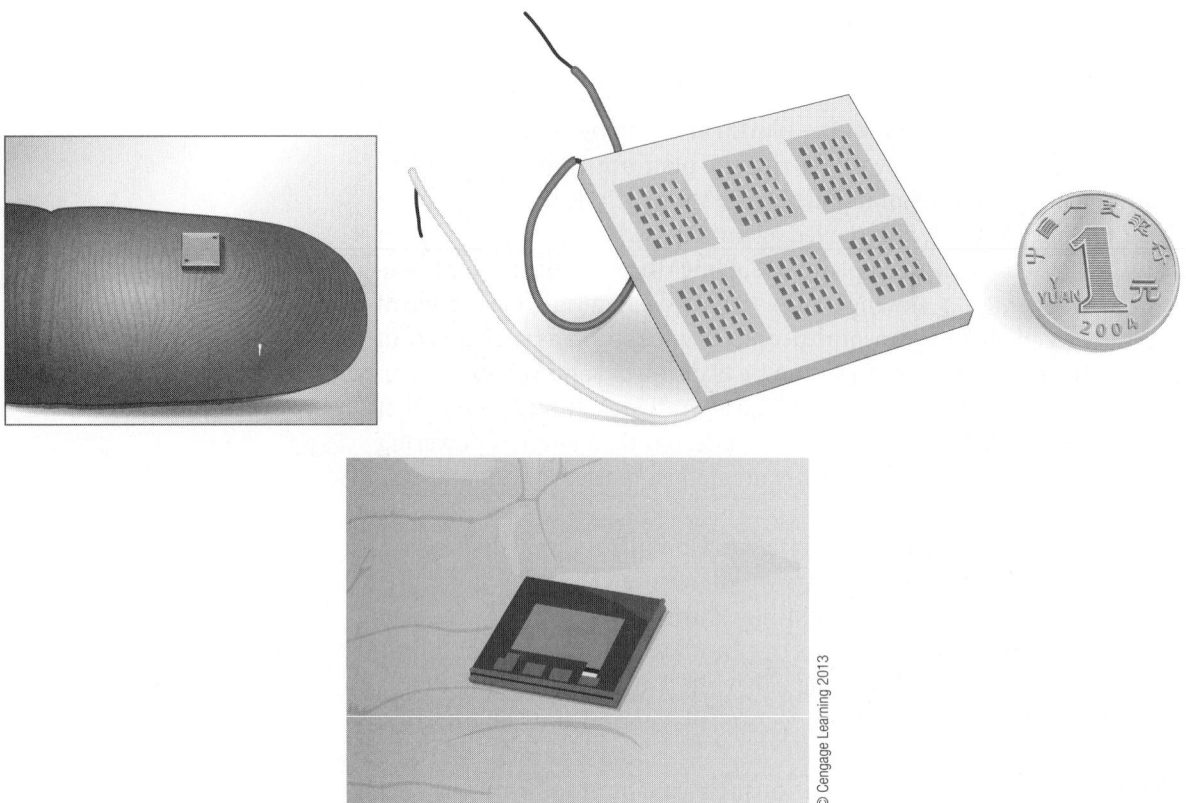

■ **Figure 10-1** Micro fuel cells.

have an entire miniature system including a fixed amount of fuel (or perhaps a microport to fuel the system) in a single package. These types of markets are not cost driven but quality (reliability) driven with cost being a secondary factor. If these units are designed for long life, they would have maintenance and repair features built in, much like their battery-powered cousins. This might include such things as **software diagnostics**, **self-repair** (reprogramming), or even **maintenance call-out** abilities, powered of course by the fuel cell itself.

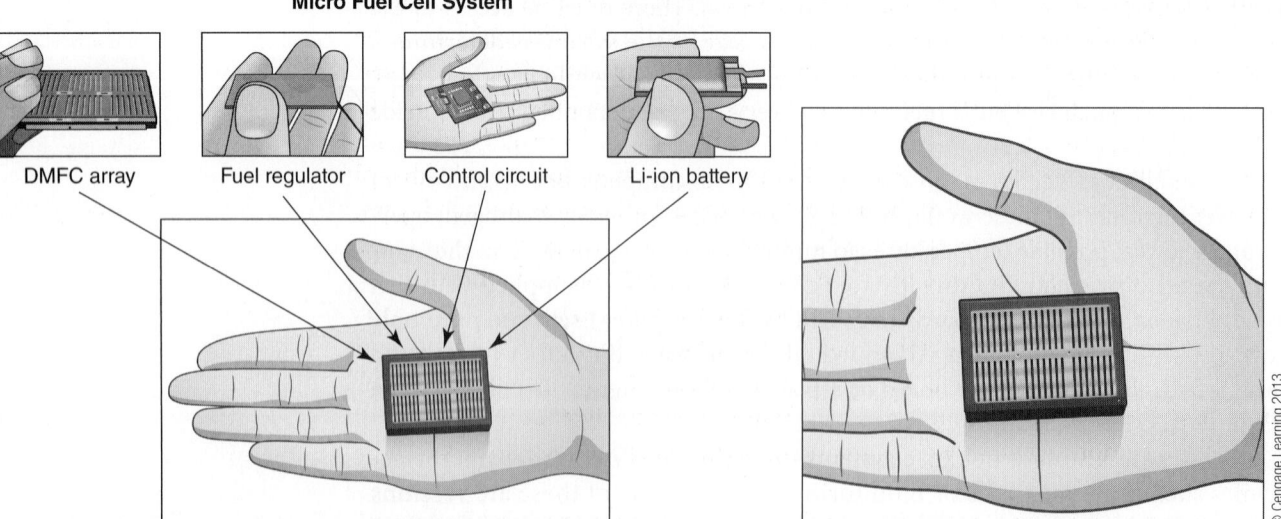

■ **Figure 10-2** Microelectronics fuel cells.

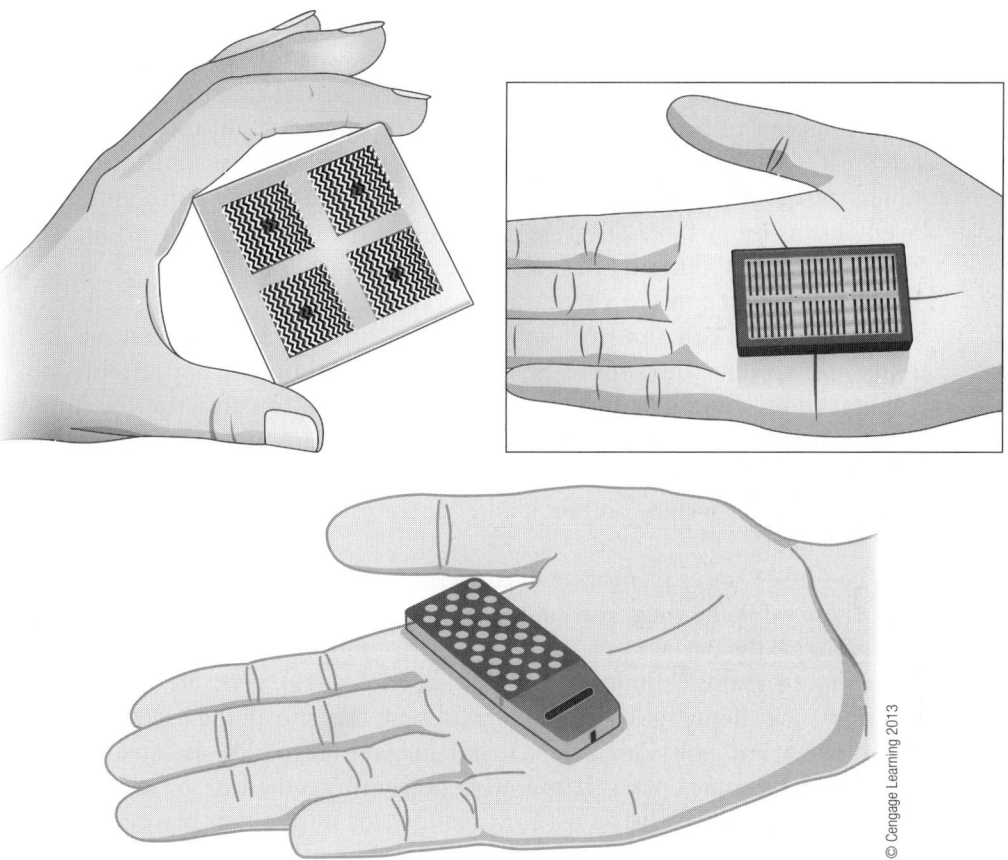

■ **Figure 10-2** (Continued)

The primary issue with these types of systems is thermal management and control so they would be complex and probably **self-reporting**, issuing system reports on a regular basis.

The size required by personal communication or surveillance devices would be the next level. While batteries are in common use over much of the world, there are applications where they are not particularly well suited to the application, such as in remote call locations lacking power or data communication transmitters. This size might be used over a limited lifetime and so be a sealed unit meant to perform a specific duty for a specific time, such as surveillance. They might also be meant for short-term operations but be used in equipment meant to be returned and refilled with fuel, such as satellite cell phones used in locations where recharging is problematical. Versions of this size are already available; usually methanol fueling a direct reforming SPFC (PEM) cell commonly termed the Direct Methanol Fuel Cell. Units like these tend to be treated like cell phones or music players, used and then replaced if damaged rather than repaired. Cost would be an issue in some of these markets since the units may well be considered as throw-away. For more complex systems produced to serve a market where power is not available but communication or computing is critical, cost would be much less of an issue and reliability the most important design criteria. These units would be used in mining locations or exploration camps, military reconnaissance, or service industries in remote areas, as well as remote imaging as either backup or primary power.

Carrying cases (suitcase units) would be used for more power hungry applications either in the field or for emergency uses. They might run power equipment such as the Jaws of Life in situations where internal combustion generators are unsafe or provide temporary power for small computer systems in emergencies. Cold starting after substantial downtime along with rapid ramping to full power would be the most important criteria for units like these. Cost would be a secondary issue in most of these applications; size, reliability, quiet operations, and lack of emissions (heat and pollutants) would be equally as important. These units represent the types of applications where fuel cells could potentially have great success, since fuel cells can be used indoors or out, can come with guaranteed supplies of fuel either as canned gas or methanol and can run with very low signatures (heat, emission, magnetic fields . . .) so that they would be equally as useful on the battlefield as in a hospital room. Emergency power generation meant to run a limited number of critical electronic devices for short to medium amounts of time is a critical application that is not currently served well by batteries or ICE generators because batteries weigh too much and generators emit too much noise, pollution, and heat.

The type of power plants able to run a small automobile is next in the progression. These would be roughly a cubic meter in size and be fully developed systems containing all the parts needed to reform fuel; control gas pressures; pretreat any of the input streams (humidify, dry, heat, or cool gases); generate and condition power; manage liquid, gas, and solid based heat; exercise all the necessary system controls; treat any emissions; and be maintained on a regular basis, repaired if need be. They would come up to **full operating power** in less than 10 seconds (probably less than 5) and be able to adjust to variable loads in fractions of a second. They would be relatively high cost, several thousands of dollars per unit at a minimum, and generally run for as long as a decade in service (but not constantly) without a statistically significant number of failures. Figure 10-3 shows a typical system that might be used in a car.

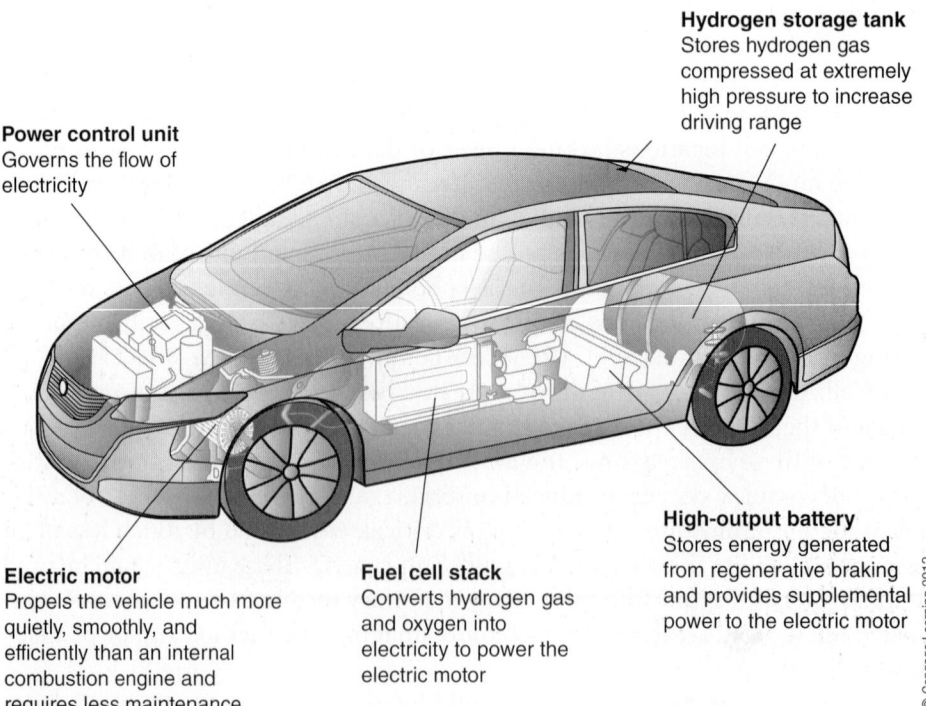

■ **Figure 10-3** Fuel cells and cars.

Fuel Cell Vehicle with Reformer

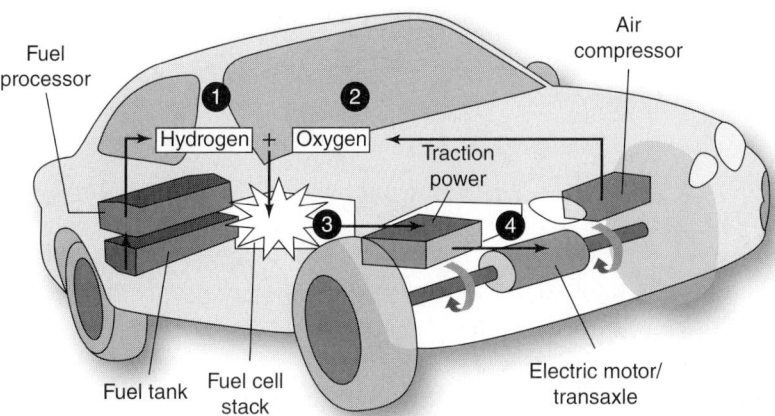

1. The fuel processor converts methanol (or gasoline) to hydrogen. Hydrogen is supplied to the fuel cell.

2. Air is supplied to the fuel cell by the air compressor.

3. Oxygen from the air and hydrogen from the fuel processor combine in the fuel cell to generate electricity, which is sent to the traction inverter module.

4. The traction inverter module converts the electricity for use by the motor/transaxle. The motor/transaxle converts the electric energy into the mechanical energy, which turns the wheels.

FCHV4 System Diagram

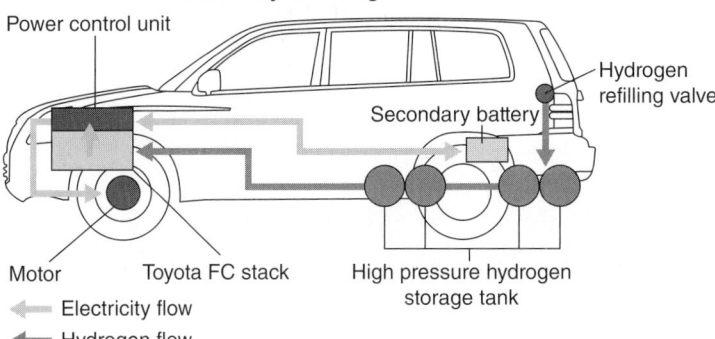

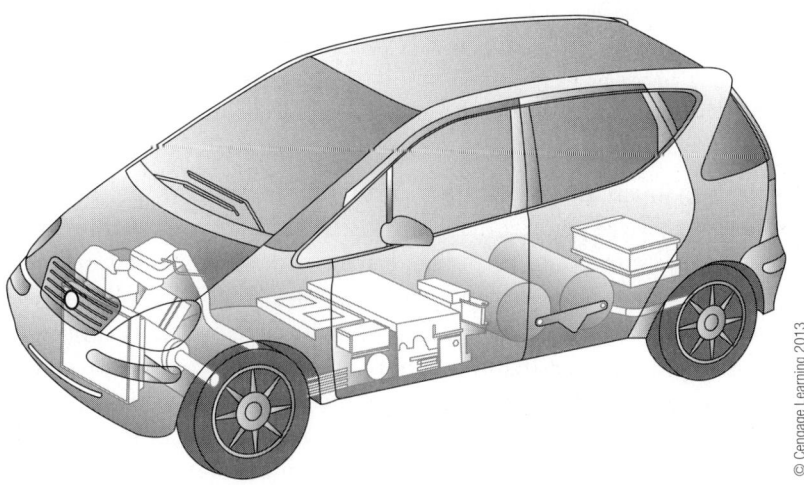

■ **Figure 10-3** (Continued)

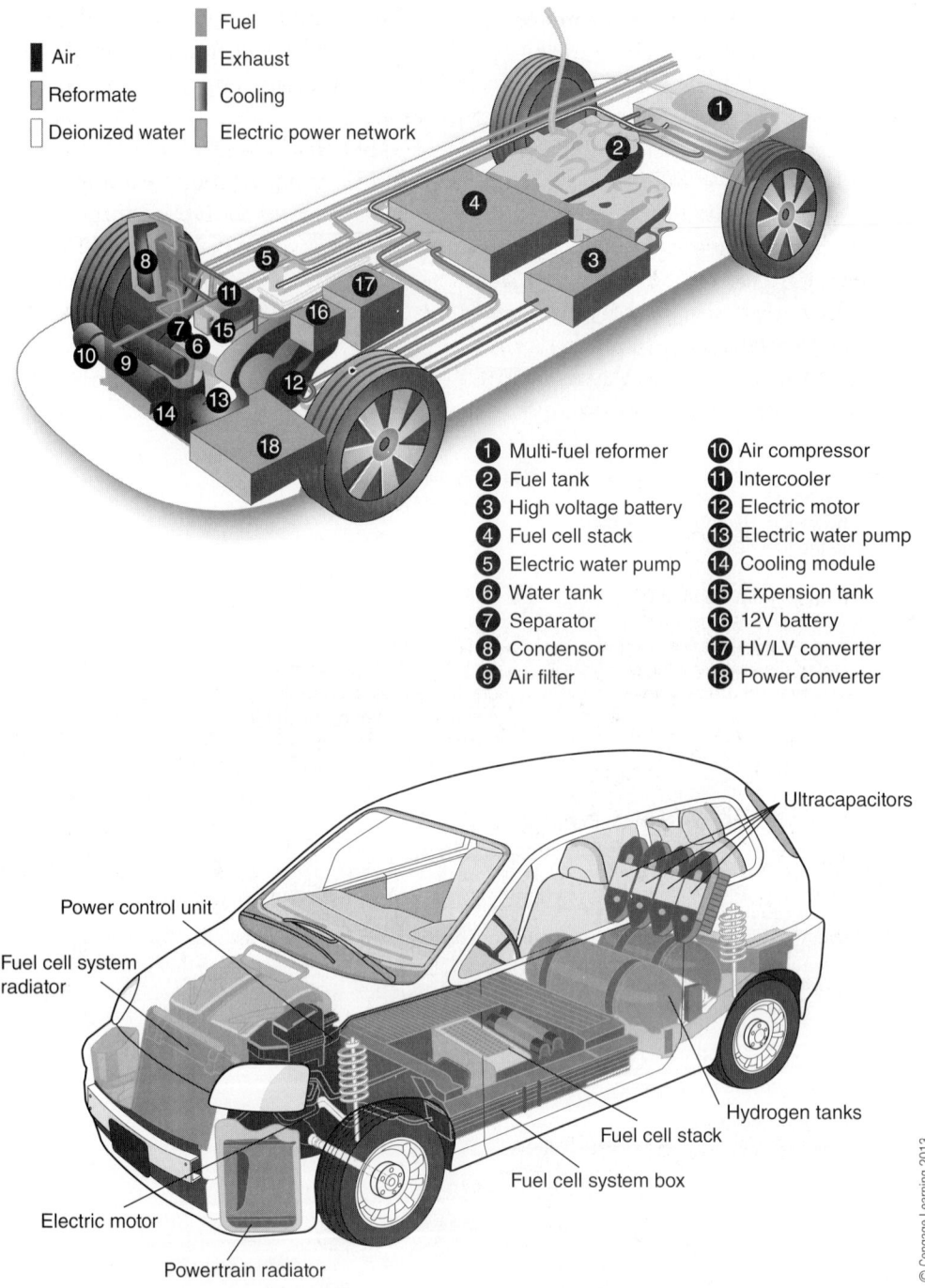

Figure 10-3 (Continued)

Although not previously discussed, this is roughly the size of fuel cell system that could act as a power generator for the grid. While not allowed in some utility districts, these systems could be plugged in at work or home, left on (better for high temperature cells anyway), and generate power that is supplied to the grid. They could also be used as generators in remote locations such as construction sites where ICE generators currently supply power. This could represent a significant amount of savings that would offset the initial cost of a vehicle. A considerably higher initial cost would be acceptable if a construction company could supply their own electricity using vehicles that now run constantly but actually work only part of that time (large diesel powered equipment is rarely shut off once it is running while on a worksite) or they could plug idling equipment in and then sell the generated electricity to the grid during part of the workday. Some of the estimates for generating power using vehicles not otherwise engaged run at over 10 times all the utility generating capacity currently in the United States. This would not be limited to electricity either. It is possible (although a bit unlikely) that your car could be plugged in while it is parked in your garage (or underground parking) but also tied into the building to supply the waste heat it generates for HVAC purposes.

Larger systems will not be the same as those used in smaller vehicles. Those that are used to power buses or trucks will be larger and more robust, but will contain the same types of sub-systems as those seen in vehicles. Figure 10-4 shows a typical fuel cell configuration that might be seen in a bus.

Truly larger mobile systems will not be the same since they will be used for a substantially different set of circumstance even though they will still be mobile. These

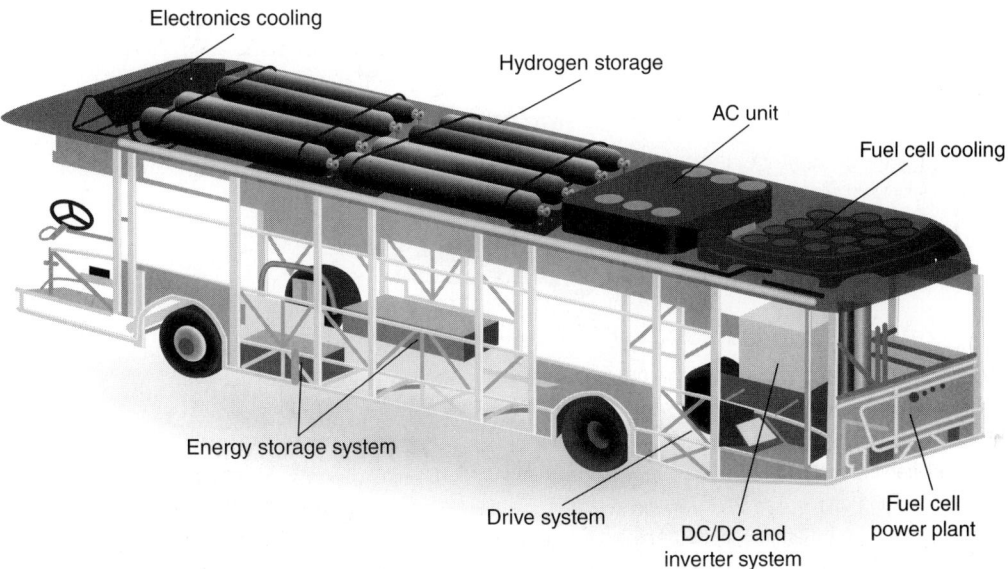

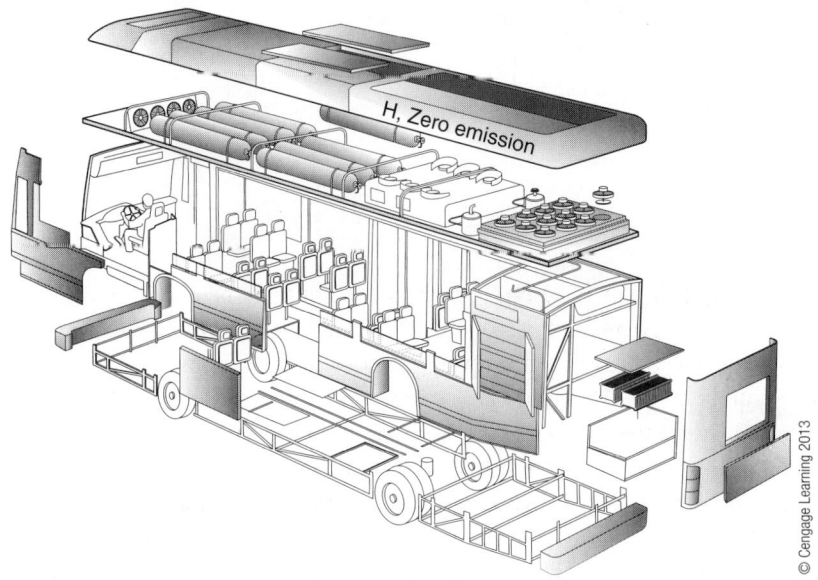

■ **Figure 10-4** Fuel cells and buses.

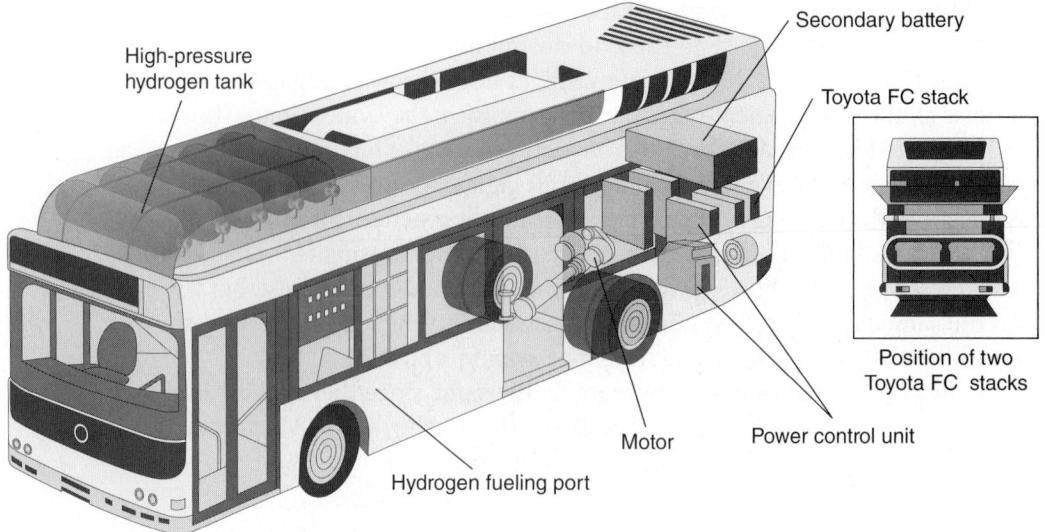

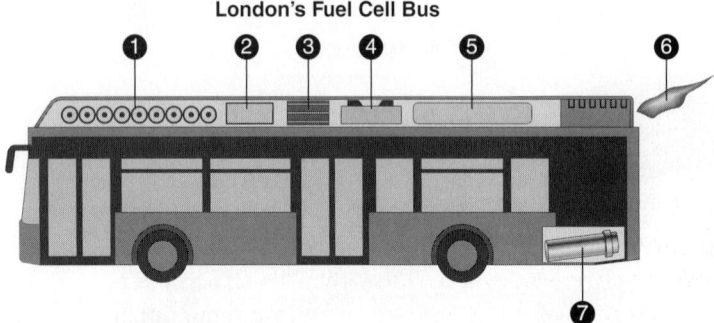

1. Cylinders hold hydrogen sourced from natural gas
2. Fuel cell supply unit - the cells have a gross power of 250 kW
3. Fuel cell stacks
4. Fuel cell cooling units
5. Air conditioning unit
6. Water vapor from exhausts are the buses' only emissions
7. Electric motor, can give a top speed of 80 kph

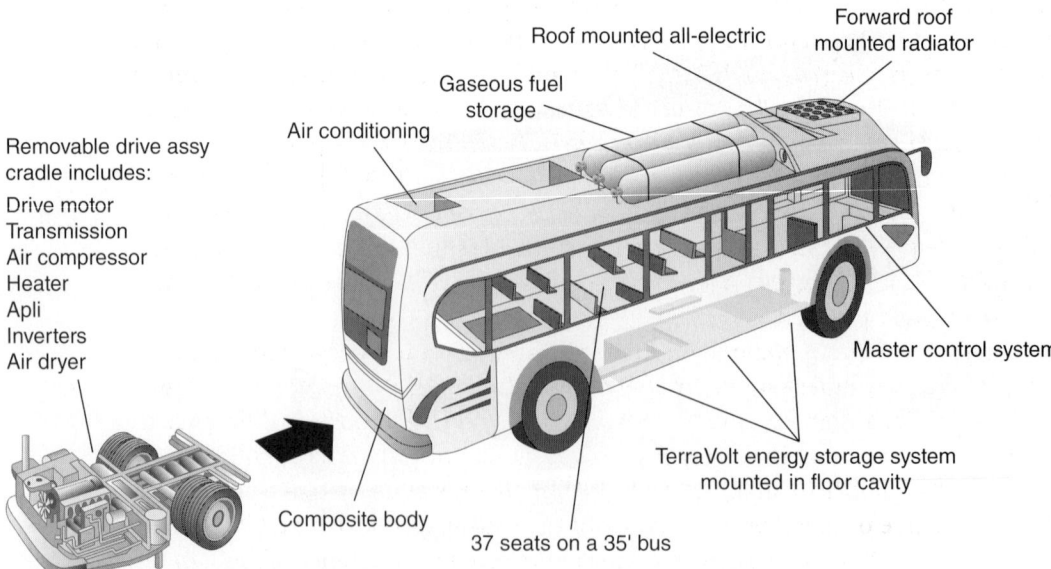

Figure 10-4 (Continued)

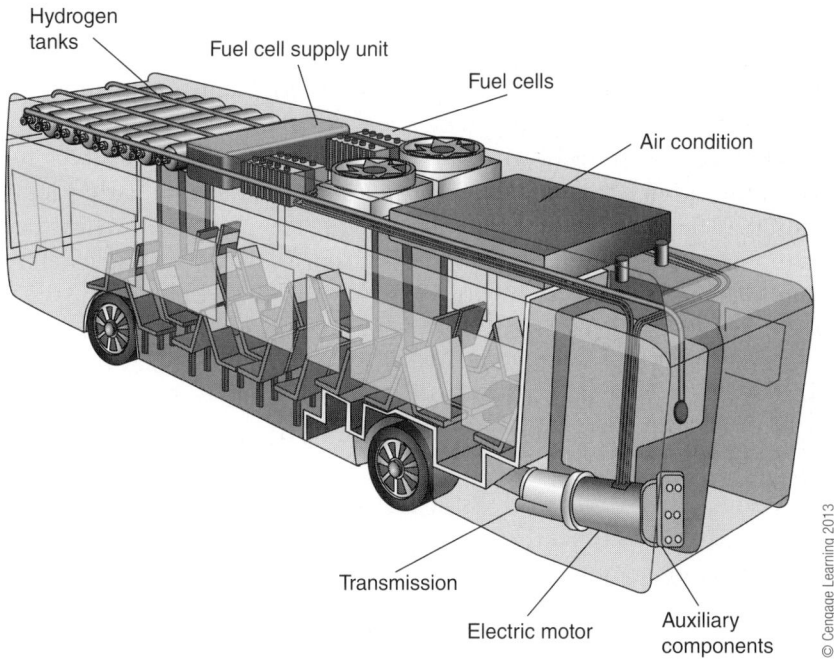

Figure 10-4 (Continued)

are units that might be set in place to supply temporary power to emergency services or used in temporary semiindustrial installations such as asphalt plants or reclamation sites. These types of units would be moved in on a trailer, then set up and operated using trained crews. They would not have to be turn-on-and-go units but rather meant to supply steady power for known functions and requirements. As such, they would not need rapid **power up cycling** or response to **variable loading**, nor the type of **control and analysis equipment** that would be needed in vehicles. When not in use as field equipment, they could also be used at office sites or at centralized locations to supply grid power directly or offset building power requirements.

What is missing in many of these mobile applications is a source of fuel. Stationary installations are having success using natural gas, in large part because the high temperature systems used can tolerate higher levels of impurities as well as internally reform the fuels used. Smaller sizes using low temperature fuel cells need a higher quality hydrogen fuel source, something that is much more difficult to supply. The direct methanol systems work well in this respect but methanol itself has both health and **supply chain** risks (it is very corrosive and has an affinity for water that makes transporting it difficult) that make it unattractive as a major bulk fuel. The fuels of choice are gasoline or diesel (JP-8 for the military) because the infrastructure to supply these is already in place. Several promising technologies exist to reform and clean these fuels but their cost is currently prohibitive.

While some technologies such as personal computers emerge without direct competition, any emerging technology meant to do what is already being done (cell phones for instance) must present a substantial and important difference to those who might adopt it. Cost, convenience, government mandate, public opinion, and even publicity can drive technological changes but only cost truly does so with any degree of speed or urgency. Until the fueling can be done easily at costs that are close to those in existence, significant consumer acceptance of fuel cell vehicles will not occur. Larger systems such as buses can be run from central fueling locations and cost offset or ignored because funding is provided by government mandate. Commercial fleet services such as service trucks or localized delivery could be fueled at centralized locations as well and their costs to a degree offset to customers and good publicity. These examples of early (higher cost) adoption tend to slowly pave the way for more generalized acceptance of new technology by paying for initial fabrication facilities and providing solid examples of successful technological innovation which in turn drive continued innovation. Early adopters are trendsetters that can point the way to emerging technologies as well as pay for that technology to emerge. Cell phones existed for well over a decade before massive adoption occurred. They were purchased by customers who profited from constant communication such as real estate agents, lawyers and sales forces; the very high cost of that initial technology was worth it to those people. They paid for cell towers that extended the reach of the technology, publicized the usefulness of the technology, and drove innovation by demanding better services until more widespread adoption occurred.

Early adoption of technology occurs several years or more in front of general adoption. It should be noted that there are a number of early adopters now using fuel cells systems, both stationary and mobile. These range from European municipalities and utilities to American corporations recognizing the independence and good publicity such systems provide. It can be stated with some certainty that the companies who initially produce the early systems many times do not live through the process (review the history of personal computers for this story) but it is equally certain that those who become involved in the early stages of emerging technologies such as engineers, operators, or technicians can develop lucrative careers if they become engaged in not only the operations and maintenance of the technology but also the efforts to improve it.

FUELING

As stated above, fueling is one of the main obstacles to the adoption of fuel cells. There are several main fuels usually mentioned in conjunction with fuel cells, hydrogen (the main and in some cases exclusive fuel used in the cell itself), natural gas, methanol, ethanol, and **transportation fuels** (gasoline, diesel, aviation fuels, and JP-8b). A number of other fuels or fueling methods are often mentioned as well, such as propane and naptha, or borohydride storage, ammonia storage, and **hybrid systems storage** (using a method wind or solar to produce hydrogen via electrolysis as a means of storing the energy produced when the sun shines or the wind blows and then using the hydrogen to produce constant power 24 hours a day by supplying fuel cells).

It should be immediately obvious that such a listing (and it is only part of the entire list of fuels or storage systems that have been investigated) points out that fueling fuel cells is no simple task and no single fuel is currently preferred or recommended. This is not a particularly bad situation since it broadens the possible appeal of fuel cells and increases their potential for expanding into markets now only minimally served by electric power. Coal has provided power for thousands of years, but not everyone has access to it. Natural gas provides a relatively clean and reasonably priced fuel but is available only where extensive and expensive pipeline systems are in place. Gasoline and diesel engines have been the staples of personal power for well over a century, but they have not supplanted horses or donkeys in many places in part because of fuel availability. The more fuels a system can utilize, the more appeal it can have in the **fractured markets** now existing when it comes to fuel and power supplies.

Fuel cells can use one of two fuels. Low temperature cells use hydrogen; high temperature cells use hydrogen and can also use carbon monoxide (CO). The fuels produced when **input fuels** such as those mentioned in the list above are chemically changed to produce hydrogen are called **reformed fuels**. There are two methods of reforming fuels, **internal reformation** where the fuel cell reforms the fuel (providing its own hydrogen or hydrogen/CO mixture) and **external reformation** where separate systems reform and usually clean the fuel as well.

Mobile applications can make use of any of these combinations but must do so in ways that minimize weight, footprint, cost, and safety as well as allow maintenance and repair access. The type of process and methods used to provide fuel is determined to a great extent by the type of fuel cell being used and the application being served. This is true even of internal combustion engine vehicles. Gasoline engines provide quicker power, lower torque, and worse mileage than diesels, so they are well suited to small passenger vehicles. Diesels better serve applications where heavy hauling is being done and so you see very few gasoline engines on freight trains. Gasoline engines work very well on small, portable generators used for supplying minimal power but larger generators are almost universally diesel engines. The same might well be true for fuel cells. Rapid start-up and shutdown cycles coupled with widely varying **power demand** might be the area where SPFC (PEM) systems are well positioned and in that case, the system to provide fuel must be capable of very high quality. Larger systems that can afford slow start-ups to supply relatively constant power outputs and do not shut down except for extended maintenance would be better served by high temperature systems like MCFC or SOFC. In that case, there might not even be fueling subsystems since internal reformation can be done. It is even possible to combine systems, so that a high temperature system is used as a reformer and fuel cleaner for a low temperature system, then both cogeneration and CHP cycling are done.

Start-up issues are particularly important in determining fueling systems. Whether water is available at start-up is an important issue. Fuel cells can dry out as has been discussed in previous chapters. Maintaining proper hydration within the fuel cell at start-up may require onboard water storage, especially in personal automobiles since they can sit for weeks in certain situations without being used. So some systems may have water stored for start-up. This water would have to be different than that used in cooling loops as well, since antifreeze would not be acceptable in a fuel cell. The very clean water produced in the fuel cell would have to be stored to supply itself, requiring a bleed-off and storage subsystem.

The amount of time available for start-up is critical but so too is the amount of energy that is needed for the start-up cycle. If the systems have to be brought to an elevated temperature then that energy has to come from somewhere outside the fuel cell, since it won't be running at the time. This is analogous to diesel engines, which use glow plugs to heat the reaction chamber so the **self-sustaining** diesel combustion reaction can begin while cars can start cold but require sparkplugs since their combustion reactions are not self-sustaining.

What happens as the first reactions start is critical, particularly in the reforming system. Carbon tends to build up on the reforming catalysts as the reactions start if the temperature is too low, there is too much fuel, or only some of the catalyst reaction sites are supplied. This is even more important if internal reforming is used because the catalysts also serve the fuel cell reactions.

Fuel itself is important, perhaps more so at the initiation of the **reaction sequences** in the reformer than at any other time. The reactions in reforming are exothermic (produce heat) and so once they start in earnest, they will continue (become self-sustaining) until something stops them. As discussed in previous chapters though, they do have **activation energies** among other things that need to be accounted for during start-up. If the fuel is not of sufficient quality, for instance, if it has too much water, sulfur, or the wrong blend of hydrocarbons (kerosene will burn but it will not combust in a gasoline engine chamber) then the reaction may not start. Once it is going, quality is a bit less critical, but initiating chemical reactions is always more of an issue than keeping them going. One of the best ways to ensure clean **initiation** of the reactions is to use pure hydrogen from a stored, high quality source rather than reformed input fuel to begin the sequences. So some systems, especially those that may be starting from very cold temperatures may well have enough hydrogen stored to start and run the systems until the operating temperature is reached.

Another consideration is the primary **fuel delivery** system. This is not the one within the reformer or the fuel cell stack but the main supply line providing the input fuel, be it hydrogen, gasoline, or methanol. Gasoline is delivered to a fueling station in trucks whereas natural gas fuel is delivered in pipelines to the site it will be used. There are well over 150,000 gasoline fueling stations in the United States and about half that many in Europe. Just as electric grid lines lose power in transmission, gasoline loses power through transmission systems. It takes a great deal of diesel to fuel the trucks delivering gasoline to tens of thousands of filling stations. It takes energy to make hydrogen and then to deliver it as well, especially since there is currently no infrastructure to do that. Trucking hydrogen is considerably more expensive than gasoline or diesel since it is so much lighter than gasoline and thus there is less energy value in each truckload delivered. Somewhere over 10 percent (depending on the efficiency and condition of the truck) of the energy in a hydrogen tanker would be consumed by the engine pulling the hydrogen over the roads with that number going up the longer the route is. Natural gas, on the other hand, suffers little transmission loss from wellhead to water heater. Minimizing the loss-on-transmission is an important consideration, and there are many who advocate hydrogen production at the filling station. It is important to remember though that it is considerably cheaper to build one very large refinery able to fuel thousands of filling stations rather than thousands of small refineries that can act as filling stations. The final form that the distribution system for fuel cell fuels will take has yet to be determined.

Internal delivery of fuel to the fuel cell has been discussed in previous chapters, but mobile applications are special cases, since the onboard fueling system must fit within the overall footprint of the system and be sufficiently robust to survive such things as crashes. There is one additional important consideration that has yet to be discussed that is a critical issue in mobile applications: if there are multiple fuel phases involved (solid, liquid, or gas) instead of a single phase fuel system, then the resulting process stream and equipment is considerably more complicated, larger in size, and more expensive. Gasoline is stored as a liquid and then delivered to the combustion chamber as a liquid, so it has a simple enough internal supply system. Fuel cells would have to store liquid gasoline, transfer that to a system to convert the liquid to a gas, and reform the gasoline gas to a hydrogen gas, all the while isolating the liquid gasoline from the high temperatures of the reformation process and keeping the gasoline gas from the gasoline liquid. Even the description is complicated and the system would be even more so. The more complicated a system is, the more expensive it is and the more maintenance it requires.

Internal reformation is the simplest method of internal fuel supply, but it requires a fuel cell running at high temperatures and some mobile applications are not well suited for that. The fueling system for internal reformation consists of manifolding, heating and control to supply the gas to the cell in a consistent manner at the right temperature. There would probably be **onboard storage** of the input fuel as well, and in some cases, a small supply of start-up hydrogen might be used. Mobile applications would not universally require onboard storage though, especially in larger mobile applications that might be temporarily hard-plumbed. The issues in systems like these have to do with gas supply, and include such things as maintaining flow and pressure consistently to the fuel cell area, expansion and contraction of gas as it heats and cools, as well as monitoring and adjusting water content to name a few. Computer control, monitoring, and evaluation would provide required information about system status and integrity so that maintenance would generally be a matter of either hot swapping failed parts or system shutdown for repair. Formal shutdown sequences, including provisions for emergency situations, would dictate the manner in which repairs began, progressed, and how operations were reestablished. Keep in mind that most internal reformation systems operate at high temperatures, so once a system is up and running, taking it back down for any reason would cause problems. Figure 10-5 shows a typical internal reformation scheme for a fuel cell.

External reformation is substantially different. There may be industrial scale external reformation where hydrogen is supplied by plants, either large or small, then delivered in some bulk amount to the mobile platform. For the mobile fuel cell system itself, onboard requirements would be much like internal reformation systems. The main issues would then be storage of hydrogen and gas supply to the cells themselves. The supply system would be even simpler, since most of these systems would operate using lower temperature fuel cells so temperature control loops would be simplified as would provisions for gas expansion and contraction. The hydrogen would be a high quality supplied gas as well, since that would be considerably cheaper than cleaning gasoline on the mobile platform, so there might be a simple **polishing step** or nothing at all.

External reformation of the input fuel on the mobile platform is by far the most complicated of the systems. A generalized system using steam reformation (there are three types of reformation methods, **steam**, **partial oxidation**, and **autothermal**)

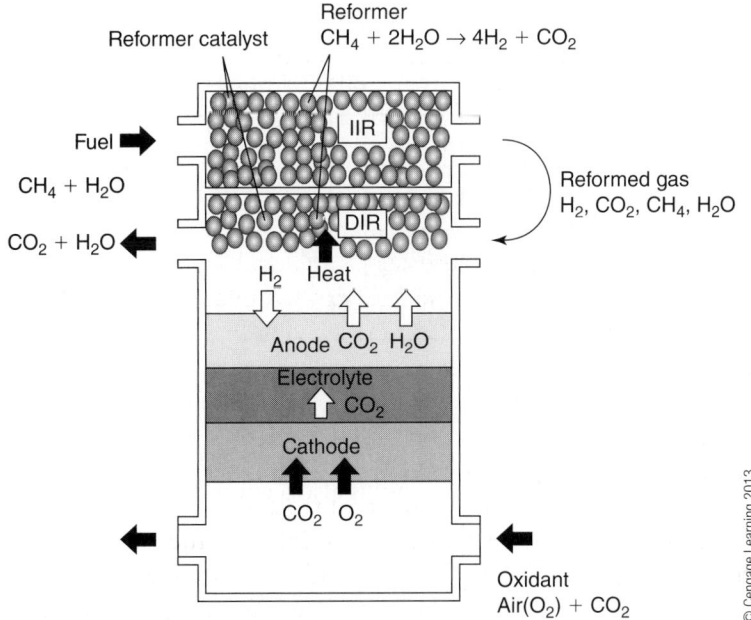

■ **Figure 10-5** Internal reformation.

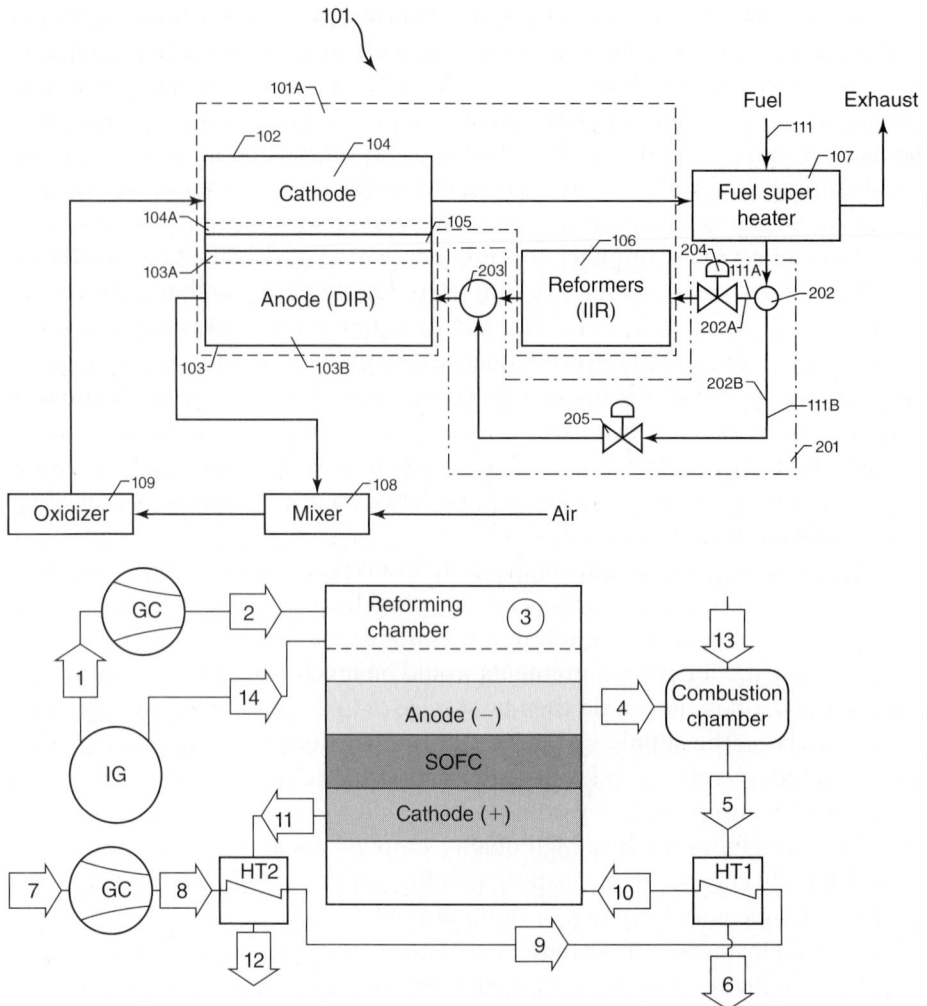

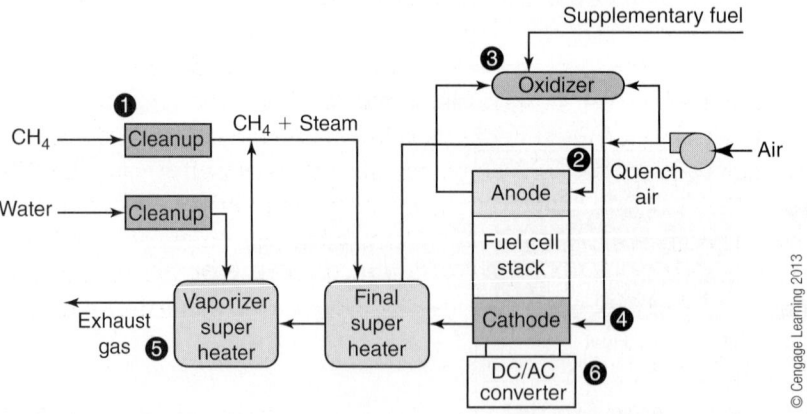

Figure 10-5 (Continued)

would require a fuel tank. The fuel of choice is gasoline, since that would mean the existing infrastructure could be utilized. Methanol is also a good choice since it can be utilized directly with a SPFC (PEM) system without reformation, but methanol has a lower **energy density** (the amount of energy in any given volume that can be used) and has higher **health** and **safety risks**. In any case, if gasoline is used, then current designs in tanks, fuel pumps, and computer control and monitoring can be used as well. That is the second part of the required system: getting the fuel in metered amounts to the gasification/humidification system. Again using gasoline as an example of the input fuel, the liquid gasoline would have to be turned into a gas. A controlled temperature/expansion chamber would heat the gasoline, allow it to transition to a gas (expanding as it does this) and then mix with steam for the input to the main reformation reaction chamber. Temperature considerations are very important in these systems, since the various components run at different temperatures, some up to 1500°C. Balancing, monitoring, and controlling the temperature requirements is complex and requires room that some mobile applications lack.

Current gasoline products are too high in sulfur. Many system developers are assuming that will change at the refinery and do not include a step to remove sulfur from the gasoline. It is important to note here that there is not one type of gasoline, or even three or four. Many states and municipalities dictate **gasoline blends** to deal with specific pollution problems and there are cases where only one refinery makes a blend acceptable to a major city. The fractured nature of the current gasoline delivery infrastructure makes design difficult but also represents opportunity because the major gasoline supply infrastructure is set up to supply multiple grades, so supplying one specific to fuel cells would not need huge investments.

Once the water and hydrocarbon source (gasoline) is available, it is supplied (again in metered amounts at the right temperature and pressure) to the initial reformation reaction chamber where hydrogen is produced. We revisit Equation 18 for this reaction:

$$C_nH_m + nH_2O \rightarrow (m/2 + n)H_2 + nCO \quad \text{Eq. 18}$$

If partial oxidation methods are used (you can review all these equations in Chapter 6), then a steam system is not needed but a supply of oxygen must be available; in either case CO and H_2 are still the products. Remember that steam systems produce more hydrogen and are more efficient but POX systems are smaller, better handle variable load conditions, and start quicker. POX systems are better suited to smaller applications or automobiles while steam reformation is more appropriate to larger applications both in mobile and transportation areas.

Note something important here: mobile fuel cell systems that do reformation on the mobile platform itself are not **zero emission vehicles**. They produce carbon monoxide (CO) as shown above. CO cannot be used in low temperature fuel cells so it is removed in the next required piece of equipment, the **water gas shift reactor** (Eq. 17) where water and the CO gas are shifted to hydrogen gas and carbon dioxide (CO_2). Some systems use both a lower temperature shift reaction and a higher temperature shift reaction to clean the CO out of the fuel gas going to the fuel cells, and in that case, two separate temperature adjusting and controlling systems would be needed.

$$CO + H_2O \rightarrow H_2 + CO_2 \quad \text{Eq. 17}$$

Once the hydrogen is good enough, it has to be supplied to the fuel cell. The process steps seem relatively straightforward: storage to initial reformation to shift reactions to cleanup reformation to final quality adjustments to fuel cells. However input flows, pressure, temperature, water content, and output flows for both liquid and gas species have to be controlled and monitored so that the system responds quickly enough to changes in operating conditions. Pumps, compressors, controllers, monitoring equipment, operating condition gauges, and data storage are all required as well as the basic process loop equipment. The more conditions a mobile unit will operate under, the more support equipment is needed to run the system—all of which has to be located under the hood of a two passenger mini or perhaps even under the seat of a scooter.

The hydrogen does not have to be used immediately although most systems are set up to do so. The types of applications where load is highly variable might be offset by onboard storage of hydrogen gas, used in conjunction with cold starting requirements. The storage and gas delivery from any storage system would increase the overall system's initial cost. That higher initial cost might be offset in part by lowering the initial cost for the reforming and start-up subsystems, and providing the ability to bring stored hydrogen in to the system whenever it is needed would mean the reformation system would not have to respond as quickly to variable load requirements and could make use of a less robust start-up system. Pressurizing hydrogen is difficult given how light the gas is, but there are borohydride storage systems that may be used in these circumstances as well. The amount would not be large either, just enough to supplement supply under heavy load and start the system until it comes to operating temperatures.

The fueling system for mobile applications has to be compact in size with a minimal weight. It has to be able to supply high quality fuel to the cell in a short time, perhaps seconds. It has to use available fuels (probably including **"heavy" hydrocarbons**) but be capable of utilizing fuels that might be available or mandated in the future. It has to be capable of responding to load changes quickly and efficiently. It has to deliver somewhere around 700 liters to 1000 liters of hydrogen per hour for a cell producing 1 kilowatt of electricity (that number depends on a number of details though, so it is only a very general number). System cost has to be roughly equivalent to the current cost of fuel storage and delivery system (tank, fuel pump, injectors, control, analysis, feedback instrumentation . . .) used in ICE systems. It has to be accepted by an extremely diverse market.

VARIABLE OPERATING CONDITIONS AND LOAD CYCLING

One of the hallmarks of mobile systems is the range over which operations occur, no matter if powered by fuel cells or some other engine. There are several aspects to this **range of operations**. The engine must operate when ambient temperatures are well below the freezing point of water to near the boiling point of water. The engine must operate when it is in the horizontal position and continue to do so when the operating position changes to the vertical (the operation must maintain stable output during the change as well). The engine must operate while it is stationary and then continue to operate up to what can be substantial speeds. It must operate when its movement is smooth, bumpy, jerky, twisting, interrupted, and stopped instantly as well as a number of other interesting choices. The engine must provide either current or **torque** under dead stop conditions, where demand and stress to the parts are very high, then transition smoothly to less demanding requirements, stop supplying almost instantly, and then repeat the cycle tens of thousands of times over the course of its operating life.

Variable operating conditions and load cycling have been worked into the designs of internal combustion engine designs for over a century. Their ability to respond to variation has changed markedly over that time, improving in some respects but getting worse in others. For instance the shift from carburetors to fuel injection has speeded up both responding to load cycling and monitoring fuel flow, but the computer-driven nature of this change means any disruption in the electronics shuts down operations completely. The nature of markets and products dictate that any replacement for products supplying mobile power will have to mimic or improve on current abilities. Mobile fuel cells systems for automotive applications will be judged on acceleration from zero to 60, how long it takes to go a quarter mile, how far the vehicle can travel on a single fueling, and how many foot pounds of torque are delivered at the driving wheels. Other mobile applications will be judged on whether instant-on electricity is supplied when power fails, how quiet operations are in residential neighborhoods, if it can be dropped out of an airplane for delivery to customers, and whether it can be supplied in Caterpillar Yellow or Deere Green.

Fuel cell systems can deal with changes in ways that current ICE engine sets cannot. This is due in part to how power is used in the two systems; fuel cells generate electric currents whereas ICE sets directly drive power shafts. Directly driving power shafts means the engines have to reduce the number of combustion reactions in order to reduce the number of turns in the drive shaft or install something like a transmission to do this mechanically. Fuel cells can be paired with batteries in hybrid systems so they run at optimal settings all the time. In these systems, not only is power available from the batteries instantly but efficiency can be increased even further by taking advantage of such things as **reverse cycling** during braking to recover some of the energy already gone into bringing an automobile up to speed but lost when it has to stop. Since fuel cells are already manifolded, they can be operated in banks once they are fully operational, with fuel gas flows sent to additional stacks when more power is needed, providing everything is kept at temperature. This is unwieldy but many ICE engines do a version of this by shutting off some of the cylinders when torque requirements are minimal, such as when cruising down a flat highway. Some of the systems now sold in stationary applications make use of centralized reforming capabilities to supply multiple stack sets and mobile systems for larger applications such as trains or ships could use versions of this to shut down excess capacity simply by shutting down some of the stacks.

Many of the cycling problems associated with fuel cells such as expansion and contraction mismatches, inconsistent gas flow, and lack of complete fuel gas cleaning are associated with the system as a whole moving from one temperature to another. As mentioned above, there can be a huge difference in temperature for fuel cell systems from room temperature up to well over 1000°C (1832°F). Most of the goals for start-up of mobile systems, automotive in particular, are in the neighborhood of 30 seconds. That means that for some of the high temperature components of the systems (actual fuel cells or gas reformation components), the material has to increase in temperature even as much as 50°C in one second. That is an extremely aggressive ramping rate for any material to endure without failing.

There are many type of failure seen in temperature ramping. If there is an inner material that expands more than the material surrounding it, then the outer material can split. If one material connected in line with another expands more, then the two can be bent or even buckle. If a single, large block of material expands, the outer part will get hotter faster, expand more, and crack away from the inner, cooler part. This is even more complicated when a gas is involved. As a gas heats, it expands (think hot air balloons) and that expansion has to be taken into account when pipes and valves are used to move the gas. Any manifolding such as that done in interconnects (bipolar plates) complicates the matter since the material is also expanding at the same time. Contraction has the same issues as expansion, just in reverse with one major exception: when gases cool, they can form a **vacuum** if conditions are right. Many things are designed to handle gas exerting a pressure but not a vacuum. This particular issue can cause substantial problems in any gas system and is especially troublesome when rapid temperature cycling occurs.

ICE systems handle cycling of loads by increasing the number of reactions occurring in the combustion chamber (increasing rpm). The chemical reactions in fuel cells happen at relatively constant rates once they start. You can increase the rate that chemical reaction happen (kinetics) by increasing temperature or by using catalysts, but fuel cells already do that from the start so handling variable load requirements cannot be done the same way ICE systems do. You can do a version of this by designing your anode and cathode catalyst loading based on expected maximum load and then pushing less gas through in normal situations. When maximum load is called for, the electrodes can handle more gas with the extra catalyst and thus develop more current. This increases the cost and lowers the overall system efficiency though. Fuel cells (all engines actually) work best when they run within optimized conditions, where the gas flow and the temperature and the number of reaction sites and reformation and thermal management system and everything else run at peak efficiency levels. Just pumping more gas through is not a particularly good solution to the problem. If a system is not going to use a hybrid battery design, then it is probably going to have **compartmentalized stacks** coming on and off line as need dictates.

Systems could well cover the entire cycling load scenario on several levels. There might be systems that combine several technologies into a massive hybrid. A ship system, for instance, could use part of the hull for hydrogen storage that is produced via electrolysis using solar cells and possibly even smaller windmills. The main fuel cell system could be used with battery sets which augment fuel cell capabilities to provide the power initially needed to get out of port and underway. The fuel cells could then be capable of running the propulsion system at cruising speed with some extra capacity to recharge the batteries. At port, the fuel cell could provide truly pollution free power, possible even selling it into a grid while it is unloaded and loaded again.

Variable conditions and loading also require thermal management systems capable of handling the variable heat. This is more critical in fuel cell systems than in ICE systems for instance, because CHP or turbine cogeneration is included in many fuel cell systems. Handling rapidly varying thermal loads or bringing turbines on and off line as conditions change requires a very high level of monitoring and control capabilities. With very powerful computers now available in small footprints, a fine level of control can be exercised over even very complex systems like fuel cells. Managing heat generation and distribution, variable cooling requirements, cogeneration, and even associated HVAC systems on mobile platforms lowers costs, improves efficiency, minimizes pollution, and provides flexibility not achievable even 10 or 15 years ago. This improvement in control capability may be the breakthrough that finally brings fuel cells into the mainstream, especially in mobile applications.

POLLUTION

The final discussion point about mobile applications is **pollution**. The internal combustion engine is a main source of pollution; even more so in developing countries where the very common small two-cycle engines used on scooters have no pollution control systems. Coal plants in countries where industrial regulation is limited are also main sources of pollution. Rapid industrialization using ICE systems for transportation and coal to generate electricity often results in massive pollution problems. Limiting or even eliminating pollution is often mentioned as one of the strengths of fuel cell technologies.

While it is not truly possible to achieve zero emission energy, fuel cells create limited pollution, primarily CO_2. Remember that even wind or solar must use materials and energy in the fabrication process that produce pollutants and in some cases toxic pollutants. Fuel cell manufacturing is no different but because the systems are generally simple, they require less material and a smaller number of manufacturing steps, thus producing correspondingly less pollution. Reforming systems can be complicated and require high temperature materials, but are in general replacing systems of the same sort, for instance the piston chamber of an automobile, so the overall pollution may be equivalent.

In some applications, governments are driving change through regulations. A good example of this is in the North American waterways, where regulations require use of electric motors on many lakes and rivers. This is due to the lack of pollution control in ICE boat engines and the usual discharge of exhaust and cooling waters into the waterway. Even larger ships may be forced into this type of propulsion system, since the regulations of a single major port could determine the type of vessel used in an entire fleet.

In mobile application uses (transportation in particular), current uses favor ICE sets using petroleum based fuels as regards total energy losses from well-to-wheel, which equates to lower emissions overall. Compressed gas delivered via pipeline also has very little loss, while methanol, bioethanol, and hydrogen have moderate loss. Current hydrogen generation methods produce significant amounts of **greenhouse gases** (CO_2 in particular) as well as NO_x and SO_x pollutants but in general are less polluting overall than ICE competitors, in part because they can produce the same amount of energy using less fuel (are more efficient).

Fuel cells may or may not reduce overall worldwide pollution. Pollution can be dealt with in two ways. The source of energy can be centralized and very expensive, reasonably effective pollution control done at that site. The source of energy can be distributed over numerous smaller sites and less expensive, reasonably effective pollution control done at all those sites. How effective those methods are depends on how effective the maintenance and operation of the energy system and its associated pollution control systems are. No pollution control system will be effective if it is broken or removed from service while no one is looking. Fuel cells have the potential to reduce pollution but only if a commitment is made to do so by society and backed by a willingness to pay for it.

KEY WORDS

Knowing the terminology used is critical when dealing with fuel cells. Following is a list of the important terms in this chapter, which are also in bold typeface within the chapter. It is recommended that students be required to submit definitions of some of these words as homework assignments in which they look the terms up in other books, articles, or on the Internet.

activation energies	gasoline blends	reaction sequences
autothermal reformation	greenhouse gases	reformed fuels
	health risks	reverse cycling
auxiliary power	heavy hydrocarbons	robust
compact	hybrid systems storage	safety risks
compartmentalized stacks	initiation	self-repair
	input fuels	self-reporting
competitive price	internal reformation	self-sustaining reactions
control and analysis equipment	maintenance call-out	software diagnostics
	mobile system	start-up issues
cooling parameters	moving platform	steam reformation
cost of disposal	onboard storage	supply chain
cost of fabrication	polishing step	torque
early adoption	pollution	transportable engines
energy density	power demand	transportation fuels
external reformation	power up cycling	vacuum
fractured markets	partial oxidation	variable loading
fuel delivery	primary power	water gas shift reactor
full operating power	range of operations	zero emission vehicles

DISCUSSION QUESTIONS

1. What are some of the difficulties mobile systems need to overcome that stationary systems do not generally have to deal with?

2. Why are low temperature fuel cells more suited to mobile applications than high temperature cells?

3. What are some of the common design elements that might be seen in mobile systems?

4. What are two major problems in transporting liquid methanol?

5. What are the two fuels actually used at the fuel cell?

6. Starting up a cold fuel cell presents many problems. One of the most serious is in dealing with any trace contaminants in reformed fuels. What is one way to deal with this problem?

7. Why is trucking hydrogen considerably more expensive than trucking gasoline or diesel?

8. What are the three types of reformation methods?

9. What are some of the problems associated with ramping temperatures up at high rates?

10. Why is there no such thing as true zero emission technologies?

CHAPTER 11

FUEL CELL SYSTEMS: PROCESS AND INSTRUMENTATION

objectives

This chapter introduces the student to fuel cells as processes flowing from beginning to end and the ways such processes are presented. Understanding how the flow of an industrial, commercial or personal system is presented on paper is critical to fuel cell systems and this chapter introduces students to several of the ways drawings and diagrams are used in detailing such flows.

INTRODUCTION

Fuel cell systems are not simple. They contain several different subsystems, use flammable fuels at elevated temperatures and pressures, and may well be the future of electrical supply. It is important that they are understood then as systems and not as a collection of individual units that just happen to be located near each other. This chapter is meant to introduce the concept of systems to the student not only as a collection of dependent units supporting each other but also as a means of seeing and understanding such collections as a whole.

Critical to such understanding is the way systems are presented on paper. Designs and processes are documented in a number of ways. This chapter will introduce several types of drawings used in industry to explain in detail what the parts of a system are, how they are connected, what is moving in the system, and a general idea of how they are arranged. There are several ways to present this information: **piping diagrams**, **flow diagrams**, and **instrumentation diagrams**, as well as diagrams that combine several of these. There are also **architectural drawings**, **engineering drawings**, and **fabrication drawings**, as well as other types used for particular reasons. Maintenance or engineering personnel will probably see every type of drawing and have to understand what they are seeing in order to do their jobs.

This chapter will discuss some of the more common drawings and present several that include fuel cells and associated systems as a whole. It is intended as an introduction to the many types of drawings and diagrams used, not as a comprehensive treatment of them.

COMMON INDUSTRIAL DIAGRAMS

Most industrial processes are complex to run, difficult to troubleshoot, and expensive to maintain. Many are sealed so they cannot be easily examined and built using proprietary methods or parts so their workings are not easily understood or even shipped with documentation. Decent training opportunities must be carefully arranged or often do not happen. When individual pieces or subsystems are installed, the plant infrastructure is often changed to accommodate the new equipment but not well documented and even major utilities are altered to meet changing needs.

It is engineering and maintenance that are left to sort it all out, often when something breaks and has to be repaired in short order. The starting point for all such work, whether repairing or installing, is the set of site drawings showing buildings, utilities, equipment, process flows, and even **material balances**. What power sources need to be deactivated, where a line can be blocked, how much material might be in a vessel, and any number of other questions about what a job might entail can be answered only from drawings like this.

Block Flow Diagram

Block flow diagrams (BFD) are the simplest of all the diagrams used to present process data in a graphical format. They are meant to provide an overall view of the process and generally show where the unit operations are, usually as blocks with the raw material streams identified. These are used less as technical data and more as informational data and might be included in information packets to local media or civic groups when information about an industry is requested. They are useful in training as well, especially during introductory health and safety training.

A simple BFD of fuel cell systems is given below. It does not include what type of fuel cell is in place, what fuel is being used, whether CHP cycles are in place, or even how much electricity is generated. This kind of information is not generally presented in simple BFD arrangements. Some basic block flow diagrams are shown in Figure 11-1.

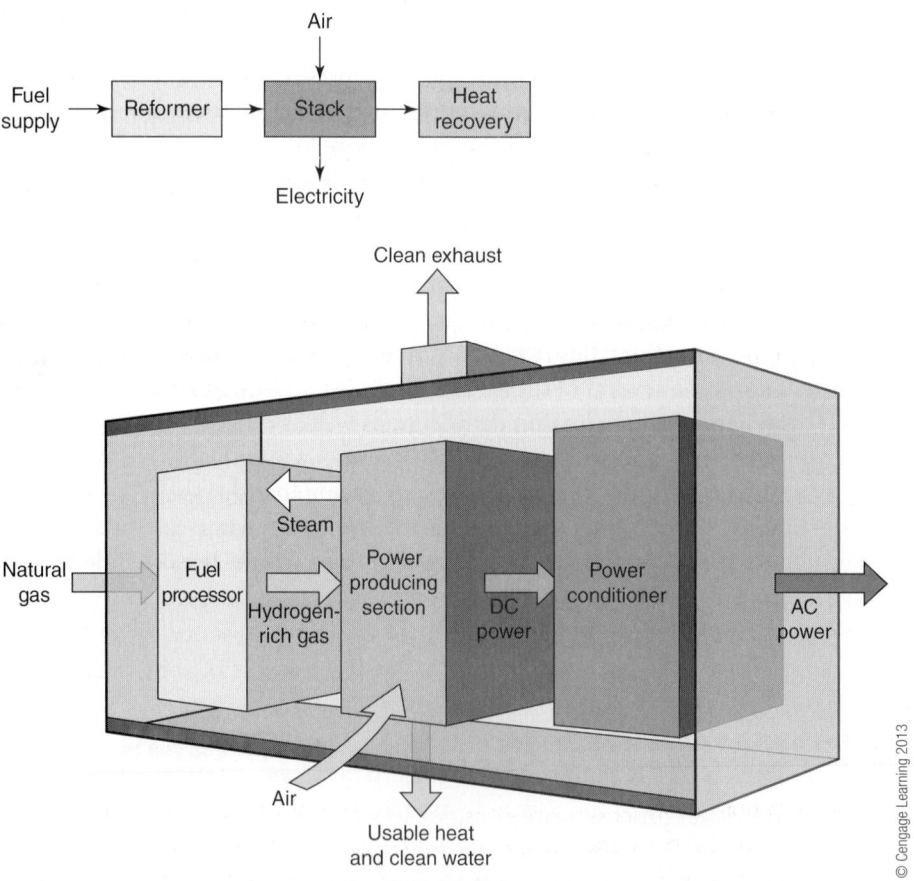

Figure 11-1 A block flow diagram for fuel cells.

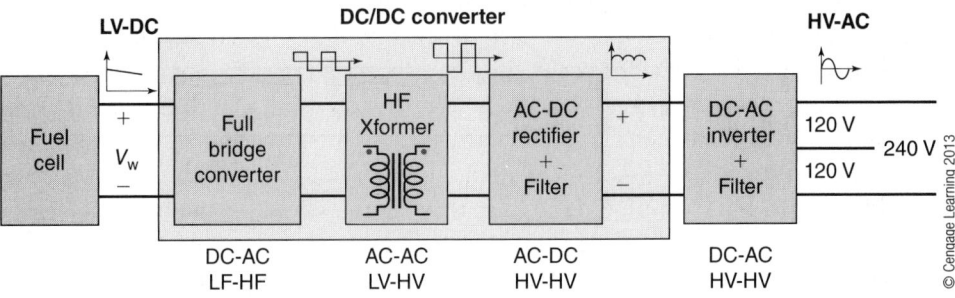

■ **Figure 11-1** (Continued)

Process Flow Diagram

Process Flow Diagrams (PFD) present the process as a whole. They will show the primary **process vessels** and related equipment as well as showing input and output flows in such a way that the entire flow of the system can be traced from a single document. Pressure, solid or liquid flows, temperatures, and even material balances (the weight and makeup of input and output streams) are included. Depending on the complexity of the system some minor flows may also be left out or merely noted. Piping is included with any heating or cooling processes that are done along with the temperature and pressure of what is in the piping along with the flow rate and composition of what is in the piping. Many times, specific safety issues for flow streams may be noted as well. Any **monitoring devices** or other instrumentation are included but usually without showing particular wiring paths.

PFDs are used to show the relationships and flow paths between the major equipment in a system and to provide information on both **standard operating values** for that equipment as well as **upset conditions** (minimum and maximum allowed). In general, any stream that must be controlled in some way is detailed graphically in its relation to other parts of the process. This can include such things as any required pressure or feed rate as well as critical temperatures or fluid levels that have to be maintained. Written data is also providing that can be used to determine whether the system is in control or not. This is the most common use of PFD in operations, so that personnel can monitor and control processes within the narrow band that returns the best (most profitable) product mix. In maintenance, PFD sets allow personnel to trace flows when some disruption, accident, or catastrophe occurs, since many times they are more familiar with the equipment than the actual process running across the plant as a whole.

Process Flow Diagrams use **symbols** to represent a number of different things within a process. There are several different standards for symbols used to represent tanks, valves, instrumentation, and other parts of a process. ISO 14617-6 is one such standard, as are IEC 61346, ANSI/ISA-S5.1, and DIN 30600. **Computer Aided Drafting** (CAD) programs may include particular sets of symbols, the standards mentioned above can be purchased or companies may even develop particular symbols based on their own needs. In any case, smaller drawings may include an explanation of the symbols used in the drawing, but many times, a separate sheet or even an attached standard itself will be included in the drawing packet for reference. Figure 11-2 shows a small sample of basic and some ANSI symbols.

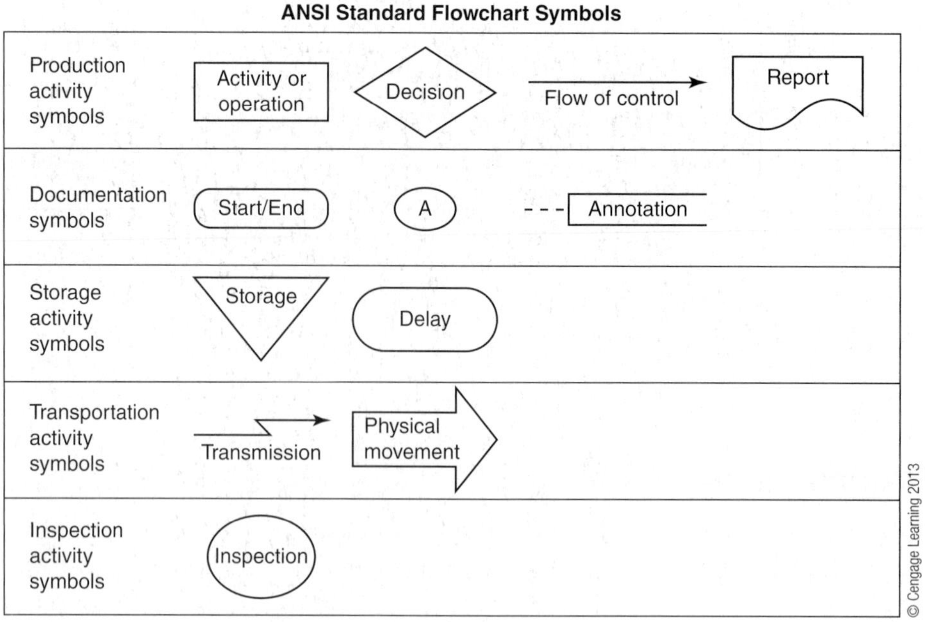

Figure 11-2 ANSI diagramming symbols.

Equipment commonly noted in PFD sets are process vessels such as reactors or mixing tanks, pumps, compressor, heat exchangers, valves (any that act to control a flow or influence the process), pressure regulators, heaters, pipes, and heat jacketing. Monitoring devices for such things as flow, temperature, and pressure are included. Ratings such as volumes of vessels, the pressures allowed in piping or vessels as well as normal, maximum, and minimum operating temperature, pressure, and flows that are allowed are included. In addition, all system connections are included, and that can mean any connection **upstream** of a particular drawing, within the area represented by the drawing and needed **downstream** of the drawing. The names and numbers of all equipment should be noted if appropriate. Company data is included, usually in a standard box that states names, dates, designers, reviewers, drawing numbers, revisions, and a number of other things.

PFD sets are used by maintenance, engineering, and operations personnel to trace raw material, **work-in-progress**, and output flows as well as to find **reference numbers** to particular pieces of equipment. They can be used as training aids and are made available to emergency services as the need arises. A much simplified drawing is shown in Figure 11-3. Students are advised to find and review fuel cell drawings available on the Internet as discussion items.

Process Control Diagram

Process Control Diagrams usually accompany Process Flow Diagrams and provide information on all **control equipment** being used and the values of the **control variables** involved in the process. These are sometimes used in what is termed **Hazard Operability Studies** where each part of the process is examined in detail to provide a guide to what can go wrong, especially as regards health and safety but also the type and extent of damage that can occur to equipment.

The main purpose of PCD sets is to identify the lines, loops, instruments, and devices used to control a process. Unlike the PFD, the flow through the process is less important than measuring the flow (what does the measurement, where it

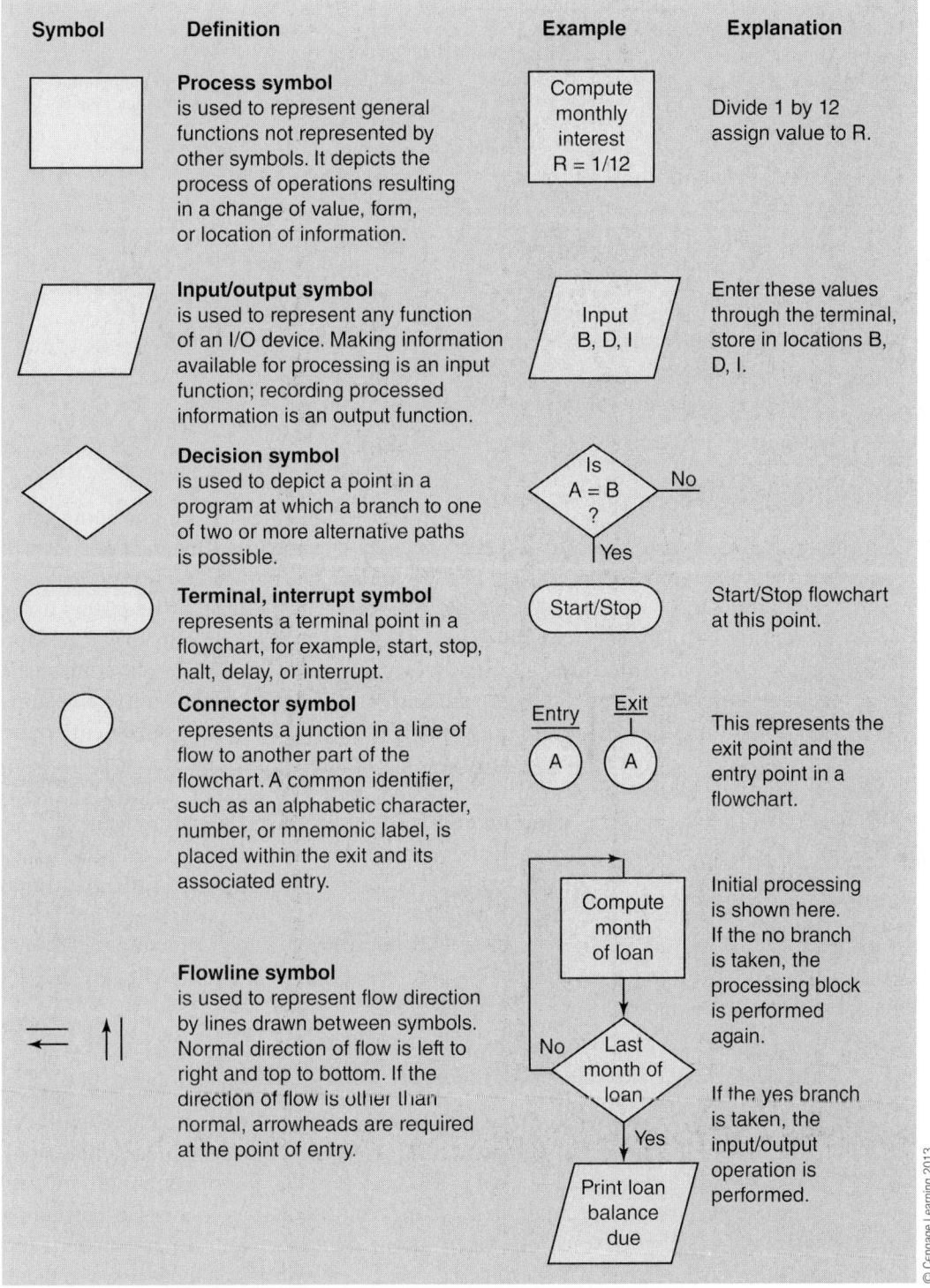

Figure 11-2 (Continued)

is measured, when it is measured, and how those measurements are used in the **control loops** running the equipment).

A PCD will identify all instruments used in monitoring, where it is located, what control loop is involved, how it is physically identified, how it is accessed, and of course what variable is being measured. It will identify the process signal being measured, the type of signal (**pneumatic**, electrical, or even software compiled)

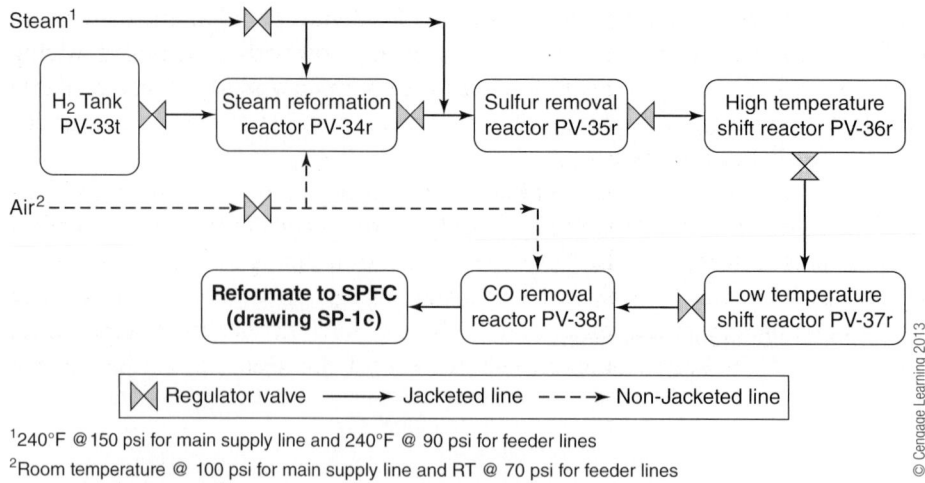

Figure 11-3 Simple PFD of reformate system.

being measured, and how the connection is established (is weight being measured directly using a scale or indirectly with **strain gauges** for instance). It will describe any loop the instrument is part of (usually a control loop but not always since some loops display information that will be acted upon by operators using separate control loops) as well as describe the functions of the loop itself (one loop might control a single pressure regulator valve while another might control the temperature of an entire low temperature shift reactor). It will identify all equipment and instruments used in any loop by unique numbers. It will identify any power sources, junction boxes, or fuse or breaker systems associated with the loop and provide **lockout-tagout** locations if appropriate. It will also identify any **sampling locations** as well as **decision points** where product quality is judged.

PCD sets are generally internal documents that may even contain proprietary information about the process and its variables. They are used for operational control, system maintenance (particularly electrical and computer), and quality control or quality analysis (QA/QC) monitoring. These drawings are rarely allowed off site and access to them is usually controlled. Some simplified PCD sets are shown in Figure 11-4.

Piping and Instrumentation Diagram

Piping and Instrumentation Diagram (P&ID) sets are the most comprehensive of the industrial diagramming methods in use. They show piping and process flow along with major and minor equipment as well as all installed instrumentation. PID sets must show all the process equipment, all the connections across the plant, including such things as piping, conveyors, mechanical loading such as front-end loaders, and any other method used to move material as well as all wiring and loops between them. All instrumentation in use to monitor or control the process is included as well as any **generated numbers** used in other plantwide systems such as stores or accounting.

P&IDs are considered to be schematics and are commonly used to understand the levels of control in the process (**manual**, **automatic**, **semiautomatic**, process, safety, quality, **stop-go**, continuous, and a number of others). They show in great detail the placement of equipment, how connects occur across an entire plant, the physical sequence of the production process and the control schemes in use. Some may provide extensive notes on the operation of individual equipment and control loops as well as other explanatory details concerning flow within the piping, wiring diagrams, and control systems. References are commonly made to any available **Standard Operating Procedures**, maintenance sequences or timing and documentation provided by vendors.

PID sets are complimentary to PFD sets, providing a more comprehensive listing of the equipment that is used to support the process and control the flow within it. For systems such as fuel cells, they will show not only piping and instrumentation in great detail but also control, shutdown, and emergency sequences in place; any safety and regulatory equipment; ties to both fuel input and electrical output lines; and basic start-up, operational, and shutdown information. So, for instance, a

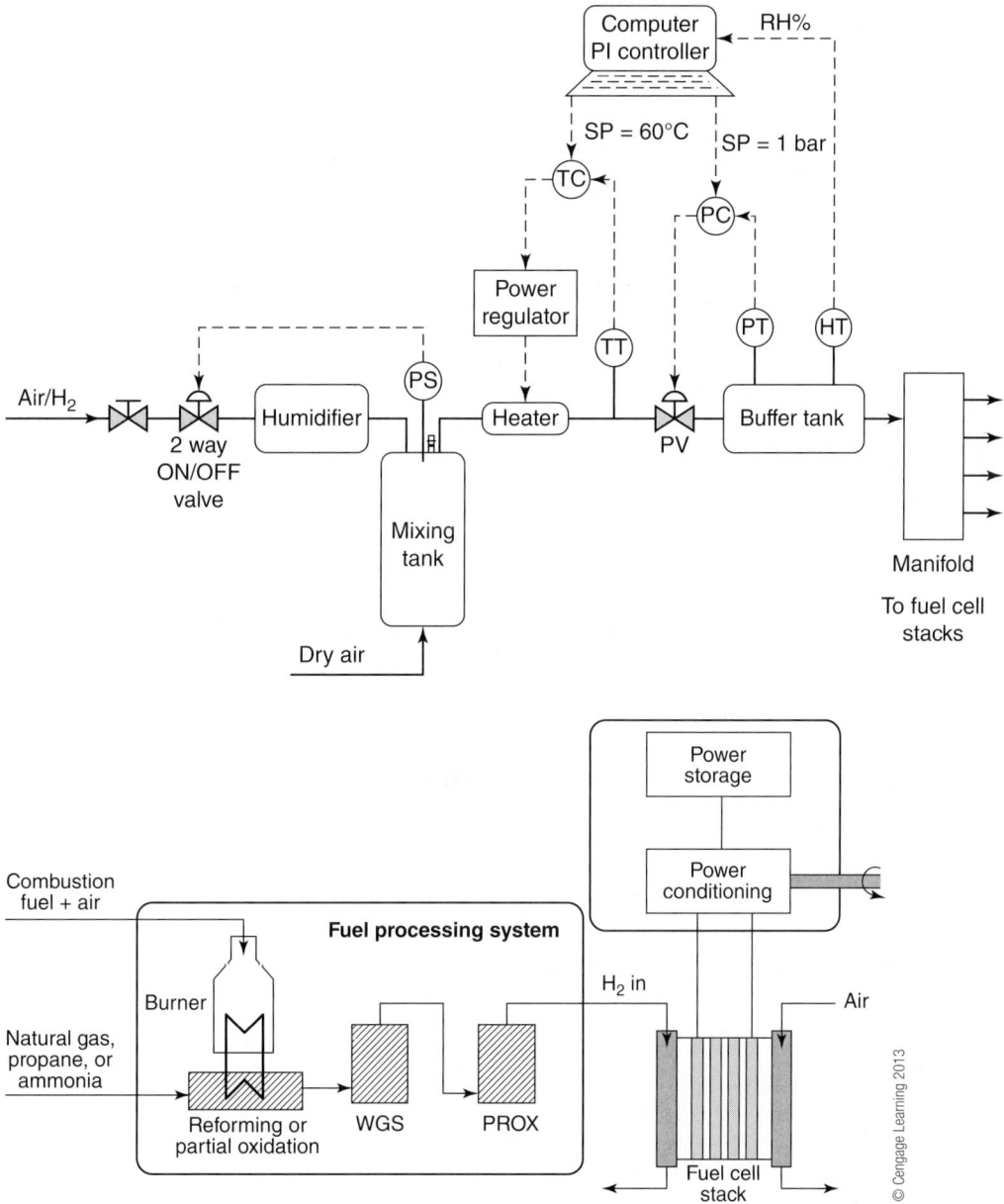

■ **Figure 11-4** Simplified fuel cell PCD.

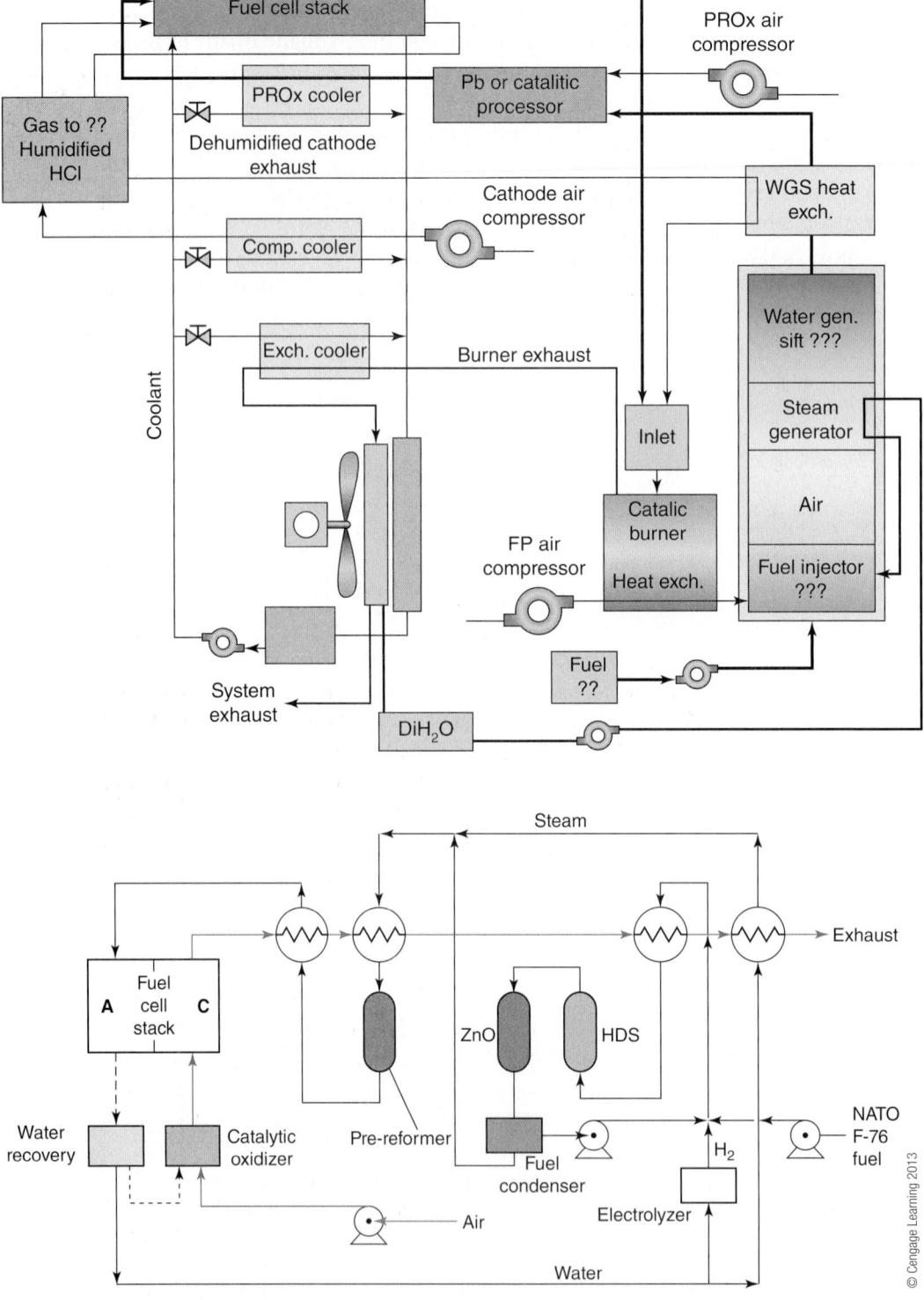

Figure 11-4 (Continued)

SOFC would start with the input gas, specifying its rate of flow and possibly general quality as well as details about supply. All the information about the interconnects from the gas supply utility would be included, such as upstream, utility-owned piping and valving, the location of **main shutoff valves**, where the utility line ends and the process line starts, as well as expected minimum and maximum flow and pressures. Utility and company input feed control would be specified including the types, designations, and identification of all valves matching one to the other, including whether manual or automatic, process or emergency shutdown, and control or flow valves. All required information to build the piping such as sizes and identification (material of construction, the unique numbers used to identify the individual parts used in fabricating the piping system, and any welded or mechanical connections), special requirements such as heating, vents, sampling lines, line reducers, swaged fittings or special connections, **flushing lines**, and flow direction (to name a few) would be included. We aren't even to the reformer on the fuel cell by the way, and we have yet to address the control issues on the main natural gas input line, which would be substantial, as some are driven by the process QA/QC issues, others by safety, and still others by regulatory requirements for high pressure gas lines.

Fully describing a PID is not really possible, which is why they are diagrams. It is important to keep in mind that these are **primary** design, construction, and operations **documents**. They are usually proprietary and rarely handed out even to company personnel unless they work directly with engineering, operations, or maintenance of the system. Unlike PFD sets, they are not used for general training or safety discussion, but are used for hazard operability studies and should be faithfully maintained whenever changes to a system occur. In some industries, they are required by federal, state, and local regulations to be up-to-date and authoritative and fully able to supply emergency personnel with critical data in emergencies.

PID layouts should mimic the process flow as much as possible, since one of their primary purposes is to provide a graphical representation of what is going on in an operating facility or **equipment cluster** (such as a fuel cell system). They are not usually to scale and often do not include physical measurements such as the lengths of piping runs since that level of detail is for other types of construction drawings. Rather, they are meant to provide the type of detail that personnel familiar with the operations and conversant with PID sets could use to quickly understand what is happening in normal operations using only the PID (and maybe an accompanying PFD). If an **upset condition** occurs or a **control interlock** is triggered, personnel should be also able to quickly pinpoint the problem simply by following the flow and control loops.

There are extensive lists of what should and should not be included in PID sets, some detailing dozens of requirements ranging from flush lines to anything at all coming into a facility that is supplied by outside vendors, up to and including cleaning supplies. As with PFD sets, standards can provide listings of what should be in PIDs as well as what figures in the drawings mean as well as a reasonable idea of how they should be produced. Including such lists is beyond the scope of this introductory discussion but the student should note that each set will be slightly different, being produced for a unique set of circumstances and equipment. Some general guidelines, however, are in order so that the student can gain an appreciation of what might distinguish a good PID set from one less acceptable. A good PID should have at least the following components:

- An arrangement that it is not crowded (not more than about half a dozen major pieces), smudged, has no overlapping lines, and has been hand erased or treated in any other way that makes it confusing.

- Company information is prominently displayed (usually in a separate block), including who produced the document, who reviewed it and all the dates associated with it, including any revisions (if it has not been revised on a regular basis, it is probably wrong).

- All equipment, vessels, and fabricated pieces are identified in full (name, size, capacity, horsepower, duty cycle, standard operating conditions, vendor. . .).

- Anything that requires **leveling** or **elevation control** is clearly identified, whether pieces of equipment or drain lines, with elevations called out.
- Anything coming in from outside the lines or equipment (power, raw material, water, chemicals, heat...) is clearly identified and quantified as well as showing where they are coming from and going to.
- All flows are identified and quantified and flow directions are clearly evident.
- All instrumentation is fully identified (line or control type, local or transmitted, any codes associated with the piece or its installation stated...).
- Anything that can act to exert control on the process, whether in regular or upset conditions, is identified and probably discussed in detail somewhere in the set.
- Anything that can act as a **safety release** or **guarantee** (relief valves, alarms, automatic shutdown sequences, emergency contacting sequences...) is identified and probably discussed in detail somewhere in the set.
- Plant site elevations are if not fully stated, are at least implied by the drawings.
- Lockout/Tagout stations are clearly identified even if they are at locations remote from any single piece of equipment.
- All insulation is fully identified.
- Any **required line of sight** is called out and detailed as important.
- The way **crossing lines** are handled on the drawings is fully explained, easily understood, and followed faithfully.
- References to PFD sets are clear and concise, matching all components.
- Quality Control and Analysis requirements are fully integrated into the set.
- Any **in-line** measurement or analysis equipment is fully explained.
- Spelling is correct; numbered references are double-checked.
- Access to critical areas for maintenance and operations is noted and discussed if needed.
- All symbols used are identified in a full listing with critical ones identified on the individual drawings.
- All **emergency locations** (safety showers, evacuation points, fire extinguishers...) are clearly identified.
- Any local **programmable device** is called out in detail.
- **Command and control centers** are clearly identified as such.
- Anything enclosed for a reason is explained.
- At least some limited design criteria are included for everything, with references to full design sheets called out.
- Fabrication and installation details are provided.
- Any conversions or calculations that are made from instrumentation to control system are explained, with examples included.
- All monitored and controlled variables are listed and explained.
- Required **calibrations** are prominently mentioned.
- Any **codes or standards** used are called out and included in the set.
- **Troubleshooting** procedures for all equipment are called out with full sequences included in the set.
- Prior revisions should be available and can be examined to track changes.

KEY WORDS

Knowing the terminology used is critical when dealing with fuel cells. Following is a list of the important terms in this chapter, which are also in bold typeface within the chapter. It is recommended that students be required to submit definitions of some of these words as homework assignments in which they look the terms up in other books, articles, or on the Internet.

architectural drawings
automatic control
block flow diagrams
calibrations
codes and standards
command and control centers
control variables
computer aided drafting
control equipment
control interlock
control loops
crossing lines
decision points
downstream
elevation control
emergency locations
engineering drawings
equipment cluster
fabrication drawings
flow diagrams
flushing lines
generated numbers
Hazard Operability Studies
in-line equipment
instrumentation diagrams
leveling
lockout-tagout
main shutoff valves
manual control
material balances
monitoring devices
Piping and Instrumentation Diagram
piping diagrams
pneumatic
primary documents
Process Control Diagrams
Process Flow Diagrams
process vessels
programmable device
reference numbers
required line of sight
safety guarantee
safety release
sampling locations
semiautomatic control
standard operating values
stop-go control
strain gauges
symbols
troubleshooting
upset condition
upset conditions
upstream
work-in-progresss

DISCUSSION QUESTIONS

1. In general, what types of things would be shown in drawings?

2. Which are the simplest of all the diagrams used to present process data in a graphical format?

3. What are Process Flow Diagrams used for?

4. Process Flow Diagrams use symbols to represent a number of different things within a process; name four standards that detail which symbols are used in these diagrams.

5. What are Process Control Diagrams used for?

6. What do Piping and Instrumentation Diagrams usually show?

7. P&ID drawing sets are almost always used along with PFD drawing sets. How should the two tie together?

CHAPTER 12

FUEL CELL SYSTEMS: POWER AND CONTROL

objectives

This chapter introduces the student to fuel cells as materials handling systems. The fuel cell and its balance of plant is a collection of different materials, other materials flow in, change form, break apart, recombine and finally exit as something different. Some of these materials change between solids, liquids and gases as they move through the systems. All these materials in different forms must be controlled as they move through the fuel cell system from input to output. Heat and the electrons that produce electricity are also introduced as materials conditioning and handling problems that have to be dealt with just as gas or liquid materials streams must be dealt with. Some of the means to keeping all these streams separate from each other are also discussed.

INTRODUCTION

Much of the discussion about fuel cells so far has been about individual parts of the cells themselves or about overall systems. This chapter will follow an entire system from start to finish to better explain how energy is changed from chemical to electrical, what is needed to make that change, and how the change is controlled to produce a stable supply of power. Recall the general flow mentioned in previous chapters: input fuel gas, reforming, cleaning, fuel cell, and power conversion. It seems relatively straightforward, and in fact is pretty reasonable when compared to nuclear power plants, but we have managed to ignore a number of systems and items that must be included for fuel cell systems to operate. This chapter is meant to fill in the smaller details and close out the discussion of fuel cells as operating systems.

PRODUCING POWER

Start with gas and end with different gas and some electricity. As discussed several times in previous chapters, input fuel gas can be methanol, gasoline, natural gas, or even methane from landfills. It is not the gas that is important so much as the amount of hydrogen in the gas and how you free that hydrogen from everything else that might be in your fuel such as carbon or sulfur. The chemical and fuel parts of the input gas have been presented but not the physical and mechanical requirement of gas itself. How do you distribute a gas, move it from one place to another, or make electricity out of it? Blowers, turbines, ejectors, compressor, heat exchangers, combustion chambers, and preheaters do something to gas to set it in a particular form so we can extract energy. What kind of energy is that and why do we use that type of energy but not a different type for this bit of work? What type of material is available to move that energy from one place to another and will that material be suitable for connecting to another type of material if we need to change the form of energy we are dealing with? How do the balances for materials and energy add up so we can sell the final product for a reasonable cost?

These are the types of issues designers, engineers, and technicians deal with on a day to day basis. Understanding some of the basic concepts involved in the equipment being used is important but so too is having a solid background in how solids, liquids, and gases differ as well as how they might be the same. This chapter is meant to introduce the student to some of the basic concepts and workings of material movements within the cell, stack, and balance-of-plant equipment used in fuel cell systems.

Pressure

There are all kinds of **pressure**. Pressure in fuel cells can be from **gas**, **liquids**, or **solids**; it can come from weight or how many gas molecules are in a certain volume or even the force exerted by a liquid as the liquid turns a corner. We will start with simple solids. Something that weighs one pound and has a footprint of one square inch will exert one **pound per square inch** of pressure on whatever it sits on. If it weighs two pounds but still has a one square inch footprint or if it weighs one pound but has a footprint of only one-half a square inch, then it will exert 2 psi of pressure on whatever it sits on. That sounds simple but it is misleading. The piece has to be exactly flat and whatever it sits on has to be exactly flat too, but exactly flat is rare in the real world, almost impossible to achieve. In fact, every surface has variation so that how things touch on a very small scale is a matter of valleys, bumps, holes, and ridges lining up. In large things like foundations or bridge abutments, these small things can be ignored to a great extent but the smaller a piece is, the more important it becomes.

Fuel cells are small and the way they are manufactured requires mechanical pressures to be carefully considered when operating and maintaining them. In SOFC stacks, for instance, the seals between the cells are glass based and won't set properly until they flow under temperature and pressure to fill all the little holes and cracks. If the stack bolt set pressure is too high or too low or uneven, then the seals may not work properly. If personnel adjust internal parts incorrectly, they can cause them to flow a bit while in service or even crack them. This is one of the important considerations that distinguish fuel cells from tightening metal bolts in an internal combustion engine. Metal **distributes stresses** like weight or bolt plate pressures pretty evenly across itself, even taking into account such things as variation in topography. A ceramic solid oxide fuel cell plate will not distribute stress the same way because of the way the atoms and molecules are connected (covalent bonding versus metallic bonding). You can view this as the difference between welding something and bolting something. Welds become part of the metals they join and pass stress just like the metals do but bolting stops the stress at the bolt. Bolts or rivets are set in metal structures so that the engineers know where the stress will go and how much it will be whereas welded joints just pass it on until it finds a weak spot somewhere else. Applying the correct **mechanical pressures** called out in fuel cell installations and maintenance is critical because the efficiency of the system requires sealed units but most fuel cell types are not well suited to robust and reliable sealing methods.

Mechanical pressures are critical but there are other pressures as well. Since fuel cells use solids, liquids and gasses, the pressure in liquids is as important as that of solids. Some cells use liquid electrolytes (MCFC) so that the pressure exerted by the liquid on the solid parts of the system is as important as that exerted by the solid components on themselves. Liquids are like solids in that they will not compress under pressure (or at least compress very little) and they pass stress just like solid, through until some **stress riser** (bolt or rivet plates are intentionally set stress risers) intercepts the flow of stress. Force moving in a structure can be viewed like liquid flowing in a pipe, where the liquid flows along the straight lines without much trouble and only causes a bit of wear as it moves on. When the liquid encounters something though, such as an elbow or valve, the water exerts force when it must turn and it is that change in the **direction of flow** which causes trouble. Figure 12-1 shows how stress flows within a solid and how it can move and concentrate much like a fluid. Where the lines are compressed means the force stays the same but is distributed over a smaller area, increasing the pound per square inch seen at the stress riser (the amount of solid may also be physically reduced if the stress riser is a notch).

Liquids coexist with metals and other than some chemical reactions (water causing rust for instance), they mostly exist without much interaction (they don't really mix together or form new substances). It is the movement of liquids that most affects solids. Think of molten carbonate flowing in a very thin channel into and out of a fuel cell, through a stack, and around a pumping and cleaning system. If the liquid pressure is too great, the force exerted as it flows around all the angles in the system can wear material faster than designs call for and can exert force against very small components that is too much for them to handle. If too little pressure is in the pumping cycle, flow does not occur evenly and nonuniform electrolyte builds up in the cell, affecting performance.

There is one other major issue when dealing with liquids in fuel cells: **gas pressure**. In MCFC systems, the molten electrolyte has to be in contact with the electrodes or ions and electrons will not flow to produce electricity. If the gas pressure is greater than the liquid pressure, the hydrogen or the air (oxygen) will push the electrolyte away from the electrode, stopping the reaction. If the liquid pressure is

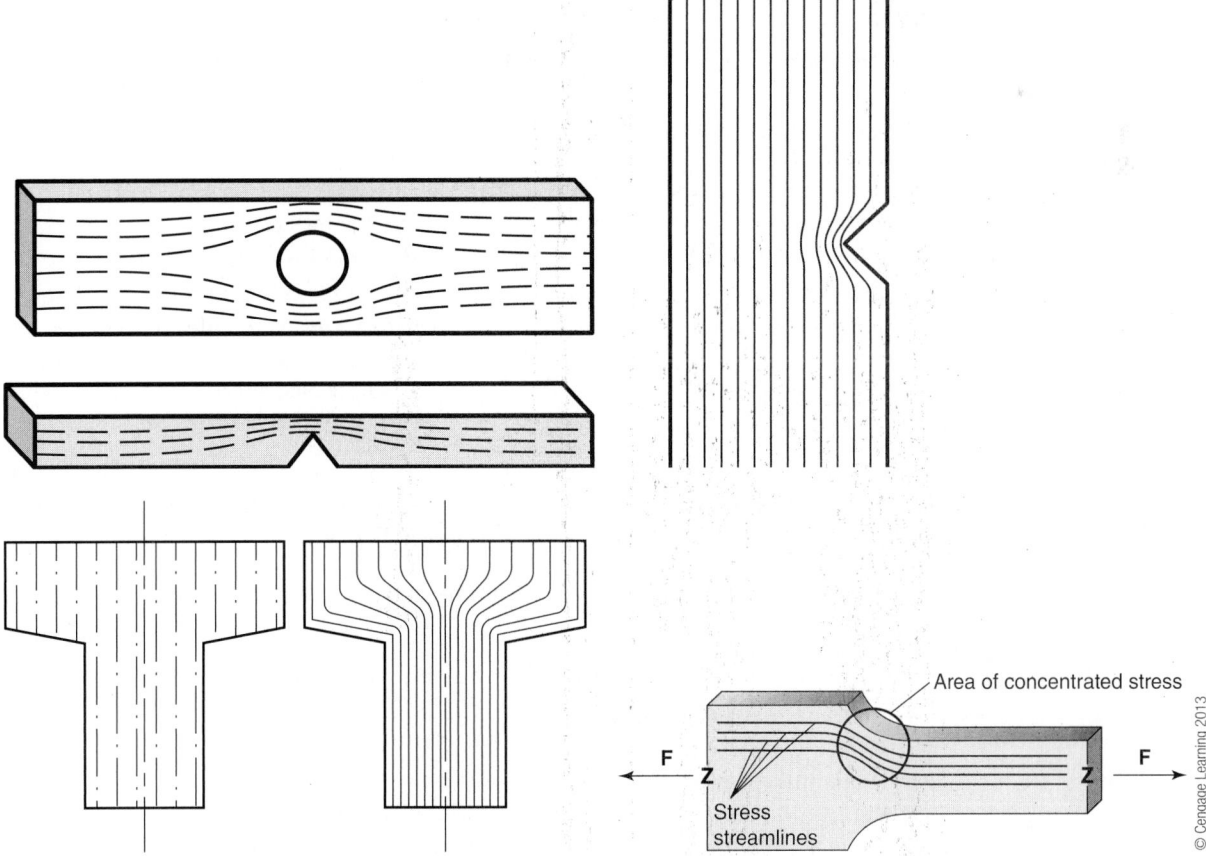

■ **Figure 12-1** Stress concentrating at a stress riser.

too great, then it will push the electrolyte too far into the electrode, flooding the part of the electrode with the catalyst on it and keeping the fuel gas from coming in contact with the catalyst; the reaction will stop.

There are other considerations throughout the system as well. The water pressure feeding the steam system determines to a great extent the amount of water still in the steam or the treatment of steam as it changes back into water can affect the humidity within the cells. Controlling the water pressure systems in fuel cells is an important aspect of controlling the system itself.

The most important pressure in fuel cells is not solid or liquid pressure but gas pressures. Two things are usually mentioned when gas pressures are discussed, pressure and flow. Both pressure and flow are important to the operations of fuel cells and both are complicated by the odd requirements of these systems. We will discuss these two properties but first we will talk about what a gas is. Solids have atoms or molecules bonded chemically together and set in place over long distances (at least if you are an atom); liquids have atoms and molecules bonded chemically together and set in place but tied together only over short distances. Moving any one atom or molecule (or a small group of them) is very difficult (not impossible just very difficult) in solids but moving individuals or small groups in a liquid is relatively easy. The relatively easy movement of small groups or individuals in a liquid is called **flow**. Metals can flow like liquid but it takes higher temperatures or pressures. Liquids have one other defining characteristic in the way they respond to **shear stresses**. Shear stresses are forces that oppose each other. Put your two palms together with one thumb pointed up and the other pointed down and then drive your fingertips toward your elbows; that is shear stress and if you do it hard enough, your palms will get hot from the **friction** of the two solid surfaces moving against each other. If you take a solid block and exert a shear stress on it (push in opposite directions from either end), the solid block resists the shear stress. Push hard enough and eventually the block will tear apart. Try that on a liquid and the liquid just moves as the small groups and individual liquid molecules flow. Figure 12-2 shows some different types of stresses such as tensile, compressive, and shear.

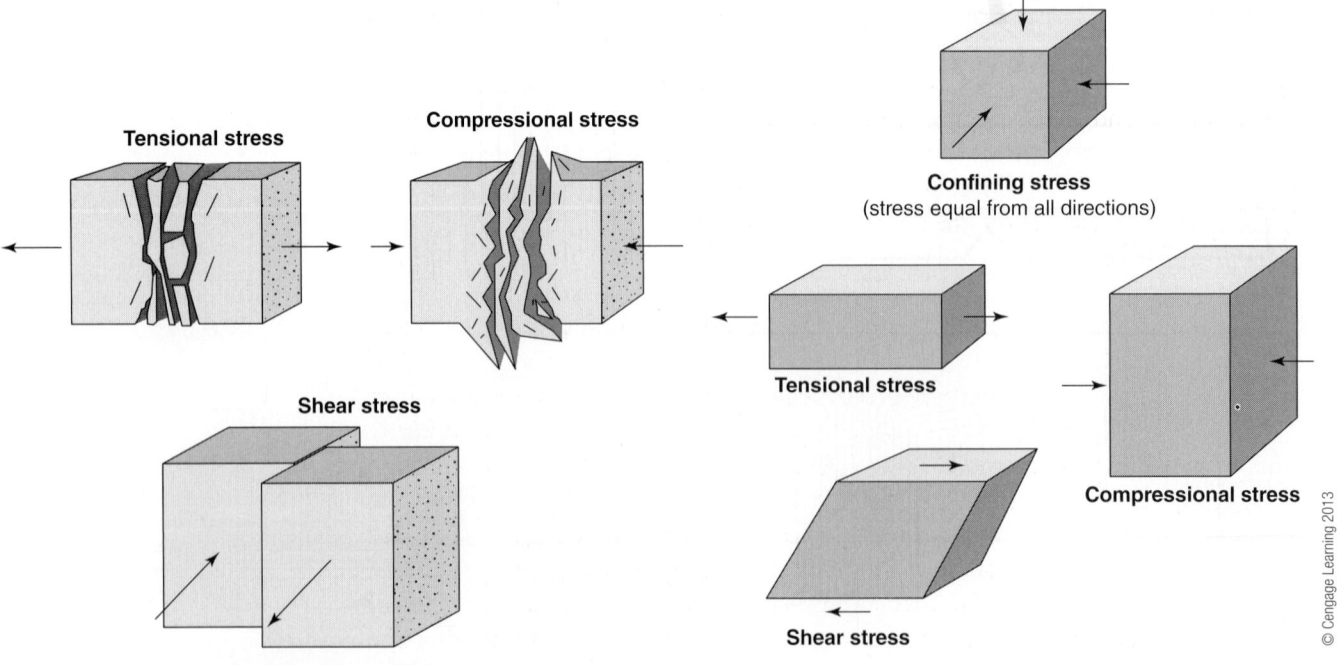

■ **Figure 12-2** Different types of stress.

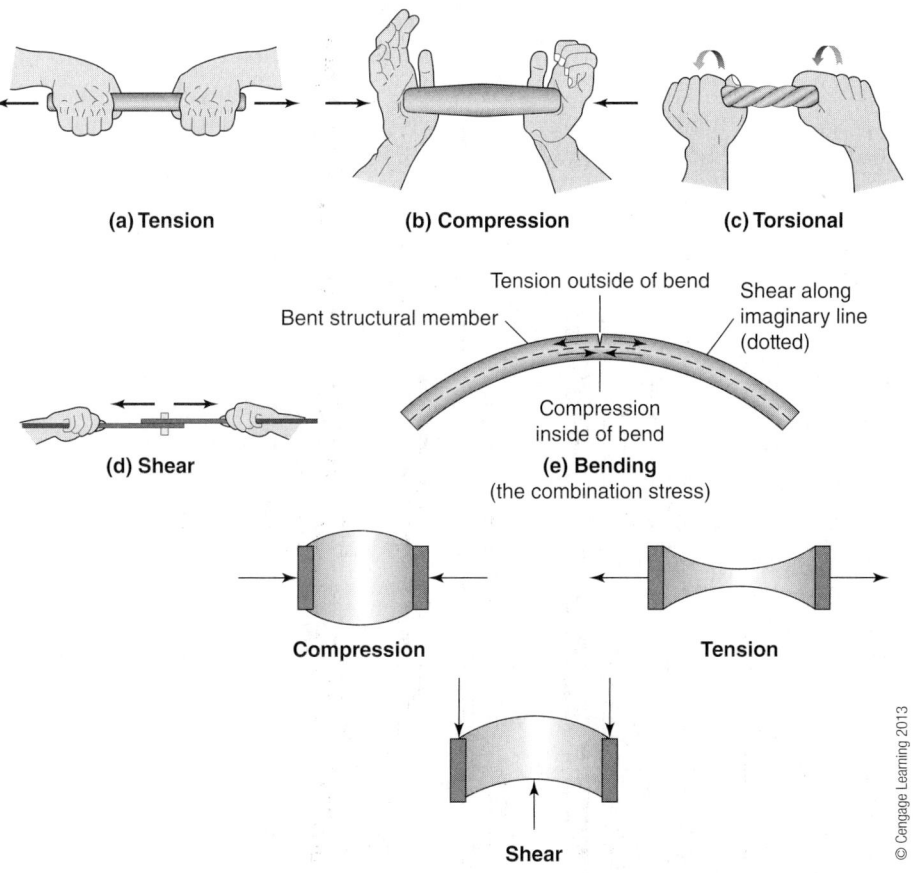

Figure 12-2 (Continued)

Some of the primary differences between solids, liquids, and gases are in those two concepts. Gas atoms or molecules are not bonded together at all but are separate and separated from each other and force exerted on a gas has little effect. You can easily distinguish the difference between the three using those principles. Walk through a gas (air works nicely) and not much happens as the individual gas molecules or atoms just slide around you or are pushed out of the way. You won't feel much until you start to go faster (stick your hand out of the window of a moving car) because more molecules are hitting you and you are going faster when you hit them. Walk through liquid (water in a pool works nicely) and it takes considerably more effort because you have to exert substantially more force to move the water molecules (cause them to flow around you). Walk through any solid; you either move it or it moves you. Thus, the important concept of fuel cells is that to move the solids in a fuel cell, you move the fuel cell, but you can move the liquid (harder) or gas (easier) within the fuel cell.

Gas pressures are stated in units like pounds per square inch, just as a force in the solid world, but the two are a bit different. As discussed above, for solids weight per unit area is psi but gases don't really work that way. Solids exert a force on this planet because gravity is trying to suck them into the center of the earth. There is **pounds force** and **pounds weight** and the two are not the same; a pound force is a pound of weight multiplied by the force of gravity trying to move it. A pound of liquid will exert the same force as pound of solid by the way. Gas does not work the same way as liquids or solids though. Put a pound of any solid or liquid in a balloon and weigh it to find a pound on the scale but put a pound of helium in a balloon and you won't see the scale read a pound, you will see the balloon float away. Gas molecules or atoms are not bound together like solids or liquids and because they are not, they try to get away from each other (like repels like). As they try to get away (they are in constant movement), they go until they hit something, and then they change directions and keep going. Remember from earlier chapters that there are a lot of molecules in even very small amounts of things (trillions of trillion in something roughly the size of your smallest fingertip) so when gas molecules go zinging all around, they hit whatever they are contained in. The force of trillions of trillion of trillions of very, very small molecules hitting the walls of our balloon is what causes the balloon to inflate or in other words, builds gas pressure within the balloon. Put more molecules in to bang against the sides more often and you get higher gas pressures; put less molecules in and you get less gas pressure.

There are many ways to express pressure. One psi is 0.68 standard atmospheres (or bars) of pressure but you can only use that to describe a gas pressure. One psi is 2.31 foot-of-head, but only if you use it to describe water pressure. One psi is 6895 newtons per square meter (or pascals) but only if you are talking force. Given the international nature of fuel cells, students are urged to familiarize themselves with the many terms used in the industry.

Flow is often stated in such terms as liters per minute or gallons per second, but for gas, you have to be careful. **Gas flow** has to be standardized so it makes sense. In essence, you have to state how many molecules are flowing past a point at any given time. Does this sound familiar? It should, and if it does not, then review the definition of an amp given in Chapter 2. This is usually done by using what are called standard liters (if you use the metric system, that is), which has a certain number of gas atoms or molecules in a certain volume. This number implies that the gas is under a certain pressure (the same number of gas molecules hitting the same amount of wall) as well. If you want more gas (individual atoms or molecules) flowing, you can compress the gas or flow more volumes by the same point.

The question then becomes, how do you **"compress" gas** (change pressure)? In most cases, this is done by either changing the volume of the container the gas is in or adding more gas molecules into the container. The opposite is true as well, in that you can lower gas pressure by putting the gas into a larger container or taking some of the molecules out of the container. This does not work for solids or liquids, by the way. Put a solid in a larger container and nothing happens, it is already bonded together and its shape does not change. Put a liquid in a smaller container and it occupies the same volume until the new container is filled, then it just spills out. To change gas pressure in a pipe, change the diameter of the pipe. The gas molecules flowing in the pipe will be forced closer together, repel each other a bit more but hit a smaller area of pipe wall, resulting in higher gas pressure. To get higher pressure in a tank, drive a piston up to mechanically compress the gas then open the piston to the tank, and repeat until the tank pressure finally equals the piston pressure (the gas pressure from the piston will go down at first by a considerable amount as it enters the larger volume of the tank, but after thousands of cycles and thousands of pistons full of gas molecules going into the tank, the two will finally equalize and the piston won't be able to push any more gas molecules in).

There are other ways to do it though, and fuel cells make use of those ways as well. You can convert either a solid or a liquid to a gas in a closed container, for instance. Since there was no gas in the container to begin with, the gas pressure will increase with every gas molecule that is created (provided the container is sealed to stop the molecules from getting out). Boiling water with a lid on does this, and once the gas pressure gets high enough, the lid will start to lift off the pot (exert pressure greater than the force determined by multiplying the weight of the lid by the force of gravity holding it down). You can create gas in a chemical reaction too, and that happens in fuel cells. Let's look at that for a moment because it is very important. In AFC systems, the anode reaction is shown in Equation 4 below.

$$2H_2 + 4OH^- \leftrightarrow 4H_2O + 4e^- \quad \text{Eq. 4}$$

The gas pressure for hydrogen (on the left side of the equation) is dependent on two molecules of gas hitting the walls of the container (the fuel cell itself). The hydroxide ion (OH^-) is not in the gas phase but is an ion in the alkaline electrolyte. After the reaction there are four water molecules (H_2O) but as they exist as steam (a gas) because of the temperature involved. Two molecules of gas exerting pressure are now four molecules of gas exerting more pressure; the gas pressure on the upstream side of the anode in an alkaline fuel cell is lower than the gas pressure on the downstream side of the anode in an alkaline fuel cell because there are more gas molecules on the right than on the left. In MCFC systems, the reaction is shown in Equation 8a and the situation is the same, lower pressure on the upstream side than the downstream, but it is even more complicated because where the left has only one gas (hydrogen), the right now has two, water (H_2O) and carbon dioxide (CO_2).

$$H_2 + CO_3^{2-} \leftrightarrow H_2O + CO_2 + 2e^- \quad \text{Eq. 8a}$$

These pressures must be balanced or the lower pressure on the anode side will allow electrolyte to flood the anode. If you generate twice as much pressure on the right side, then you have to equalize the system by taking twice as much gas out as you put in. Wait though. Do not forget the cathode reactions. For MCFC systems, the cathode reaction is (Equation 9):

$$2CO_2 + O_2 + 4e^- \leftrightarrow 2CO_3^{2-} \quad \text{Eq. 9}$$

So on the upstream side of the cathode (coming from the electrolyte), two carbon dioxide (CO_2) gas molecules arrive and at the downstream side comes one oxygen molecule (O_2) with the two meeting at the catalyst site. Again, there are more molecules on one side than the other with the chemical equations having to balance as well as the gas pressures. Notice, however, that the water is now gone. That is one way to balance pressures, by using the water gas molecules (steam) as the balancing agent. The hydrogen gas pressure fed to the anode is measured along with the air (oxygen) gas pressure fed to the cathode with the steam molecules generated being drawn off so the pressures are kept in line. There is another simple way as well, by changing the chamber volumes of the electrode and electrolyte. If the left side has a lower pressure than the right, you can have a larger volume on the right so the greater number of gas molecules are just in a larger chamber, effectively reducing the pressure. Search for MCFC system piping and instrumentation diagrams on the Internet and look for pressure control loops or chamber sizes as well as pressure drops. You will find several different kinds of examples showing how pressure is handled, with pressure drops across electrodes running in the 300 millibar (4.3 psi) range.

The input pressures are generated by tanks with regulators or piston pumps or turbine blades compressing gas as they spin or any other mechanical system used to either decrease volume or increase the number of gas molecules in a set volume with those systems matched to the gas generated within the fuel cell itself. There are also blowers used to supply a sufficient number of gas molecules to a **compressor**, **recycle gas loops** to build pressures in steps, and **heat-aided systems** where the expansion of gas due to heat is used to build pressure in pretreatment steps (hot exhaust gas can drive a small turbine which in turn drives a compressor used on the input fuel gas stream, a method commonly called **turbocharging**). A simple turbine with compressor is shown in Figure 12-3.

Smaller systems producing less than 5–10 kW may use a blower or just rely on volume changes and heat to balance pressure and mass in the system. Much like ICE systems though, higher pressures mean higher efficiencies (diesels get better gas mileage than gasoline engines in part because of this) so many smaller higher temperature systems make the trade-off in initial cost for cheaper kilowatts produced over the life of the system. Higher pressure can also mean smaller designs, since more gas molecules are supplied to the electrodes in any given block of time.

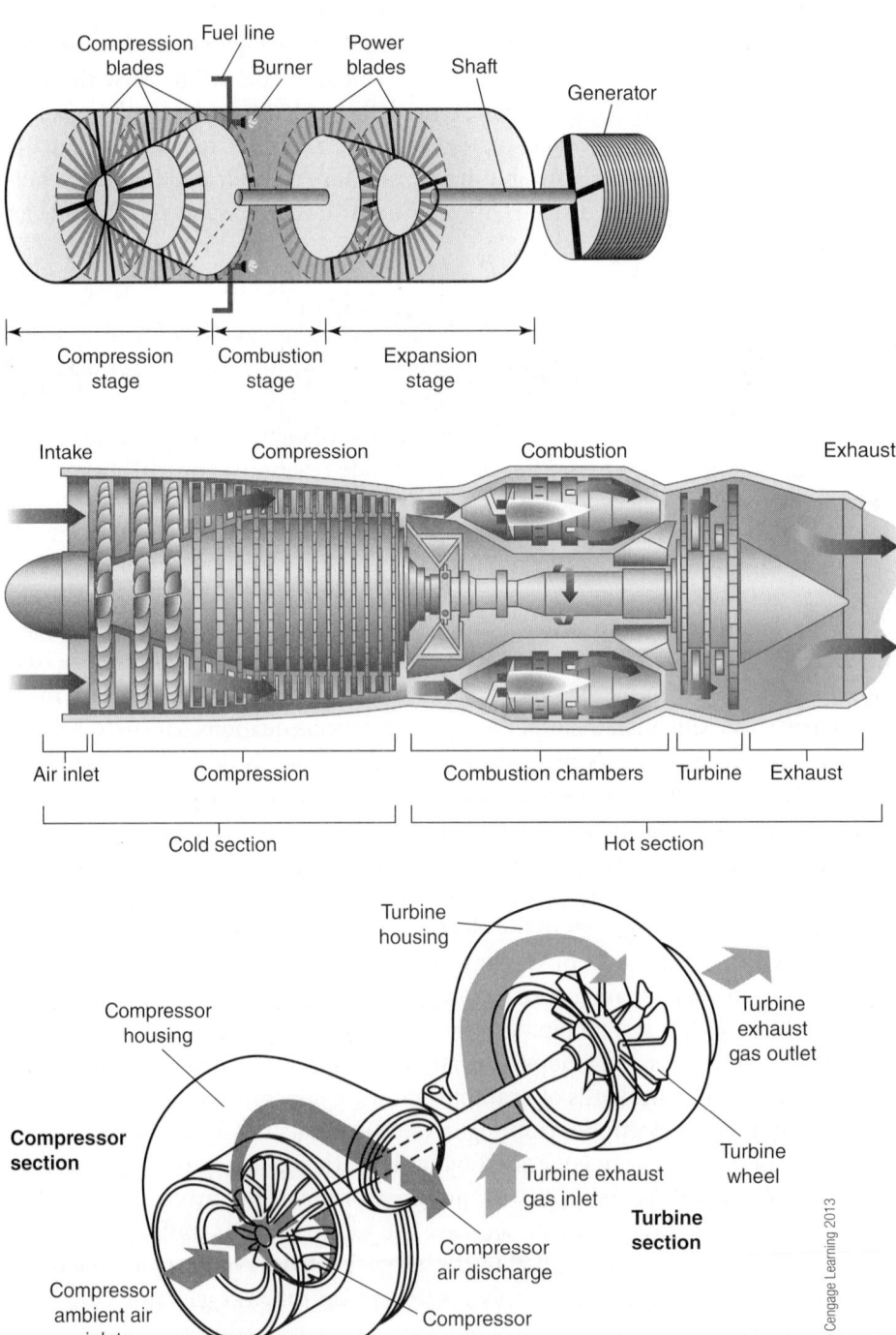

Figure 12-3 Simple turbine.

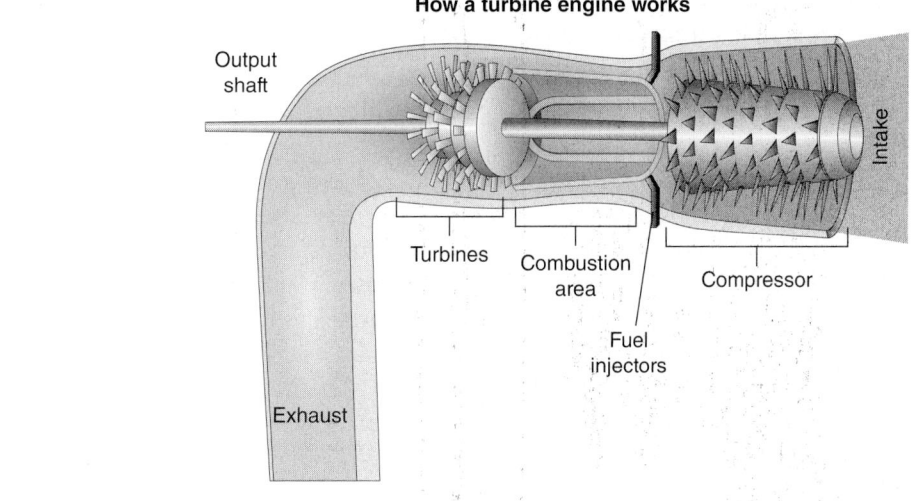

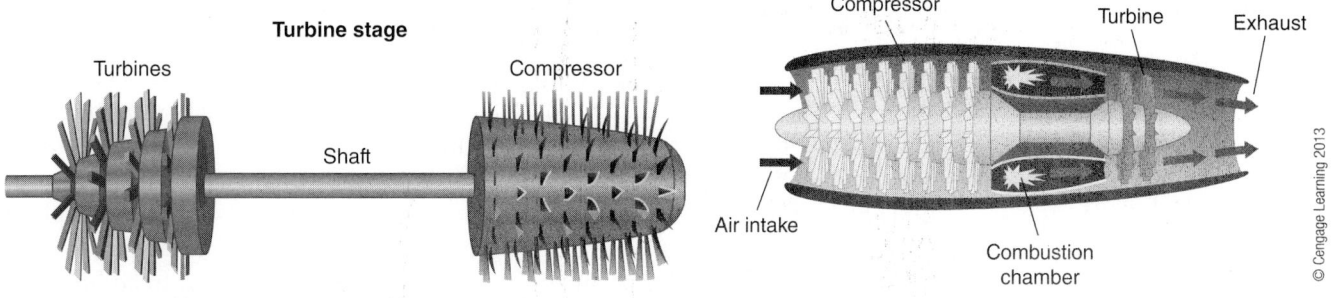

■ **Figure 12-3** (Continued)

Each add-on raises the cost not only of the equipment, but of the control system and of operations and maintenance. In high temperature systems, where heat is readily available, secondary systems like turbochargers can essentially run for free, getting the energy needed from the heat produced by the chemical reactions but in lower temperature systems, the energy usually comes from the cell itself in the form of electricity, reducing the output current accordingly. Some of these **parasitic losses** can run up to 20% of rated output and the gas handling systems are one of the main consumers of power. Even systems that use compressed tanks must pay for gas pressure, since compressed gases are not cheap.

Whether systems use elevated gas pressures within the fuel cell is many times dependent on how the fuel is obtained. External reformation is done at elevated temperatures and most of these systems will then use the energy supplied by the exothermic reformation and shift reactions to drive compression systems. Keep in mind that if a high temperature reformation system is used with a low temperature fuel cell system, then using temperature to compress gas means that the stream must be cooled down before it enters the actual fuel cell, a process that alters pressures yet again. In any case, there is always a point of diminishing returns where more pressure causes more problems in other places, such as humidification requirements (high pressure gas is always hotter and this changes humidification requirements) or equalizing the supply to the electrode catalyst sites (high pressure gas might not stay at a catalyst site long enough for a reaction to occur) or needing even higher pressures to drive the gas through narrower channels resulting from making the cells smaller (which requires stronger internal parts) or even corroding the cells faster (in MCFC systems, increasing pressure increase the amount of carbon dioxide which raises the acidity of the electrolyte and corrodes the cathode quicker).

Each of these components is matched in a system so that mass and energy amounts going in equal that going out. Gas molecules are small, but they each weigh a certain amount and it takes energy to move them as well as pack them in tighter (increase gas pressure). Good designs will use these balances to predict the behavior of systems in a number of different circumstances to maximize the efficiency and thus minimize the costs, many times using advanced **computer modeling**. Since the actual fuel cell itself is relatively small, balancing mass and energy can be difficult, even on a basic scale such as specifying associated equipment. Optimizing a pressure design so that the size of the system is minimized may result in having to use blowers or compressors that are small and specialized as well and thus very expensive. As mentioned, smaller systems may just eliminate the entire issue by not pressurizing at all, and even large systems tend to operate at midlevel pressures. A 1000 kilowatt MCFC systems might run around 3 atmospheres of internal pressure if it uses external reformation while internal reformation systems generally run without pressurization.

Fuel Distribution

Fuel cells will not work unless gas (hydrogen or carbon monoxide for fuel and oxygen) is consistently delivered to the catalyst sites on the electrodes. It is a bit hard to visualize this, but consider it along the lines of a stadium and hot dogs. There are one hundred thousand spectators in the stands and they will be eating hot dogs continually throughout a 4 hour event. Each person can eat one hot dog every couple of minutes and all are allergic to mustard but not ketchup, so any mustard-containing hot dogs will disrupt the system and must be removed beforehand. In addition, all the empty wrappers must be removed before the next hot dog can be handed off. Ignoring the facilities needed to cook that many hot dogs (produce hydrogen gas), the question of importance becomes how to move the three million hot dogs an hour in a steady stream to individuals while also picking up their garbage.

Fortunately, gas is not hot dogs so that the individual molecules will line up naturally and move to catalysts sites one after the other. That is one of the hallmarks of a gas, that molecules are in constant motion and will fill any space they have access to. Of course, not all the space is filled like in a solid or liquid since the molecules are each separate, but the odds of a molecule occupying any particular space at some time are very good. Unfortunately, a couple of million molecules are not what is required, but somewhere closer to a couple of million trillion molecules per second, supplied constantly over years. Designs for fuel cell electrodes all mention one thing prominently and that is **porosity**. The electrodes will allow gas to flow and since gas will move to occupy the empty space in the electrode, part of the problem is already solved. Gas moves into the electrode space and reacts at the catalyst site. Electrons move through the internal wiring and the ions left after electrons are removed transfer into the electrolyte, opening up the space around the catalyst for another gas molecule to take its place. Chemical reactions occur very quickly so the trillions of reactions that must take place every second can happen provided enough catalyst sites are available. Millions of trillions of reactions require a comparable number of catalyst sites be available and that is one important aspects of fuel distribution at electrodes. The electrodes have to be porous for gas to move but a huge number of catalyst sites are also needed so each site should be big enough for a gas molecule to find but not so big that it blocks the interior pore channels of the electrode. The catalyst sites are not six thousandths of an inch in diameter or half a millimeter but only several hundred molecules if even that, with the smallest clusters now at around 10 molecules. The process used to set catalysts must be capable of putting down individual sites that contain enough catalyst molecules to guarantee reasonable odds for a gas molecule to contact the site but not so many that the cost is too high or pores are blocked. This is particularly critical in low temperature cells where precious metal catalysts are used.

With the electrodes arranged for gas flow and trillions upon trillions of available reaction sites, all that is left is to get fuel gas to each cell in a stack and to each stack in a system. Gas flows best in sealed channels like pipes. Pipes small enough to go into individual fuel cells are too expensive to produce so channels are used instead. Interconnects and bipolar plates have been mentioned several times, and these are the names used in fuel cells for **gas diffusion plates** that act as **electron channels** as well (**conductors**). Gas will move no matter what, so all that is really needed is to provide a path. The supply system can also assist in gas distribution by pushing the gas molecules (pressurizing the system) but this is a matter of speeding movement up rather than getting a better distribution. The critical part is in providing pathways for gas to move (**diffusion**). If gas is not pressurized, it will flow more or less evenly through whatever paths are available, since no particular one will be preferred (the **path of least resistance**) but pressure can set up preferred paths, especially in electrodes, where the pore structures are not all uniform but more random. In nonpressurized systems, gas flow can be blocked if the channels are small enough in comparison to the initial chamber or pipe where the gas enters the diffusion plate. A very large volume leading to many channel openings that are much smaller can cause trouble in a several different ways, such as causing pressure differentials (large leading to small increases pressure since it moves the gas molecules closer together so more of them hit the channel walls) or if the channels are very small, the gas molecules have to find the opening and move in, bouncing of the walls as they do and slowing down considerably so that flow is restricted. Pressurized systems speed things up but can also cause **back-pressures** to build up which can restrict flow, especially if channels are small, such as in the electrode itself. This occurs essentially because gas molecules are forced together and then forced to move in unison (this is not good for gas molecules since they repel each other) so that when they come to a change in diameter in the flow channel, they can back up like a traffic jam. Some of the designs used in interconnects and bipolar plates can be seen in Figure 12-4.

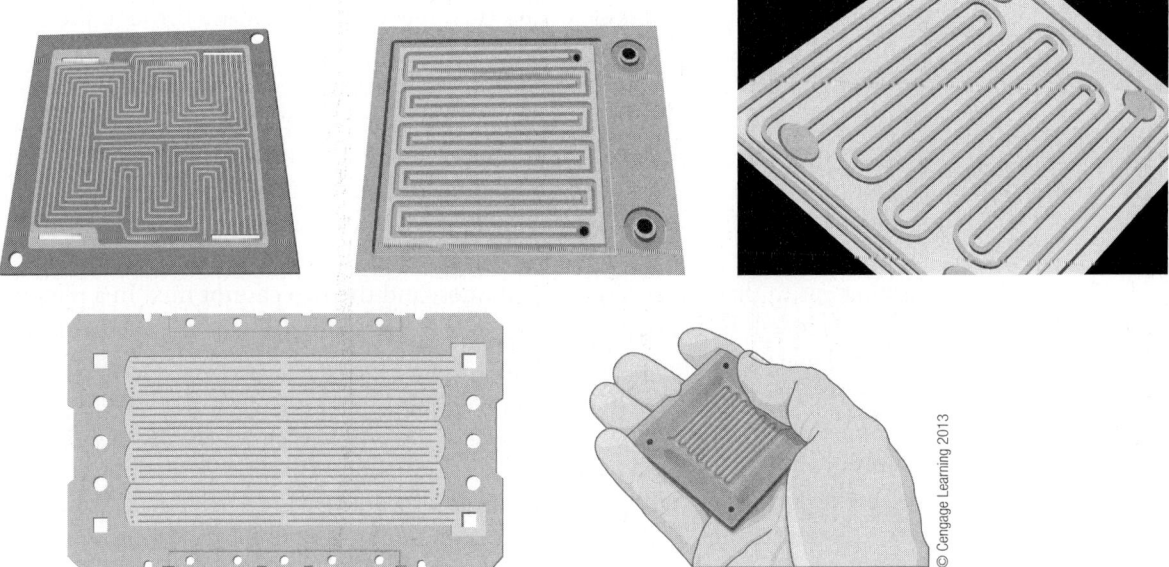

■ **Figure 12-4** Interconnects and bipolar plates.

Designing the fuel gas delivery system is done much like designing a road. How many molecules will be moving per unit of time is considered, as is the size of the molecule (hydrogen is very small) along with how much space it needs around it (since molecules of like kind repel each other, they can only get so close before they are forced back apart). The drag along the sidewalls is determined, since this causes wear as well as reduced velocity along the walls and adjacent to them. As molecules hit the wall, they lose energy which translates into slower molecules. This reaches out several layers, since other molecules will also "hit" the slower molecules (not actually touch but come close, lose energy, and then change directions), slowing down themselves. The center flow will go the fastest, with flow just along the wall going slowest, much like cars in lanes moving at different speeds. This is called **laminar flow**, while **turbulent flow** occurs when the width of the flow channel is so narrow that all the moving molecules are either bouncing off the walls or each other and flow is random. Turbulent flow in fuel cells is very bad so the channels have to be matched to the rate of speed (pressure) as well as the size of the molecule. The higher the pressure, the wider the channels have to be.

Even flow has to be supplied to the diffusers (interconnects/bipolar plates and sometimes called **flow field plates** as well) which in turn have to move the gas at a steady rate to the catalyst sites. There are a number of ways to do this but we will only look at a couple of simpler systems. In one system, the gas is piped to each diffuser plate for distribution and in the other the gas is fed to a plenum with outlets to each diffuser. Keep in mind that the diffuser plates must also conduct electricity so they must have enough physical contacts with the electrodes in order to provide paths for the electrons to move. This is a critical point for plates; many of the actual sites will not have direct contact to an electron conducting path since there are so many catalyst sites and there must be open space around enough of these sites to move gas. It was mentioned above that catalyst sites must be available for hydrogen but cannot be so big that they block gas flow but they must also provide a path out for the electrons to move into the wiring system. If the sites are set predominantly along the gas flow paths, then they won't have ready access to the electron flow channels, and if they are set to maximize access to electron flow channels, then they won't have easy access to gas. However, the electrodes are porous, so gas can and will flow through them, while conduction paths require an initial physical **contact site** and a complete contact path out to the wiring. It is usually better to guarantee good electrical contact and then rely on gas to diffuse through the electrodes. Figure 12-5 shows simplified versions of internal and external manifolding.

As discussed in previous chapters, the **Membrane Electrode Assembly** (MEA) consists of a diffuser/conduction plate (interconnect/bipolar), anode, electrolyte, and cathode. The MEA then ties into the next MEA to form a repeating stack that starts with a plate to supply the first anode and ends with the plate supplying the last cathode. Twenty-five individual MEA sets in one stack is a common number seen in literature. One gas has to be supplied to the diffuser for the anodes and another gas to the diffuser for the cathodes, and the two cannot mix. In a plenum system, one will be set on top and bottom to feed into diffuser channels running up and down while another plenum set will be on the two sides and feed diffuser channels running across the cell. At the beginning of the stack, each plenum will have a feed connection (maybe more than one) and at the end of the stack, each plenum will have an excess gas bleed-off connection (used to control pressure as well). This type of system will in essence totally enclose the stack and so must also provide some means to control temperature, such as cooling (or heating) channels on the outside. In a manifold system, a larger flow channel will connect to each of the diffuser plates which in turn supplies each of the diffuser channels. This series

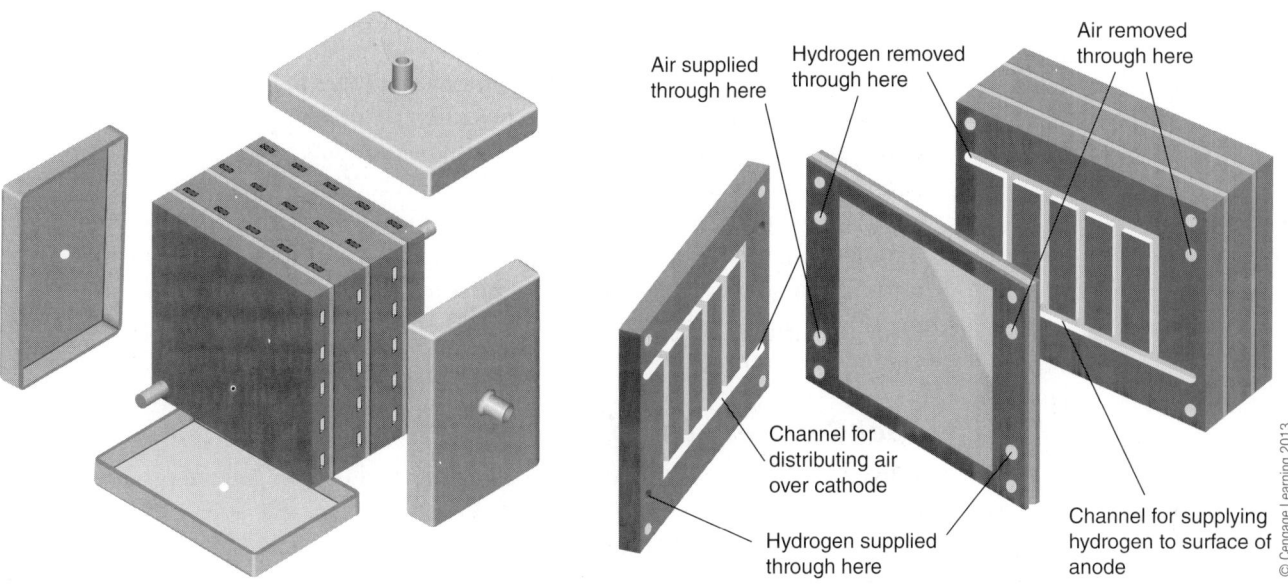

Figure 12-5 Internal versus external manifolding.

type of connection from the first channel to last channel is more expensive to produce but can provide control characteristic plenums cannot. It too has to include provisions for thermal management, but this can be done on the plate itself by setting channels in to be used exclusively for heating or cooling, usually with a cooling gas that is then recirculated to act as a preheater.

One of the important considerations for stacks is along the edges of the electrodes, since they are porous. If they are not sealed, they will conduct gas out around the edges of the assembly. Each MEA and the stack itself has to be **gastight** to prevent mixing of fuel and oxidant gas and to control pressure. This is further complicated in a cell where the electrolyte is liquid, since that has to be sealed as well. Making fuel cell MEA and stacks gas- or watertight is one of the main problems in producing fuel cell systems. There are probably as many ways to seal fuel cells as there are fuel cells since many manufacturers use proprietary methods. Maintaining these system seals, whether in input or output piping, reformers, or the cells themselves is a major operations and maintenance concern.

Much is made of the fuel and air used in the various reactions, but it is important to remember that the trash is probably more important to the stadium crew than the hot dogs, since the hot dogs remove themselves at the end of the day while the trash does not. Any fuel gas that is not reacted has to be removed, and many systems run excess amounts to provide follow-on fuel to CHP or cogeneration systems. Remember that in chemical reactions, one of the ways to hinder them is to build up too much of one of the parts in the reaction. If you do not remove the products, then the reaction goes backwards and what were supposed to be the products might become reactants. In that case, you can reverse a fuel cell system and then it generally fails catastrophically. There are also parts of the various input streams that do not get involved, like the nitrogen in air that just goes along for the ride, and since air is mostly nitrogen, most of the air introduced just has to be moved back out again. Many contaminants that are present in very small amounts cannot be allowed to build up and must be removed. What goes in must also go out whether it is in the same form or a different one and as much thought and effort has to go into the output channels as is put into the input channels.

Temperature

Fuel cells run in relatively narrow temperature ranges to achieve the best possible efficiency and lowest possible cost per kilowatt produced. In general, AFC systems run below 250°C, MCFC systems run below 650°C, PAFC systems run below 250°C, SOFC systems run below 1000°C, SPFC (PEM) systems run below 100°C, and PAFC systems run below 250°C. While the reactions will occur over a range of a couple of hundred degrees, output will drop and parts of the system will degrade if the temperature is too high, while the reactions may well stop if it is too low, especially in higher temperature systems. **Thermal management** is one of the main operations occurring in fuel cells.

Fuel Cell Heat Balance

While not strictly considered "**heat engines,**" fuel cells in general produce heat as part the reaction cycles. In higher temperature systems like SOFC, the heat generated can be substantial since the system runs at temperature up to 1000°C. In these cases, handling the heat requires equipment but the cost of such equipment can pay for itself by increasing generating efficiency or providing the heat to complimentary systems. In lower temperature systems like SPFC (PEM), the heat generated is not enough to do much with but is enough to disrupt operations if it not managed. Each fuel cell type uses different schemes to manage the heat produced and while a full treatment of these various methods is beyond the scope of this book, some basic issues can be discussed for each system.

In AFC systems, the heat generated is minimal but if the electrolyte is maintained as liquid, it can either cool or heat the inner workings of the cell stack. There are some problems with this type of system since stacks would use a common electrolyte supply and that moving electrolyte would also provide a path for the ions all around the system, which can result in a form of short-circuiting. Since water is produced at the anode of an AFC, cooling can also be done using excess hydrogen fuel, which becomes humidified, is recirculated out to a drying unit where the water is recovered and the hydrogen sent back into the fuel loop. This type of cooling does require a match with the gas pressure system as well as circulating equipment such as an ejector circulator or air pumps. One issue of importance to remember about liquid electrolyte cells is that if the temperature gets too high, the liquid will boil, eliminating its usefulness as an electrolyte and probably damaging the cell from the resulting gas expansion.

In MCFC systems, the high operating temperatures means the heat generated is used in CHP or cogeneration processes. The main issue with these cells is to maintain the operating temperature, an issue made even more critical because the difference between the pore structures in the electrodes and that in the matrix holding the electrolyte is what keeps the electrode gas and electrolyte separate and in place. If the temperature varies too much, the area where the separation is maintained can vary and by moving back and forth even a small amount, cause corrosion or even mechanical damage. The heat generated by the cells is generally used in a **hot zone design**, where the input fuel can be heated to the operating temperatures (MCFC systems can utilize internal reformation) before entering the cell stack, the entire unit maintained at operating temperature with excess heat moved out of the hot zone unit to be used in complimentary processes. In many cases, this system is modified somewhat by using an excess of fuel gas supplied to the anode and then recirculating the hot, unreacted gas to mix with the incoming fuel gas, minimizing the equipment needed to preheat fuel. This type of system does require monitoring and control equipment to be separate with any devices inside the hot zone capable of operating at the elevated temperatures. Once the units come to operating temperatures, thermal management is primarily an issue of removing excess heat, accomplished using heat exchangers, where even an entire hot zone can be used in the exchange.

In PAFC systems, the lower operating temperatures mean that the system has to be cooled. This is generally done using the gas diffusion (bipolar/interconnect) plates where separate channels are arranged within several plates (not all of them) of a stack to provide a flow of either gas or liquid to cool each stack individually. It is important to note that specific stack thermal management schemes can be used in fuel cell systems or more general system schemes. Cooling in many cases is done using water to minimize the size of the channels although this does make the overall system more complicated. Water (or liquid) cooling can be done using either a **boiling water method** or a **pressurized water method**. In the BWM method, water is raised to temperatures 30 or 40 degrees below the operating temperature to prevent thermal shock and boiling within the cell. The hot water does not turn to steam because it is within a fully enclosed and contained system that does not allow either the opportunity or the expansion required for the transformation. A version of this type of behavior can be found in microwaves, where high quality cups can contain water well above the boiling temperature but will not actually boil until they are disturbed, scalding the unfortunate person who does not notice the trouble looming. In PWM methods, the entire system can be pressurized, which raises the temperature needed to boil the liquid (the higher the pressure, the higher the temperature needed to boil a liquid, which is why water boils at lower temperatures in the mountains than at sea level). In both cases, the slightly lower temperature liquid will pull out enough heat to maintain steady operating temperatures as its own temperature is raised. The water is then cooled slightly using a heat exchanger or by just running coils through ambient air. Some cooling may also be done using the fuel gas input, although that is generally done only in smaller systems.

In SOFC systems, the high operating temperatures means the heat generated is used in CHP or cogeneration processes. A SOFC has the advantage of being a solid system needing only gas to be delivered but that tends to make thermal management a bit more challenging. Heat has to be moved through the solid to some edge where it can be removed. Heat does not move through solids very quickly. Liquids and gases can deal with heat in a relatively straightforward way, transferring the energy of heat to the molecules (or atoms) individually, which results in those molecules moving a little more, but since they are already in constant, or near constant, movement, the deal is not so difficult to arrange. In solids, atoms and molecules are pinned in place and cannot really move. That is one of the things that differentiate solids, liquids, and gases, in that melting solids is really just freeing the molecules up to move in the restricted way liquids move (in groups but not individually) and turning a liquid into a gas is really freeing the molecules from one another so they can all move individually. As a result, SOFC systems transfer heat differently than other fuel cell types and so require different means of thermal management. The methods used can be the same to a great extent, such as running channels through the stack gas diffusion plates (interconnects/bipolar plates) but the heat itself will not move as quickly nor distribute within the cell the same way. For SOFC systems operating at the highest temperatures (~1000°C), heat transfer is mostly a matter of moving enough air to bring the waste heat into the associated CHP or cogeneration system, but for systems running at the lower end of the temperature range (~700°C), then a more aggressive system has to be in place to manage the heat, since the materials of construction will not tolerate temperatures much beyond the operating temperature. Liquids may be used in some cases, but given the very high temperatures involved, they are not ideal unless kept under high pressure, adding substantial cost to the systems.

Recovering Heat

Several terms have been used in the preceding chapters such as combined heat and power and cogeneration to describe what is done with the heat generated by fuel cells. One other may be encountered as well, referred to as a **bottoming cycle**, where excess heat is recovered from any heat producing system as part of its process but where the heat itself is not used in the process itself. Most fuel cells make use of bottoming cycles using the heat produced in preceding or following systems rather than within the fuel cell stack itself. External reformation, HVAC, or turbines are all examples of bottoming cycles while preheating incoming gas is an example of internal use.

There are several different ways bottoming heat can be utilized. In the simplest form, any heated air can be used as either a primary heat source or to supplement an existing heating system. Many of the smaller stationary systems designed for residential or small commercial applications make use of this form of heat recovery. It adds a small amount to the overall system efficiency, but because the air itself is not very hot, the increase is minimal. Some of the smaller systems for transport application, particular the auxiliary systems, would also provide either heated or cooled air to mobile platforms while the main internal combustion systems were turned off to limit pollution.

If the temperature of the waste heat is above that needed to produce steam but is more or less at ambient pressure, the heat can be used to either directly produce steam or to preheat water going to a conventional boiler. The steam can then be used in a number of ways within a building or complex, such as laundry services, HVAC systems, or radiant heat loops. The steam can also be used to drive engines such as turbines, although those systems are much more expensive and considerably more complicated to operate and maintain. In mobile applications, steam is of limited use, and even in smaller stationary applications, steam does not fit well into existing infrastructure. In those types of systems, if the heat cannot be used in HVAC type applications, it will probably be recycled internally in preheating cycles or discharged into the atmosphere as waste. Larger systems operating at high temperature will probably all make use of bottoming cycles to not only lower the cost of power produced but to position the products as more versatile and useful than simple boiler/grid connections are.

Turbines are also common bottoming cycles used in follow-on processes, especially when combined with high temperature units. **Turbine systems** are simple enough technology. In a gas turbine, air is first accelerated by a compressor and then slowed in **diffusers** (this combination raises both temperature and pressure of the gas and is often referred to as a turbocharger). The hot, compressed gas is then mixed with a fuel and fed into a combustion chamber (or other heat adding chamber if combustion is not used) that is held at constant pressure. If you raise the temperature of a gas but hold the pressure constant, then the volume has to increase (Pressure, Volume, and Temperature of a gas are related by what is called the Ideal Gas Law $PV = nrT$ where "n" and "r" are constants). The gas is allowed to expand through nozzles aimed at vanes attached to a shaft. As the gas pushed against the vanes, the shaft rotates. If the shaft is attached to an alternator, electricity is generated. All of this can be set together on a single unit or broken up. Combining it all together acts as a control on the turbine, since the cycle of compression combustion and expansion can runaway if there is not some type of limit imposed, usually the amount of fuel supply. In fuel cells, there is no need for either compressor or combustion chamber as the gas is already hot, under pressure and directed as it exits the stack and can go straight to a turbine. In an interesting note, the smaller the turbine, the faster the system runs, with some **microturbines** (the types that would be used in smaller fuel cell systems) running at over half a million **revolutions per minute (RPM)**. Figure 12-6 shows a general microturbine arrangement.

The combined systems mentioned above still retain enough heat in the gas exiting the turbine that it can then be used to generate steam. **Steam turbines** are even simpler and in fact the first one described (and probably built) was by Hero of Alexandria in

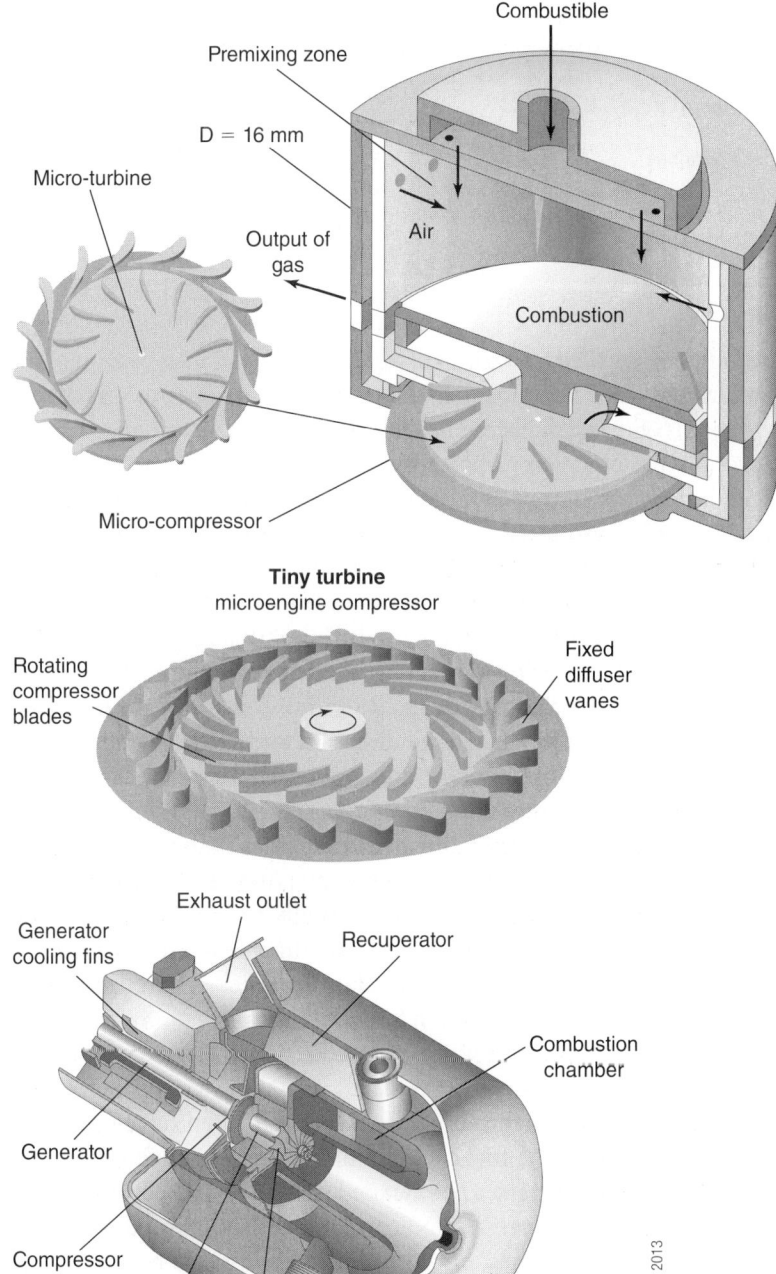

■ **Figure 12-6** Microturbines.

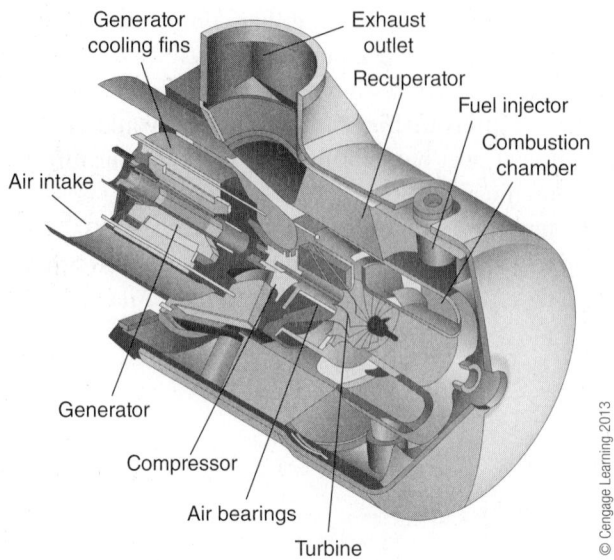

■ **Figure 12-6** (Continued)

ancient Greece, where steam exited a nozzle to drive a paddlewheel. This version of the turbine could not turn the wheel fast enough to do much work, but once the gas stream was aimed along the axis of the shaft instead of at right angles to it, that changed.

There are several types of steam turbines such as condensing, noncondensing, reheat, extraction, induction, impulse and reaction, but they employ roughly the same principles. A rotor is fixed in place to act like the nozzles mentioned above. Further down is another rotor, attached to a shaft which is free to rotate. Steam moves from piping into the main turbine chamber to expand in volume and is then directed at the **fixed rotor**. Much like a gas turbine, this results in directed jets of high velocity steam aimed at the next set of **rotating rotors** to turn the shaft. Many of the variations mentioned above have to do with the shape of the fixed and rotating rotors and how they act to change the direction and velocity of the steam jet or the state of the steam jet (condensed or noncondensed for instance) as it exits the turbines. A simple steam turbine is shown in Figure 12-7.

One issue in particular that is critical in steam turbines is that the steam be "dry." Since water absorbs heat extremely efficiently, it is difficult to transform all of it into a gas. Steam is rated on how much water (usually as very fine droplets) it still contains. Any water in the steam used to power a turbine causes mechanical wear on the turbine rotors and can destroy them relatively quickly. Large industrial steam generators can produce over 1.5 million kilowatts each.

In general, bottoming cycles that produce more electricity increase overall system efficiency the most since they use heat otherwise lost. Taking a larger SOFC system (>2000 kW) as a basic example, the air and fuel (nonreformed since the SOFC does internal reformation within the fuel cell) come into preheaters and then into the fuel cell (the fuel side gas may have to be cleaned first). The DC current generated by the cell goes to a DC/AC converter and conditioner. Both the exhaust gases (from the anode and cathode) feed first into the preheaters used to heat the input gas. Then, since both gas streams have excess amounts of fuel and oxidant (air), they are fed to a common excess fuel afterburner. In a typical SOFC high temperature system, the gas exiting the afterburner would be around 800°C. With an in-line boiler, this temperature is sufficient to produce a good quality dry steam of about 550°C, which drives a steam turbine that turns an alternator to produce more AC

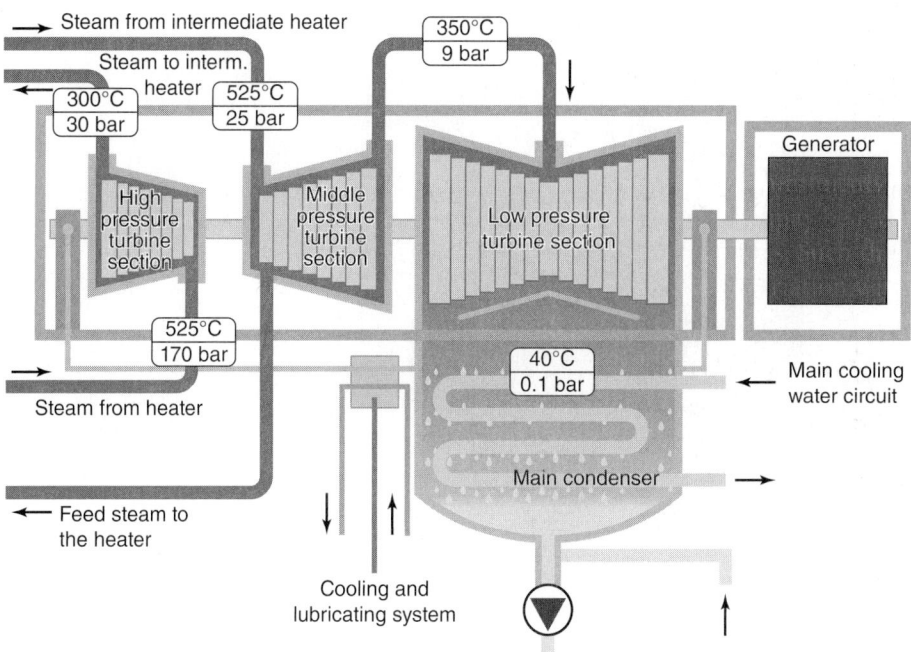
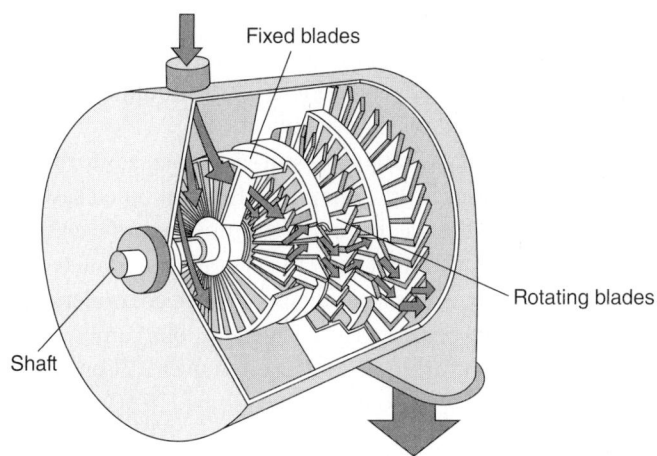
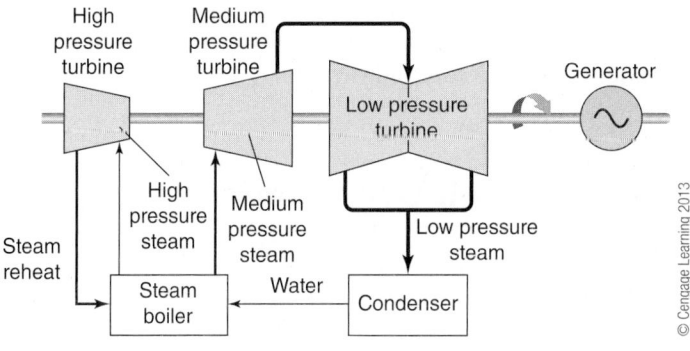

Figure 12-7 Steam turbine.

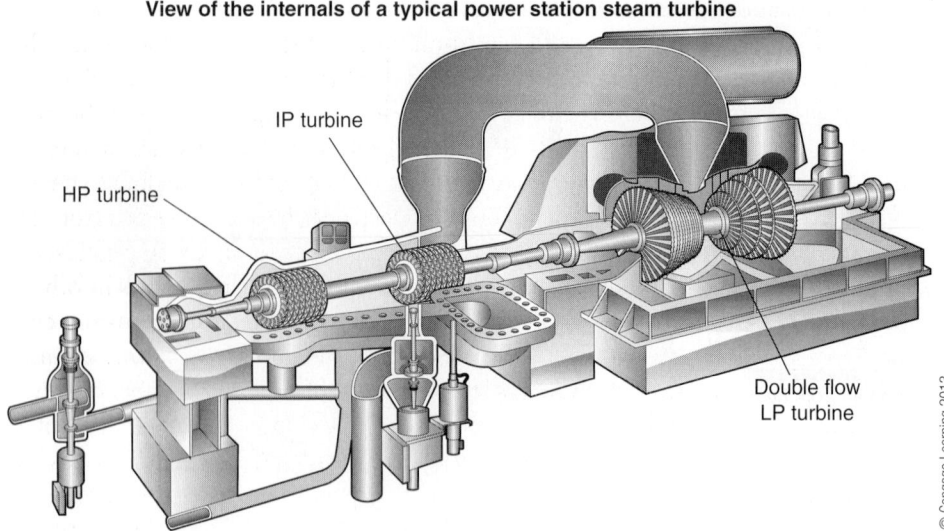

Figure 12-7 (Continued)

current. The exhaust gas from the boiler goes out at just above the boiling temperature of water and can be used with a heat exchanger to provide local hot air or hot water. The excess water from the boiler goes through a conventional condensate circuit recycle system back through the boiler. Thus, in a system like this, more than one bottoming cycle is being used to increase overall efficiency. It should be noted, however, that bottoming cycles are not suited for many applications.

If the system is pressurized (or if output gas can be pressurized), then gas turbines can be used, resulting in substantial increases in efficiency. Ideally, a combined system that generates steam and gas would be used, since that offers the most versatile configuration as well as the most overall efficiency. In that case, the system is much like the one described above but a pressurized gas turbine is used before the steam boiler produces steam for the steam turbine so three separate electric generations can be done (first fuel cell, then gas turbine, then steam turbine). Any heat left over will probably be used either to heat incoming gas or liquids or go to a conventional HVAC system.

Keep in mind that even in high temperature systems, there is a limit to what can be done based not only on the temperatures involved but also the amount of hot air produced and the laws of **heat transfer**. In many cases, hot exhaust heats some other cooler air and is then **exchanged** for the other hot air (using heat exchangers and thus the name). The exchanged air can be used as heat source (exhaust from internal reformation units have carbon and nitrogen compounds that have to be discharged under permit, not cycled through a building), exchanged again for cool air, or used to heat water.

Fuel Cell Stacking and Cell Connections

The most important part of the fuel cell stack and the way it connects together is the bipolar (interconnect) plate. This critical piece of equipment is what allows the electrons produced in the fuel cell reaction to gather and then move into a wiring system to produce electric current. They can take the forms mentioned in the gas diffusion discussion above but many times are separated from this form for more efficient gathering and set between the diffusion plate and the electrode as a mesh, screen or expanded metal. If used as part of the gas diffusion system rather than a screen, they can also supply the fuel and oxygen (air) needed for the reaction and can be used to cool (or heat) the stack, serve as the bolt plate set for holding the cell together, and provide sealing surfaces for the gas seal. The plates differ in each type of fuel cell, mostly because the conditions within the cells themselves are different while the basic designs (except for types like tubular SOFC) are similar.

They also serve as a **sealing surface**, one of the most important, if not the most important, design issues in fuel cells. Sealing fuel cells is a comprehensive issue, not merely a matter of sealing around an edge. Even a partial list shows sealing to be one of the critical issues in fuel cell fabrication. The electrodes have to be sealed from the electrolyte to prevent flooding (or drying out), which require seals keeping solids, liquids, and gases in place for some types of fuel cells and gases from gases in others. Seals must be set in place but also allow "sliding" to accommodate expansion, contraction, and operating movement within the stack. The anode gas has to be sealed from the cathode gas. The input fuel gas has to be sealed from the atmosphere as well as from the fuel used at the anode. The cooler gases have to be kept from the hotter gases, in the fuel cell as well as in reformers and heat exchangers. The steam gas has to be kept from the water in a steam turbine. The expanding gas in a gas turbine has to be kept within the turbine and not allowed to explode out into the atmosphere. Much of the work in designing, operating, and maintaining fuel cells is in making sure everything stays where it should and seals are a critical part of this effort.

In AFC systems, the plates are commonly nickel, stainless steel, or steel that is nickel or even gold plated. Carbon sheet or fleece, milled graphite and even polymers are an alternative, but do require a more complex tie-in to the wiring system than if metal is used. Polymer or carbon based types are much less conductive that metals (gold for instance is somewhere on the order of 45,000 times more conductive than graphite) but since the electrons are produced at separate sites and then move into the conduction medium in relatively small amounts, at least when compared to industrial generation, the issue of conduction is less important than it might be in other systems. The only real criteria for AFC systems are that they be resistant to the alkaline electrolyte and match the coefficient of thermal expansion of any gas diffusion plate if the two are separate. Some designs eliminate current collection layers and gas diffusion layers entirely by moving the electrons to the edge of the MEA and collecting them there, but this does lower system efficiency because the current path is longer and thus offers more **resistance** to electron flow. The trade-off between that loss and the elimination of collector and gas diffusion plates (the entire MEA is porous so it acts as its own gas diffuser) can be acceptable, especially in smaller systems. A typical assembly is shown in Figure 12-8.

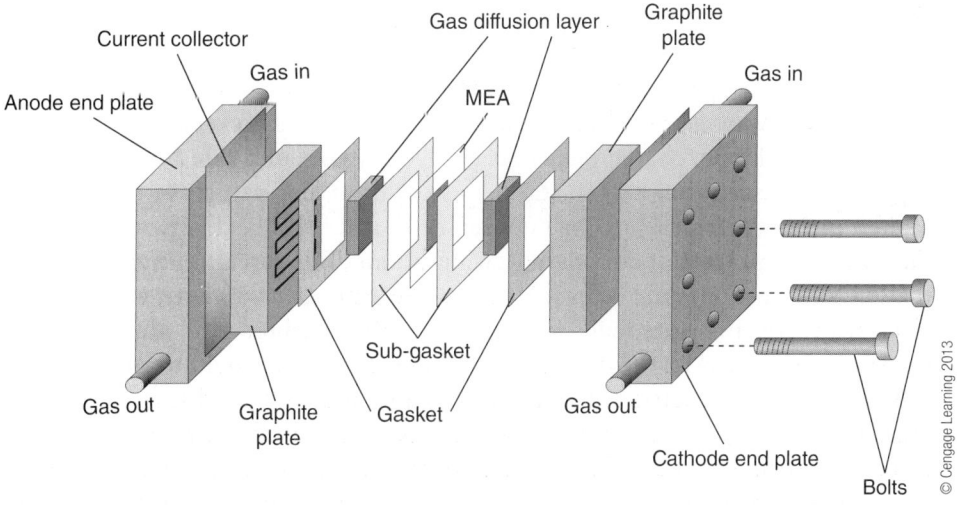

■ **Figure 12-8** Fuel cell assembly.

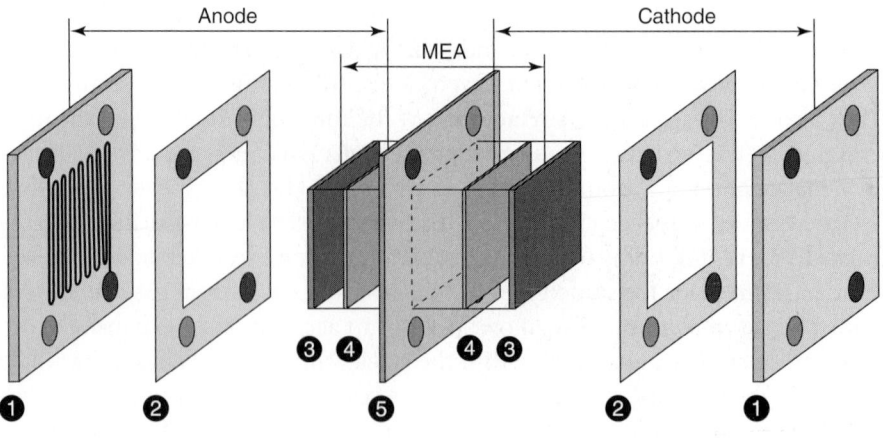

① Bipolar plate (gas flow channel)
② Gasket
③ Electrode backing layer
④ Catalyst layer — Membrane Electrode Assembly (MEA)
⑤ Polymer electrolyte membrane

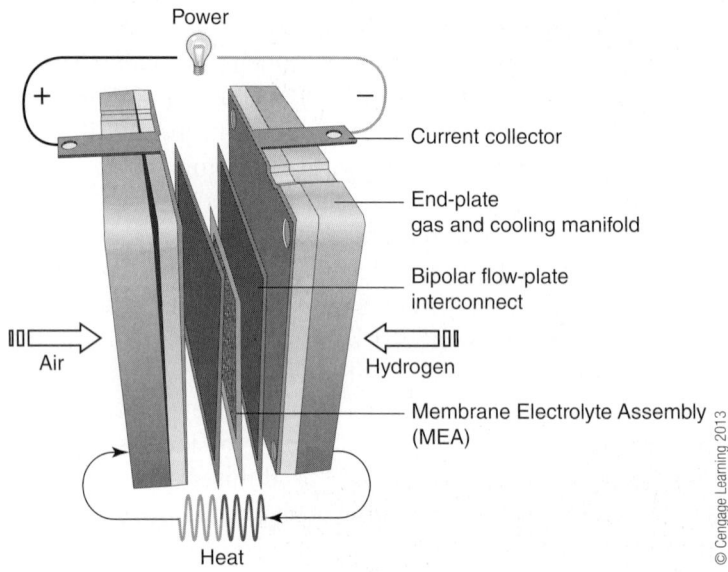

Figure 12-8 (Continued)

In MCFC systems, plates can be stainless steel that is nickel coated. No matter the fuel cell type, coated plates generally only coat one side (that facing the electrode) to save money. Thin stainless steel sheet also can serve to block the molten carbonate from bleeding between cells, since the electrolyte will not wet **porous stainless steel** under operating conditions. This can also improve overall sealing, since any contact between electrolyte and plate will then act as a seal. Any area where the plate will act as a seal is usually coated with aluminum (an oxide coating of $LiAlO_2$ is formed upon contact with the molten carbonate) to prevent corrosion between the electrolyte and the stainless steel. Common stainless steels are relatively cheap but only marginally acceptable at MCFC operating temperatures and in the atmosphere within the cell, so some designs will use more expensive, high temperature alloys such as Inconel® or Hastelloy® that are

CHAPTER 12 • FUEL CELL SYSTEMS: POWER AND CONTROL 263

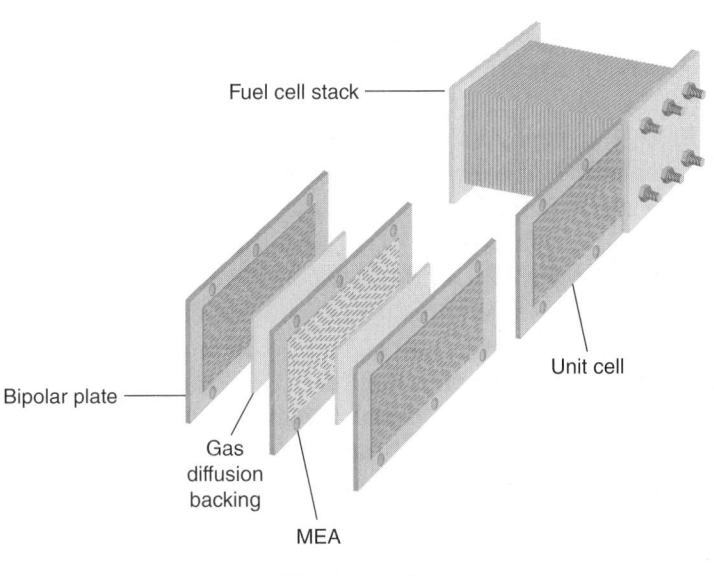

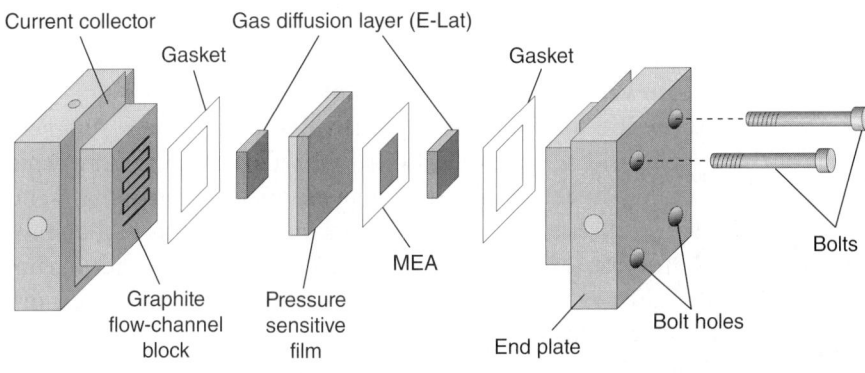

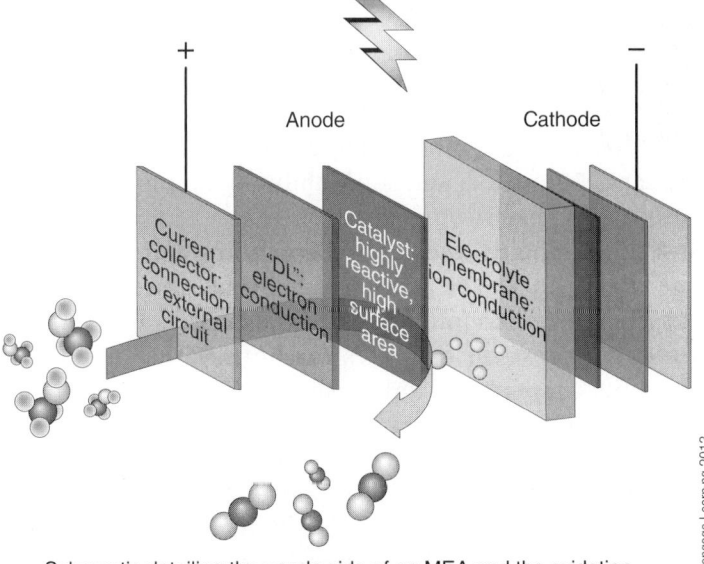

Schematic detailing the anode side of an MEA and the oxidation half-reaction of a direct-methanol fuel cell (waters omitted).

■ **Figure 12-8** (Continued)

also coated with nickel where a seal involving the electrolyte is needed. One of the primary issues with these plates is the degradation of the oxide coating at the sealing interface over time.

In PAFC systems, the plates are graphite, since that material better withstands the working environment. There are several different types of graphite plates, ranging from machined solid graphite to layered hybrid systems, and each has strengths and weaknesses that must be considered based on the size of the fuel cell system, the type of fuel, the internal pressure and temperature of the system, and the cost of the overall system. Originally, most PAFC systems used machined graphite. This approach is effective and easy to deal with but not very cheap since machining is costly, especially for small, tightly controlled pieces of relatively expensive material. Machining can produce parts that have consistent size, reliable channeling and perform well in the corrosive acid atmosphere of the PAFC but mass production is difficult. Newer multicomponent designs use several layers to make the plates, allowing additional design criteria to be used when producing the item. The base layer is a thin carbon plate that acts as a gas separation membrane between the individual fuel cells in the stack. The base carbon membrane plate in a PAFC cell serves to keep gases from mixing but it also serves as a solid platform to work from when building the rest of the plate itself. It provides structural stability, mechanical strength, and chemical resistance to the assembly. On that base is set the channels to direct gas flow to the electrodes. These are made of porous graphite and resin mix that are formed and then heated in a furnace that has a carbon containing gas at very high temperatures of around 2700°C. This process is called **carburization** and is done in controlled atmosphere furnaces since most of the gasses used are hydrocarbons that will combust (very rapid combustion, usually called an explosion) if oxygen enters the furnace. It forms a thin layer of carbon on the surface and provides both strength and chemical resistance to the piece; machining tools or chisels are many times carburized so that the outer surface is hard but the interior is softer and can absorb stress without shattering the entire piece. Recall from earlier chapters that one of the issues with PAFC cells is that the phosphoric acid (H_3PO_4) will vaporize over time so that some way of supplying a reservoir inside the sealed cell is needed. That is why the channel plates are porous, to hold extra acid and in larger systems another backing plate of the porous heat treated and carburized graphite resin mix is used for additional storage. By using both impermeable and porous materials in the plate assembly, sealing between cells can be accomplished, reservoirs can be provided, and some ability to accommodate the liquid volume fluctuations during start-up and shutdown can be provided. The entire stack assembly of solids, liquids, and gases is expanding and contracting as temperatures change and that has to be accounted for. It becomes even more important as the cells age and begin to foul for various reasons.

There has been some discussion of plates for both SOFC and SPFC (PEM) systems in previous chapters, so only a short review is given here. SOFC system plates depend on the operating temperatures. Higher temperature (>900°C) SOFC systems use plates made of chrome based alloys that are stable at high temperatures and have thermal properties that match the stabilized zirconia used in the electrode/electrolyte assembly while lower temperature systems (<700°C) can use much less expensive stainless steel plates. Between the two, high temperature alloys are commonly used. Chrome in any elevated temperature is undesirable, since it tends to **diffuse** through other solids, contaminating some areas with chrome and leaving others deficient in a principle alloying element. Any metallic plate used in fuel cells must have protective oxide coating of some kind to reduce corrosion, and SOFC plates are no different. Some SOFC systems use ceramic plates but ceramics are not good conductors, so proprietary methods of increasing this property usually accompany any such plate set, most having to do with introducing metals into the molecular structure (**cermets**). In SPFC (PEM) systems, the plates are generally high purity graphite (with the same issues as PAFC), although metals such as aluminum or stainless steel are also used. As in other cells, the gas diffusion plate can be separated from the current collection plate to alleviate the restrictions of graphite conduction, with many being porous carbon paper treated with a hydrophobic polymer such as PTFE.

DELIVERING POWER

All of this leads to the main point of fuel cells, delivering current. Fuel cells produce Direct Current (DC) as do many primary generating installations. DC has limited used for one main reason: it does not transmit over long distances at all well. While fuel cells produce what is considered a high quality current output, this is the result of a separate system **conditioning** the output rather than some basic result from the fuel cell itself. All methods to generate electrical power have variation, especially when comparing the relative amounts of volts compared to amps. Fuel cells in fact have rather steep volt/amp curves, so supplying constant current is not done at the fuel cell or stack level but with secondary conditioning units.

One of the main uses of the balance-of-plant sets in fuel cells then is to convert DC to AC and condition it so it is suitable for whatever application the electricity is being created for. This is an important issue in all power generating schemes. Previous chapters discussed the possible markets for fuel cells and all these markets already exist. The type of current used in these applications is uniformly AC (at least the power going into the application unit is AC although it may be reconverted to DC there) but the current itself can differ by large amounts. Such things as the overall consistency of the voltage and amperage feeds, the variation within very small timeframes of those feeds and the time cycling of AC power all vary but whether that matters or not depends on the equipment and applications involved. Running a timer for watering your lawn does not require consistent, high quality delivered current but powering a massive data center does; flickering lightbulbs in a garage are annoying but flickering equipment during open heart surgery is dangerous. This section talks about the type of equipment used in fuel cell systems to deliver power appropriate to the circumstances.

Current

Some fundamental definitions in electricity were presented in earlier chapters and they must be understood in the following discussion so a review by the student would be appropriate. Electricity is electrons moving in a wire and where the electrons come from is not really important to electricity. Electrons from a coal-fired power plant are indistinguishable from electrons produced by rubbing a balloon on your head so you can watch your hair rise up. What is important are the wires the electrons move through, and the quality of the connections made between wires, the metal the wire is made of, and a number of other issues based on materials, system maintenance, care in manufacturing, and excellence in design.

The two basic currents are DC (Direct Current) and AC (Alternating Current). And the difference between them is substantial. As mentioned, electricity is electrons traveling in a wire. In DC power, the electrons go from a negative **pole** (electrode) to a positive pole. In AC power, the poles (electrodes) reverse so the electrons first go in one direction and then reverse flow and head back in the other direction. The 60 Hz cycle used in the United States tells how often the poles reverse and the electrons change direction. A directional cycle represents a full movement back to the original direction though, so each cycle has two directional changes. In the first half-cycle, pole 1 starts as negative and the electrons flow away from it, and then it changes to a positive pole so the electrons flow back, but to complete the cycle, pole 1 then has to return to its original negative value in the second half cycle so the electrons once again move away. In a 60 Hz system, there are 120 changes in direction each second.

DC power is relatively constant because the electrons flow as a constant stream from one pole (electrode) to another. AC power is not constant because the electrons are particles with mass and size and so have to come to a stop, turn around and start flow up again. Remember that electrons do work like heating a wire so it glows or turning a shaft. To do work, they must be moving, but when electron flow reverses, there comes a point where the electrons do not move at all (just before they reverse directions). You can look at this like sweeping where you can push a broom in a straight line to move the dirt or you can sweep the broom back and forth to move the dirt. When you stop the broom at the end of the sweep, it is not moving any dirt. You accelerate the broom but then have to slow it back down to stop it at the other end of the sweep. It does the most work (moves the most dirt) in the middle of the swing, where it is going the fastest. If you just push a broom in a straight line, it moves dirt the whole time.

DC power is relatively straightforward but AC power can get quite complicated because of what is involved in starting, stopping, and reversing. Electrons themselves are very complicated since they can behave as both a particle and a wave so anything other than simple, **unidirectional flow** introduces complications well beyond the scope of this book.

Conditioning

It can be said in general that if you want quality electricity, you deliver it over the shortest possible path using the highest quality materials made by the best possible workers. That is unfortunately not possible in **centralized generation** models, where you are producing power at a dam in upstate Washington but using it to power a banking computer in California. Instead of trying to guarantee high quality in a delivery system encompassing millions of miles of various types of wiring, grid power systems provide reasonably consistent, low quality, low cost power to millions of customers. If those customers need higher quality power, they are required to buy additional equipment to condition the power in-house.

The **distributed generation** model used in fuel cells, solar energy and some wind generating schemes differs from the grid-supplied model in that respect as well as a few others. In distributed generating, the system generating the power includes the conditioning needed for the application. Because fuel cells can employ a distributed generating model, the power outputting from the cell has to be produced, conditioned and delivered essentially all in the same unit. The steps used are conventional and do not rely on fuel cell technology but rather on electronics that have been developing for well over a century. The primary means of conditioning power consists of conditioning the DC power, changing the current from direct to alternating, and finally transforming the AC power to what is required by the final application.

To understand why electricity is not uniform, we need to go back to two basic definitions, volts (the difference in electrical potential between two points or the amount of energy available to move electrons some distance) and amps (6.25×10^{18} electrons moving past a point in one second). If you look at what is commonly called a volt/amp or polarization curve for fuel cells (each type is a bit different) showing the output voltage compared to a change in the load amperage, you will see that the potential between the electrodes goes down as the number of amps (electrons) flowing between them goes up. While this may seem backwards, you can look at it as the amount of energy available to move each electron. The more electrons (amperage) there are flowing in the circuit (the higher the amperage is), the less energy there is available to drive any individual electron (the lower the voltage).

This is why AC power is used in the grid. If DC power was to be used, the electrons would have to travel from generating site to where the lightbulb actually is located, which might be hundreds of miles away. The energy needed to move the electrons that far in the wire is lost, and at some point, the energy needed to move the electrons is greater than the energy generated. In AC, the electrons move only as far as 1/120th of a second will allow them to travel, and that does not take much energy.

Fuel cell DC voltages are generally low and vary quite a bit (about 50 volts for a small system up to several hundred for large systems) because the number of chemical reactions that happen in any given amount of time (say 1/120th of a second or 60 Hz) is not absolutely uniform. Remember that each reaction only generates a couple of electrons and you need trillions of trillions of electrons coming from each cell every second. Grid systems generally run very high voltages (systems transmitting at 110,000 volts or more are common), because the electrons are moving in hundreds if not thousands of miles of wire (although they are not going very far, an enormous number of them are being moved back and forth). The energy needed to drive the movement (the electrical potential between the pole where the electrons enter the grid and the pole in some other state where they are used or more commonly called voltage) through such a system is substantial. Fuel cells move electrons over smaller distances and so the amount of energy built up to move them (voltage or the potential between the positive and negative electrodes) is not large. Remember that the electrons have to go out of the cell, do some work, and then come back to the cell, so larger systems will have somewhat higher voltages because the path out and back is longer and thus more energy is needed to go the distance. Even in AC systems, where electrons are moving a small amount in space, the longer the wire, the more electrons there are to move and so the more energy is required.

Volt/amp fluctuation within a cell is also understandable if you consider that the chemical reactions producing electrons don't run like clockwork machines, but rather surge as hydrogen gas molecules react at an anode while oxygen and ions react at the cathode. There are a number of things that can change the chemical reactions and how fast they proceed, in particular the temperature and the gas pressure not only at the anode or the cathode but at the individual catalyst sites within the anode and cathode. Pressure and temperature vary, as we have discussed numerous times in this book, but also, given the huge number of molecules and electrons involved, there is no way that the number of atoms and molecules reacting at the anode is exactly the same as at the cathode, nor is the number ever going to be exactly the same over any two given periods of time. Variation can also be caused by catalyst sites being fouled and taken out of business or electrolyte moving a bit and changing the movement of ions or any number of other things. In low temperature cells, the **reaction kinetics** (speed of chemical reactions) are slow enough that monitoring, measurement and control lag the actual reactions by a reasonable amount, making the units either **underperform** or **overperform** on a regular basis while high temperature kinetics respond so rapidly they tend to induce cycling.

A fuel cell specification might call for a $+/-$ 6% **variation in line voltage** across a supplied range but what does that really mean? Let us take a 10,000 watt system. One watt (volt-amp) is one amp of electrons moved between two electrodes held at one volt potential. So 10,000 amps ($62,500 \times 10^{18}$ electrons moving past a point in one second) will flow between anode and cathode if the potential is held at 1 volt. Not many things run at 1 volt, so we will set our example at 110 volts which means 568×10^{18} electrons moving past a point in one second (91 amps in the 110 volt circuit). Recall that most residential services are in the range of 200 amps supplied to a whole house while individual circuits within the house are rarely over 30 amps and these are generally for circuits where larger items such as stoves are used, so a 90 amp service would work but only for a smaller house. Our allowed 6% variation then means that the number of electrons moving past a point in one second can vary from 534×10^{18} electrons to 602×10^{18} electrons. If we are talking about an AFC where each molecule of hydrogen gas (H_2) produces two electrons, then the number of hydrogen atoms reacting in one second over all the cells and stacks in use for a 10 kilowatt system (a "smaller" system) runs from 267×10^{18} to 301×10^{18}. Seen from the standpoint of the cell itself where individual atoms and molecules must move and find catalysts sites and react to change into ions and flow across electrolytes and find another catalyst site and recombine, the numbers involved are huge but seen from the standpoint of volts and amps, they are not. The student should remember that every number encountered in this business is relative. A 6% variation in residential service will flicker your lights but the flow variation in the lightbulb itself is millions of trillions of electrons. Industrial generation deals with numbers of electrons moving that are so large they cannot even be described without **scientific notation**. Fuel cells operate on a smaller scale so we can talk about such things as millions and trillions to make them more "user friendly" but to truly grasp what is important in fuel cells systems and why, the student must understand that each electron has to be produced individually in a chemical reaction rather than produced wholesale in generators as big as houses in plants bigger than some small towns. That is both the appeal and the weakness of fuel cells.

If the number of electrons moving in the circuit (amperage) changes, then the voltage must change as well. The actual relationship between volts and amps is determined by the resistance (to electrons flowing) between the two poles over which the **potential** is measured or stated somewhat differently, volts are amps times resistance ($V = IR$). Since the internal resistance in a circuit does not generally change except for over long periods of time, then if amps (the number of electrons) change, volts (the energy available to drive each individual electron) must also change. Change is bad in electricity. For some equipment like a hair dryer or a toaster, if there is a variation of a couple of million electrons out of several million trillion electrons moving in a one second time span, then not too much is at stake. Other types of equipment like computers or medical devices are much more picky about the number of electrons moving in and out every second, so a 6% variation is not going to be acceptable.

Fuel cells must then guarantee a steady supply of amps and set the volts in the system, both with minimal variation. While fully describing power conditioning is well beyond the scope of this book, the methods used are not. The first order of business is to stabilize the DC current output coming directly from the cells and this is done by chopping the output using switches such as

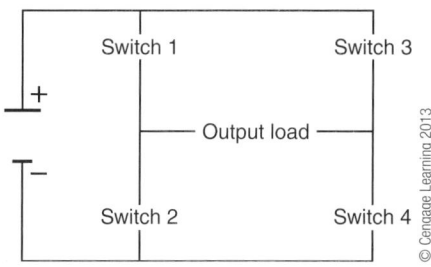

Figure 12-9 A simple DC to AC converter.

thyristors, metal oxide semiconductor field effect transistors (MOSFET used in low voltage systems), or insulated gate bipolar transistors (**IGBT** used in systems up to several hundred kilowatts). These are commonly called **DC-DC Converters** and are used to stabilize the output (they damp "rippling" or variation in electrons produced per second) and in fuel cells at least step-up the voltage as well (120/240 AC volts systems typically use 400 volt DC input before the conversion). In essence, these units act like a bucket brigade for regulating water flow, where a steady output rate is set by filling thousands of buckets and then dumping them all at a steady rate. Rather than controlling flow, they chop it up into a series of timed sequences and then add them steadily back together to make a steady output.

Changing DC to AC is next. The fundamentals of this require a drawing, shown in Figure 12-9. If Switch 1 and Switch 2 are open but Switch 3 and Switch 4 are closed, no current flows through the Output Load channel (the wiring sending the current out of the fuel cell) located between the switch sets. If Switch 1 and Switch 4 are open but Switch 2 and Switch 3 are closed, then the current flows through S1 and left to right through the load. Open S3 and S2 but close S1 and S4 and the current through right to left through the load. That is AC current with the sine wave output and the hertz cycling determined by the timing of the switches opening and closing. There are some issues that involve other electronic devices to better control current, but in essence that is how DC is changed to AC using device with names like pulse width modulation or tolerance-band pulse inverters.

Many systems will insert a battery between DC-DC converter and DC-AC inverter to supply peak loading for such things as motor startups as well. Other devices such as isolation inverters, LC filters, and transformers can also be included in these circuits depending on the applications and even whether the output is to be single phase or triple phase. Monitoring and control circuits, of course, are included as well. The actual configuration of the conditioning set will depend on the application being served, whether it is supplying a dedicated load, backup, or auxiliary load or gird service supply and whether generating current part time or full time. There are also government regulations involved that set the "purity" of grid supplied power to protect the overall system, so grid connections can have additional requirements beyond dedicated servers.

BALANCE-OF-PLANT

It should be clear by now that fuel cells are not much use by themselves. What is called the balance-of-plant (BOP) determines much of the usefulness of fuel cells and whether they can be used in most applications at all. Even the applications themselves are only a small part of the equation, since roughly half of all the electricity used worldwide does nothing more than power induction motors of one kind or another (the dynamo of Chapter 1 used in reverse as a motor). The ability to use off-the-shelf items in the associated BOP sets will to a great extent determine whether fuel cell systems enter the mainstream as primary generating installations rather than niche players.

The importance of BOP sets can be seen in the automotive industry, where there is some degree of difference between a Ford Model T and a new Mustang 302 Boss; both are mobile power platforms to transport people and goods, but they are not the same because their BOP sets are profoundly different. Even in these early stages of the fuel cell industry, differences are immediately obvious not so much in the fuel cells but in the BOP sets. SOFC fuel cells, for instance, all operate at higher temperatures, can take advantage of internal reformation, and use both hydrogen and carbon monoxide as fuels; however, a mobile SOFC used to supply power to a small army patrol will use more or less the same actual fuel cell configuration as a mobile SOFC used as an auxiliary power supply for a bus in Germany but their BOP will not be the same. The bus will be heated but the patrol unit will have a minimal heat signature to prevent detection. The bus will have access to a ready supply of uniform fuel but the patrol unit will have to rely on field supplies of uncertain quality and amount. The list of these differences is large and growing larger as more people and industries begin to recognize the versatility, size, and quality of power supplied by fuel cells systems.

The size and weight, as well as the all important cost of fuel cells are determined in great part by the BOP. It follows then that the maintenance and engineering requirements for fuel cells will also be driven in large part by the BOP. Using standardized pumps, compressors, valves, heat exchangers, piping, and inverters will not only lower system costs but minimize maintenance costs since anyone with knowledge of pumps will be able to fix the pump on the fuel cell unit. It should be noted that in many installed system studies, the maintenance cost for fuel cells was more about the BOP equipment than about the fuel cell itself and in many cases system life was determined not by whether the fuel cells still produced electrons but whether the BOP still worked.

Rather than calling out the BOP as a separate category, we have discussed most of the BOP and related equipment in context with the fuel cell because they are in fact inseparable. What is important to maintenance and engineering in fuel cells is not the specifications and hose requirements of a compressor but rather that it is being used in the fuel preparation system as well as the cooling system and might even run a few valves as well. If that compressor goes down, fixing it is the easy part while understanding what else is being affected within the fuel cell system and the associated consequences of the failure are what is truly important.

KEY WORDS

Knowing the terminology used is critical when dealing with fuel cells. Following is a list of the important terms in this chapter, which are also in bold typeface within the chapter. It is recommended that students be required to submit definitions of some of these words as homework assignments in which they look the terms up in other books, articles, or on the Internet.

back-pressures
boiling water method
bottoming cycle
carburization
centralized generation
cermets
compress gas
compressor
computer modeling
conditioning
conductors
contact site
DC-DC Converters
diffuse
diffusers
diffusion
direction of flow
distributed generation
distributes stresses
electron channels
exchanged
fixed rotor
flow
flow field plates
friction
gas

gas diffusion plates
gas flow
gas pressure
gas-tight
heat engines
heat transfer
heat-aided systems
hot zone design
IGBT
laminar flow
liquids
mechanical pressures
Membrane Electrode Assembly
microturbines
MOSFET
overperform
parasitic losses
path of least resistance
pole
porosity
porous stainless steel
potential
pound per square inch
pounds force
pounds weight

pressure
pressurized water method
reaction kinetics
recycle gas loops
resistance
revolutions per minute (RPM)
rotating rotors
scientific notation
sealing surface
shear stresses
solids
steam turbines
stress riser
thermal management
thyristors
turbine systems
turbocharging
turbulent flow
underperform
unidirectional flow
variation in line voltage
volt/amp fluctuation

DISCUSSION QUESTIONS

1. Liquids and solids generally do not mix together. Why is not a good idea then to put a piece of steel in water?

2. Why is balancing the liquid pressure and the gas pressure important in a fuel cell like an MSCF system?

3. Of the three types of pressure, solid, liquid, and gas, which is more important in fuel cells?

4. Search for MCFC system piping and instrumentation diagrams on the Internet and look for pressure control loops or chamber sizes as well as pressure drops. Pick one such system and detail it as a homework assignment.

5. What is one of the main issues in low temperature systems where a pressurized cell is going to be used?

6. What are 3 atmospheres of standard pressure expressed as pounds per square inch (psi)?

7. Why do electrodes have to be porous?

8. Why does transferring a gas flow from a large volume to a smaller volume increase the gas pressure?

9. Why are the edges of the electrodes important in fuel cell design?

10. Why is thermal management usually more important in liquid electrolyte cells than in solid electrolyte cells, no matter which operating temperature is used?

11. How would a fuel cell use excess heat in a bottoming cycle?

12. Can only one bottoming cycle be used with a fuel cell?

13. What functions do seals perform in a fuel cell?

14. If high quality electricity is needed, what are the three general requirements for the delivery system?

15. How does the conditioning of grid-generated electricity differ from distribute generation electricity?

16. Why are grid systems operated at high voltages?

17. Do all fuel cells use the same power conditioning systems?

CHAPTER 13

FUEL CELL SYSTEMS: ENGINEERING, OPERATIONS, AND MAINTENANCE

objectives

This chapter introduces the student to issues that may arise when installing, operating or maintaining fuel cell and associated systems. Some general issues reported by those installing and operating fuel cell systems are discussed and specific guidelines given if available.

INTRODUCTION

Fuel cells can be solid oxides or Teflon-based polymers. Fuel cell stacks can operate at low temperatures or high temperatures. Fuel cell systems can be portable or stationary. How does a company or a maintenance group or the engineering staff deal with technology that is so broad yet at the same time changes almost month to month? This chapter discusses some of the issues staff may face if (or perhaps when) their company decides adopting fuel cell technology is an economically advantageous, politically astute decision and sends word out to install a couple within the next quarter. It is not meant to be comprehensive but rather a guide based on a review of articles detailing the topic and the experience of the author.

ENGINEERING

One of the primary issues with fuel cell systems is in integrating whatever fuel cell system and design is proposed into an existing infrastructure. The first order of business is the same no matter what devices are proposed, operated, maintained, or even imagined: is it plugged it? What energy source is feeding the system? How is it doing so? Where is it located in regards to the proposed site? What codes and standards are involved? The list of questions is long and composed mostly of common sense items having to do with matching existing site infrastructure. Common sense many times has little to do with long term planning though, and so decisions are often made without asking maintenance if the mains can handle more power or if the existing natural gas line coming into the plant is already at maximum flow feeding the furnaces. Without a solid understanding of existing infrastructure, adding something to it can be an unpleasant experience.

Fuel cells in particular require a comprehensive understanding of what is in place since their needs can vary widely. A good example is in the ongoing effort by many utilities to bring fuel cells into waste water treatment facilities (**WWTF**) where the production of methane gas is common and can be used to supply fuel cells. Much of the methane generated in treatment plants or landfills is wasted, either dumped or flared directly to the atmosphere so there have been many efforts to harness this source of power. The initial attempts discovered some fundamental issues having to do with basic infrastructure and operations. The fuel cell system pipes were too small since they had been designed for a much higher quality fuel such as natural gas and this resulted in insufficient fuel supplied and excessive pressure drops within the cells. Larger piping systems and added compressors were needed but since the units were skid mounted as plug-and-play systems (an option used to reduce costs through **mass production**), changing the input lines to the reformers was difficult and costly. The incoming fuel itself contains contaminants not usually seen in fuel gasses (organic halides in particular, which are carbon bonded to either fluorine, chlorine, bromine, or iodine) as well as contaminants normally in fuel gas (mostly sulfur compounds). These are usually present at elevated levels beyond those called out in the fuel cell system design. External gas processing was required to deal with these contaminants, something understood before the system came online but still requiring extra equipment that had to be included in both funding and time-to-completion planning. These systems included an initial gas dryer stage, a chiller unit to condense out some gasses and an adsorption stage for final cleanup. Since the gas produced by the **anaerobic digesters** (systems where organic matter is digested but with no oxygen present) used in waste water treatment also varied with the waste composition, fuel variation was beyond original design parameters and control schemes had to be adjusted accordingly. Excessive water also occurred in the digester gas, causing substantial problems, and in one such Washington state test excess moisture meant the system only ran on digester gas for a short period of the 2 year test although it did run on natural gas during most of the period. Permitting the units proved to be an issue given the limited number of installations and the variability in fuel cell unit production. To ensure ultralow emission, extra catalytic reduction units were added to some units, raising costs and increasing the complexity of the turbine cogeneration systems. Other systems that add some percentage of pipeline supplied natural gas found they could use smaller external reformers and more cells per stack but the higher initial fuel pressure and increased corrosive nature of the anaerobic digester gas degraded the gas manifolding much quicker than designs called for. The added units and additional problems then required extended monitoring and control as well as more scheduled maintenance.

Any fuel cell installation requires a thorough understanding of not only the numbers supplied by manufacturers, plant facilities personnel, utilities, and accounting but also a working knowledge of how much difference in these numbers can be tolerated and what numbers are critical. Planning for and installing fuel cells systems must match the planned application to the available systems with more care than is normally used in bringing a new process into an existing facility. In particular, it is important that staff remember the importance of pressure and temperature. The pressure in the example above had to be changed because the energy value of the incoming fuel was lower than the design called for. The same issue may arise for ambient pressure units located in different places, such as the mountains of Switzerland versus the riverside of London. The atmospheric pressure differs from the mountains to the sea and that may affect fuel cell or reformer performance (it definitely affects the boiling point of water in any associated steam systems). The difference in ambient temperature is substantial between Chicago in the winter wind and Phoenix in the summer sun. It is important that any operational or durability information provided by vendors be suitable for the planned installation and not just general information. Heat to power ratios should also be examined, since this will be critical to achieving stated goals. If the primary purpose of the installation is to provide power with heat recovery being a secondary issue, then that will require different installation requirements than an installation where heat and power share importance. Some larger industrial installation may even see heat generation as the primary goal with power generation being secondary since the cost of conventional ICE or turbine industrial **block heating stations** (any installation which generates power and heat) is relatively high (the smaller the system, the higher the relative cost) and it may turn out that midrange fuel cells can effectively compete as heating stations that also provide electricity. Block heating is also less efficient when partial loads are used, since heat engines operate most efficiently at set values where fuel cells can operate efficiently over a much wider range, as discussed in previous chapters.

It is best to ask questions about temperature and pressure before installation occurs. It is best to review any calculations a vendor might provide before the system is purchased. It is best to have a solid understanding of operational variables before a system is recommended. Maintenance may need special tools, for instance, and those costs should be included in the capital spending program rather than part of a yearly maintenance budget. Computers may need to be upgraded for the control software or certain programs purchased, system monitoring and control may not be compatible with existing displays in control rooms or require alternative wiring sets. This type of detail is critical in analyzing and recommending a fuel cell system in particular. Many, if not most, fuel cell system designs and fabrication methods change regularly since in essence the systems are produced one by one and in many respects resemble research projects more than industrial product lines.

There are some general issues that should be addressed as well as the specific ones mentioned above. The codes, standards, and regulations involving fuel cells are not yet set very well so particular attention should be paid to these as they evolve to make sure planned or existing installations are up to date. There is a good chance that the codes for pressure piping and boilers, fire and insurance requirements, and local, regional, and federal regulations will not be the same between the time initial discussions are begun and final installation takes place. The buildings involved need to match. Most fuel cell systems are skid type assemblies that may have a gas treatment system, a reformer, a fuel cells system, associated thermal management systems, and monitoring and control equipment, all possibly contained in separate shells or even their own small buildings. Up to half the weight and possibly more than half of the actual footprint may be associated systems, and it is highly recommended that more than adequate room be made available. Equally as important for many installations are the tie-ins to the local grid. Both the fuel cell installation and the local grid must be protected and this is relatively straightforward given the use of inverters on the fuel cell, which generally come with appropriate sensors but may also require special software requirements. It should be understood that protecting the fuel cell from grid disturbances is more difficult than protecting the grid from the fuel cell. Protecting fuel cells requires much faster response time and is done using **subcycle response** schemes (capable of responding to transient fluctuations of shorter duration than the hertz cycling used in the grid). The local utility must approve all in-place protections schemes before installation with most of the problems encountered having to do with the embedded protection in the inverter control.

Installation operations and maintenance access for the entire fuel cell system and all associated subsystems must be guaranteed. Units cannot be subjected to excessive wind or dust (air filters are commonly disregarded in these systems and this can cause substantial problems during service), nor can they be fully enclosed due to heat buildup. Plumbing, electrical, sidewalks, catwalks, fire suppression, and a number of other systems already in place at the site have to be matched to the incoming fuel cell systems. In particular, any water to be used must meet temperature as well as both particulate and contaminant requirements (fuel cells require high quality water, usually deionized since ionized water may interfere with the electrolyte operations). Maintenance routines must be provided with any specialized training done on site with an operating unit. Fuel requirements must be strictly defined. **Grid feeds**, buyback potential for excess power, **load balancing** requirements, and cost contracts must be understood well in advance. Repair procedures, spare parts lists, replacement warranties, and service contracts should be in place before installation occurs. Finally, site preparation should be inspected by the manufacturer's representative well in advance of installation, even if only done remotely.

Installation itself should be done by qualified personnel who are familiar not only with fuel cells but also pressure piping, grid connections, and gas delivery systems. If contractors are used, then maintenance and operations personnel familiar with the existing structures and equipment must oversee installation. Engineering support is required. Substantial testing should be done, including such things as pressure testing, emergency venting, fire suppression, emergency response using both site teams and local Emergency Response teams and in fact a full HAZOP study should be done before the installation is deemed complete. ANSI Z21.83/CGA 12.10 provides a standard for equipment design manufacture and testing and ASTM Standard PTC 50 gives performance standards that should be consulted prior to any installation.

And of course, cost. For residential or small scale installations, if the initial cost of the system is roughly equal to the cost of a competing systems within 10% (plus or minus), then it will be considered by contractors in major developments. Keep in mind though that costing fuel cell systems requires careful consideration of what to compare the fuel cell to. In residential areas, the cost may include hot water and heat as well as the cost of extending electricity to a new subdivision. The cost of extending a grid line is somewhere between $4 and $10 per foot, at least in developed countries with existing grids in good shape. In undeveloped or underdeveloped countries, this cost may be substantially higher. It should also be noted that in rural areas of the United States, many smaller grids were created as cooperatives in the early 1900s and are not in particularly good shape. The cost of upgrading or replacing these smaller grids-within-a-grid are enormous and in all probability may not occur, opening up additional opportunities for fuel cells. For residential developments consisting of apartment or co-op blocks, a slightly higher cost may be acceptable if the social or reliability aspects of fuel cells allow for a slight increase in rent or purchase cost. For industrial costing, the usual analysis is based on return of investment based on a 3 to 5 year model. Again, this can get complicated. If the system is meant to replace an aging boiler (or allow it to be downsized), then reduced maintenance and operating costs have to be factored in as well as any additional piping or site preparation that might have to be done, along with any reduction (or increase) in the cost of electricity that occurs.

It should also be understood that while incentives may exist to make use of fuel cell systems, using these in a costing model will skew any later analysis done when incentives change or disappear. It is also difficult to gauge the economic value of reducing reliance on grid supplied power (where prices can fluctuate), the value of environmental awareness (it should be noted that the first push toward a so-called "hydrogen economy" was in the 1970s and it soon fizzled so caution is advised on this point) or the need for truly uninterruptible, high quality power for the growing number of electronics in modern society. Costing fuel cell installation has not been particularly successful if the literature is any indication and great care should be taken when it is done.

The system in question can also have significant impact on all of the above. If, for instance, a hybrid system of photovoltaic, wind, and fuel cells are to be employed, then design loses its modularity to a great extent, costing becomes almost impossible to do and installation is much more complicated (solar energy for instance requires a tracking system and power point tracker to follow the sun). These systems are considered mostly where power interruptions are completely unacceptable and locations somewhat remote (or major urban areas where grids are stressed) and are currently served by diesel generator and battery sets. One of the major advantages fuel cells have in this area is in their monitoring and control capacity since both diesel generators and batteries have limited control capabilities.

Finally, it must be stated again: fuel cell systems are not mature product lines and should not be viewed as such. PAFC systems are probably the most mature large scale systems. AFC systems given their history with NASA are relatively mature but only at considerable cost. SPFC (PEM) systems are probably the most mature small scale systems. SOFC and MCFC systems are still dealing with high temperature issues.

OPERATIONS

As in all system start-ups, it is advisable to test feed systems in advance. Since fuel cell systems make extensive use of tie-ins as part of their economy of operations package, this is even more important. Air or water circuits, steam recycling or feed loops, electrical tie-ins to the grid or to other parts of a plant, remote monitoring, and any discharges should all be tested. In addition, all pressure piping or additional gas storage systems should be tested as per code requirements.

It should also be well understood that start-up and response times differ greatly not only among fuel cells systems but among the associated systems required by fuel cells. Even if the fuel cell stack has rapid cycling abilities, that does not mean a reformer or gas conditioning system has the same cycling or start-up abilities. In general, external reforming takes more time to come to operating conditions than fuel cells and does not respond well to partial load or rapid cycling operations. Start-up issues are not limited to the fuel cell but to the location as well so that the system must be able to cold start from the lowest temperature seen in the winter to the highest temperature seen during the summer at the installation site.

If the system is part of a campus type installation and will feed a small grid with both electricity and hot water, lines travelling distances of more than about 150 feet will require supplemental heating to maintain temperature. This is similar to boiler requirements and generally involves insulation but can also be heated lines if high temperature steam is involved. This type of installation may also be viewed as being scalable, so care should be taken during start-up to ensure any lines to be used for future **tie-ins** are adequately prepared.

System start-up should begin with recommended ramping but before the system is turned over to operations, both lower and upper limit **ramping** should be done with the manufacturers having full access to the data. Following that,

load cycling should be done to mimic the worst case scenario. The recommended operating parameters should then be followed for some period of time followed by **optimization testing** to dial the system in for the particular site rather than using general guidelines.

System operations should be relatively automatic once the start-up is completed. Monitoring is required for flows, pressure and temperature, but industrial settings should need little attention other than a routine tie-in to existing control rooms, suitable limits being imposed and QA/QC protocols established. Smaller systems such as those that may be in residential areas or campus type settings like hospitals may need no attention beyond that typically given to appliances. In fact, there is a trend already growing toward contracted operations for the larger sizes and sealed units for the smaller. This type of arrangement is already in place to a great extent for hot water and steam systems. Residential units are run until they fail and then replaced by contractors in many cases, while steam boiler may have an operator working part of the time, with maintenance and inspection done by others.

Shutting down the system is probably the most complicated of the operations expected for fuel cells systems. In mobile systems, shut downs will occur on a regular basis, but installed sites will limit shutdowns to as few as possible, especially for high temperature systems. Formal procedures for shutdowns are required and training is called for. Even low temperature systems that can tolerate cycling will probably have reforming units attached that have a lower tolerance for cycling. It is critical that operators and maintenance personnel understand the consequences of shutting down the fuel cell system, given the damage that can occur. Some units will also require a protective gas such as nitrogen to be used in the event of a shutdown so the cells themselves are protected from unwanted chemical reactions.

MAINTENANCE

Maintenance of fuel cell systems requires trained personnel. This is true of both mobile and stationary systems. While it is not possible to cover the requirements of even a small number of systems, it is possible to give the student a reasonable look into the type of issues that may be encountered on a routine basis.

Typically, fuel cells claim long scheduled maintenance intervals and high reliability, and the more experienced users tend to agree, in particular NASA, where only one systems failure occurred in the over 80,000 hours of orbiter operations. The initial commercial installations report scheduled maintenance intervals somewhat better than internal combustion machinery but also report substantial problems with unscheduled maintenance arising from fabrication, manufacturing, and installation issues that are still being resolved as systems mature. Some PAFC manufacturers, the most mature of the industrial stationary systems, report 2500 to 7000 operating hours for **mean-time-between-failures**, a respectable number at the top and a barely acceptable number at the bottom. Many reports detail startup issues and early operations problems that once resolved lead to stable operations and reasonable maintenance requirements, although most are firmly in the too-much-maintenance-required camp. Since that camp is always full of industrial representatives who feel that any maintenance is too much and would like to see maintenance entirely eliminated, such reports should be analyzed carefully by those expecting to see fuel cells become part of their responsibility.

As for actual maintenance, there are relatively few reports of that since most researchers have little if any involvement with what must be done to ensure long term operations. Some basic details emerge however for those who understand both how fuel cells operate and why they stop.

Of particular interest is the balance of plant equipment. Much of this consists of off-the-shelf items used in fuel cell systems to minimize cost, even when the system has been designed for something else. This is not unusual in manufacturing or fabrication but does tend to come back on many of the fuel cell manufacturers more often than not. One of the first things a maintenance staff should do is investigate where the individual components used in fuel cell systems come from. If they are commercially available units used primarily in other applications, staff should mark them as suspect and pay particular attention to them during initial operations. The next thing that should be done is to find out who is fabricating the various parts of the units. Since fuel cells are relatively new, fabricators in particular are not used to their demands. It is recommended that maintenance personnel have a solid understanding of the quality of fabrication, since a conspicuous lack of that shows up in many of the test reports. If possible, review of fabricator certifications and quality test results should be obtained for the particular unit purchased.

All the consumables to be used need to be well catalogued. Since many fuel cells, whether stationary or mobile, will spend the bulk of their lifetimes outdoors, it should be determined whether these items are specifically meant for outdoor use and whether planned replacement schedules take into account the added problems associated with outdoor use. This is particular true for air filters, which should also be rated for variable temperature use. These filters can cause substantial problems if they become plugged (or even dirty in some cases), ranging from local overheating to creating short circuit pathways within the system. Water treatment systems and resins are critical in fuel cells since the quality of humidification water must be very high, and these systems must be rated for industrial use with hot replacement possible. Gas filtering media such as charcoal should also be closely examined, since many filtering media are meant for emission control systems and not for input to units that can be damaged if levels exceed specifications. Attempting a record for longest time before changing filters is not recommended unless staff already has substantial experience with the individual unit in another application at the specific location of installation. If a manufacturer suggests a replacement schedule for any consumable, it should be viewed as just that, a suggestion. Given the dependence of fuel cell operations on pressure and temperature, replacement scheduling should be done at the plant using recommendations as a guide, with inspection done during initial operations to determine if the recommendations are suitable. Electronic items that are in the **reaction zones** of reformers or fuel cells such as thermocouples also seem to degrade quicker than expected in many systems.

Although fuel cell systems are being presented as skid sets ready for production as soon as they are plugged in, that should not be attempted. It is highly recommended that maintenance staff become well acquainted with the list of reasons why units will trip to hot standby before operations commence and what is supposed to happen when the unit does trip to hot standby. Most of the reports of such tripping are caused by BOP units, and since fuel cells systems can have a substantial number of these, trying to track down the reason for a hot trip during initial operations can be both frustrating and dangerous. The limits of cell and stack operations should also be well understood so that upset conditions can be rapidly identified in order to limit possible damage should they occur. Specific mention of compressors as ongoing problems is made, possibly due to the units not being rated for outdoor use, and maintenance and engineering staffs are encouraged to inspect compressors even before installation and routinely after operations begin.

The effect of seals failing should be well understood by personnel monitoring the installation. Internal fuel cell seals as well as conventional gas system seals can be rapidly destroyed if temperature or pressure cycling occurs, not only from the cycling, but also from the mechanical stresses induced in the seals during cycling. This seems to be a common problem in systems operating at both low and high temperatures. Seals themselves may require special handling and storage, especially if they are rated for high pressure or temperature and can be easily damaged if treated poorly.

Starting and stopping capacities of the entire system as well as any connected system should be well understood. High temperature systems can take several hours to reach operating conditions, and unless "**standard operating conditions**" are well understood, damage can occur if ramping is terminated too soon. As mentioned previously, different units within the system will also have different standard operating conditions, especially in low temperature systems and maintenance staff should be aware of these differences and the possible monitoring, failure, or control reports they might cause within the computer systems. Shutdown sequences, lockout/tagout requirements, hot swapping capabilities, and how each unit in the system responds to catastrophic shutdowns should be adequately identified. In many cases, gas purges are done to protect fuel cell components and these should also be identified and understood as well as included in routine maintenance checks. The use of nitrogen purges appears to be an issue in some systems and there are several reports that hot swapping abilities for the nitrogen bottles had to be retrofitted into the systems.

Since the system will become part of the electrical grid within the installation site and probably within the greater surrounding grid, the ability to isolate it from other problems both locally at the plant and regionally at the grid connection is recommended. Maintenance staff should be well versed in how to isolate the system (or know if it cannot be isolated). Problems associated with this have been reported in several cases even during planned shutdown events as well as during emergency power outages when systems were forced to quickly and fully shut down because they could not be isolated.

Instrumentation also appears to be an issue. Circuit boards, **programmable logic controllers**, and various other electronic devices are reported as failing in many cases, possibly due to the outdoor sites most systems are set in. Access to electronics is mentioned as being too restrictive for many of the systems due to the way the BOP units are arranged. Interfaces with existing computer monitoring are also problematic and some installations apparently were not able to achieve true **systems integration** but had to rely on web access to the fuel cell site. This may represent a considerable problem and should be taken into consideration by maintenance personnel even before start-up.

Care should be taken when dealing with the fuel cell and some associated items. Some warranties expressly forbid using controllers not specifically designed and built for the specific fuel cell installed or tampering with them in any way. Other practices that might seem commonplace are also specifically mentioned as voiding warranties, such as changing valves, tampering with the stack or even any disassembling of the fuel cell unit. Operations outside of those clearly specified in the literature accompanying the fuel cell system such as those that might be done to troubleshoot a system might also be forbidden. Again, since these products are still maturing, there may be some limits imposed on operations and maintenance that are not normally seen in industrial settings.

As mentioned in the sections detailing fuel cell fundamentals, proper humidification of the cells is critical to their operations. The same can be said of the reformers and the heat recovery systems if they are using steam. Water is one of the most critical components of the fuel cell system. The fuel cells must be oriented correctly so flow is not compromised, cells often need to be humidified or purged with inert gasses during shutdown (and then re-humidified upon startup) and equipment used in thermal management will vary in operational aspects by a considerable amount depending on the humidity of several different streams. In addition, stacks and membranes will not tolerate solvents or soaps since they tend to absorb contaminants from these items and degrade performance or even destroy equipment. Understanding and managing the water, both in BOP and the fuel cell system itself should be one of the key components of a maintenance program. One thing of importance to note: never introduce anything into any water system associated with a fuel cell to stop leaks. At the risk of endlessly repeating the same issue, note that all water is not water. Any call for deionized or distilled water should be strictly enforced since introducing other water into systems calling specifically for deionized or distilled can result in substantial damage across the system.

Care should also be taken to monitor load conditions for the system. This includes the batteries used when start-up occurs, hot standby conditions, as well as standard and out of control operations. Excess load can cause overheating and even feedback into the cells that turn the chemical reactions in reverse, creating hydrogen gas instead of electrons (electrolysis). If this happens, the fuel cells are destroyed. Both voltage and current limits are usually in force so during any maintenance done while the unit is still operational, extreme care must be exercised to ensure those limits are respected.

Fuel cells will have any number of alarm conditions. Some are general to fuel cells such as temperature limits, start-up battery voltages, operational voltages, and currents, as well as thermal management limits (blowers, water, or cooling loop flows...). Each system will have specific alarms as well that maintenance should be aware of. Given the general habits of programmers and engineers to use odd and arcane alarm messages (or even different languages), it is recommended that maintenance personnel review alarm conditions prior to start-up in order to understand what the alarm means. Systems will also have preprogrammed routines that will occur such as periodic short circuiting to adjust cell humidity. Maintenance should be well acquainted with any such routines.

Make sure that emergency shutdown switches are well marked and accessible, but be aware that some aspect of fuel cell systems cannot be shutdown instantly or even all that quickly. Each system should specify the time needed to guarantee safe access; assuming that the system is ready to work on just because it is turned off and isolated is not a good idea. Pay attention to the recommended time limits. This is particularly important when dealing with anything that might trigger the fire suppression system since people and fire suppression rarely go well together. Some activities may trigger fire suppression even if the fuel cell is shut down and care should be exercised. Along the same line, never rely on set meters associated with the system. Use hand held equipment to check electric current gas, and fluid flows and manually vent all lines.

The fuel cell stack itself is generally fragile. Anytime it is accessed, it should be treated with considerable care and respect. Even the oil on clean hands may be enough to disrupt its operations and some types such as SPFC (PEM) cells can be damaged by even a small slip of a screwdriver. Make sure all bolts are carefully torqued and any recommended tightening progression followed. If a line or fitting leading to the cells themselves are opened, make sure they are immediately covered to ensure that not even dust gets in.

Any replacements made within the fuel cell assembly must be approved. Gaskets, lines, valves, and the like must be made of material capable of withstanding conditions within the assembly itself. These can be quite expensive and require longer lead times than might be expected. Some swaged pieces for use in high pressure and temperature applications in particular may be difficult to get on even long notice. Even if stored in sealed containers, make sure to handle and clean all such replacements with great care since even small mistakes can have disastrous consequences within the cell stack. There is a long list of things that can damage the cells themselves, including most oils, any particulates at all such as dirt, sand, rust, or dust, metal shavings, fingernails, paint thinner, hand cleaner, gasoline, blood, pipe dope, and sweat. In particular, anything containing **rare earth elements** (there are many of these) must be avoided, since these elements will react vigorously with hydrogen. Care should also be taken when dealing with any lubrication oils since many systems use specific oils that will not vaporize and thus contaminate any air stream within the fuel cell or reformer.

Most systems will have routine maintenance tasks called out. These can include checking vents, filters, water traps, oil levels, hydrogen detectors, periodic leak tests, external fuel to internal fuel supply line checks, cable connections, monitoring equipment tests, circuit tests, inverter and chopper tests, and ground integrity tests. A fuel list of all such tasks should be supplied by the manufacturer and followed carefully, especially in the first several months of operations.

Never weld or use any type of torch on any part of a fuel cell system unless certified and approved to do so. Many of these systems will qualify as **pressure vessels** and must be dealt with accordingly.

Check and verify all ground connections whenever dealing with the system.

Like any industrial operation, fuel cell systems can kill you in a number of ways and quite quickly too. Always follow proper procedures. Bleeding and blocking lines, lockout/tagout procedures, proper grounding techniques, and awareness of gas issues is critical to the safety of maintenance personnel. Given the relative unfamiliarity that exists with these systems as well as the constant refinement manufacturers' are putting in, it is highly recommended that maintenance is done using more rigorous safety precautions than might normally be used.

KEY WORDS

Knowing the terminology used is critical when dealing with fuel cells. Following is a list of the important terms in this chapter, which are also in bold typeface within the chapter. It is recommended that students be required to submit definitions of some of these words as homework assignments in which they look the terms up in other books, articles, or on the Internet.

anaerobic digesters	optimization testing	standard operating
block heating stations	pressure vessels	conditions
grid feeds	programmable logic	subcycle response
load balancing	controllers	system start-up
load cycling	ramping	systems integration
mass production	rare earth elements	tie-ins
mean-time-between-failures	reaction zones	WWTF

DISCUSSION QUESTIONS

1. What would be done if the incoming fuel for a fuel cell contained contaminates beyond the level called for in the system specifications?

2. Why would the codes for pressure piping and boilers, fire and insurance requirements, and local, regional, and federal regulations change the time initial discussions are begun and final installation takes place?

3. Why should government incentives to encourage the use of fuel cell systems not be included in long term cost estimates?

4. Which fuel cells are the most mature of the large scale systems?

5. When considering start-up times for the group of systems making up a fuel cell installation, which would be the most critical?

6. What is the most complicated aspect of systems operations in fuel cell systems?

7. Why is it best to detail all time requirements of shutdown sequences in fuel cells?

CHAPTER 14

FUEL CELL SYSTEMS: END OF LIFE

objectives

This chapter introduces the student to the end of life. Life cycle assessments are introduced as tools used to follow equipment through its operating life and then to arrange final disposal of the associated parts. The costs of making, using, and discarding items is presented as a comprehensive approach to understanding modern technologies such as fuel cells.

INTRODUCTION

Forty thousand hours. That is considered a reasonable life expectancy for a fuel cell. The system should last a bit longer than 4.5 years though, so the fuel cells will be replaced at some time, perhaps with a system upgrade included or balance of plant retrofitting for newer technology. If the technology proves itself, the system may well be in place for decades just as process equipment in major facilities is. While several years is a relatively short time for the guts of a system to last, it is not all that unusual. Boilers, furnace linings, and dies are just a few of the industrial items that have to be replaced on regular intervals. In residential applications, hot water heaters fail and dishwashers go bad almost as often.

Fuel cells are complicated mixes of materials though, and they cannot just be thrown away. **End of life** for industrial equipment is a matter of planning and paperwork as much as dis-assembly and recycling. **Life cycle assessment** (LSA) is done on almost all major equipment now, and in some countries is even required. It examines the **environmental impact** from raw materials extraction to final disposal of all parts. This final chapter will examine the final end of fuel cells and fuel cell systems, using as examples some of the LSA available in literature for several fuel cell types. Again, this is not intended to be an assessment itself but rather an introduction to what is expected when equipment is decommissioned. A generalized LSA is shown in Figure 14-1.

END OF LIFE

End of life for residential equipment is not the same as the end of life for industrial equipment. **Consumer items** are often just thrown away, ending in the local landfill or dumped in the ocean. While **recycling programs** exist in cities, smaller communities can rarely afford the cost of running them, especially in the United States, where the cost of hauling to centralized locations can be enormous

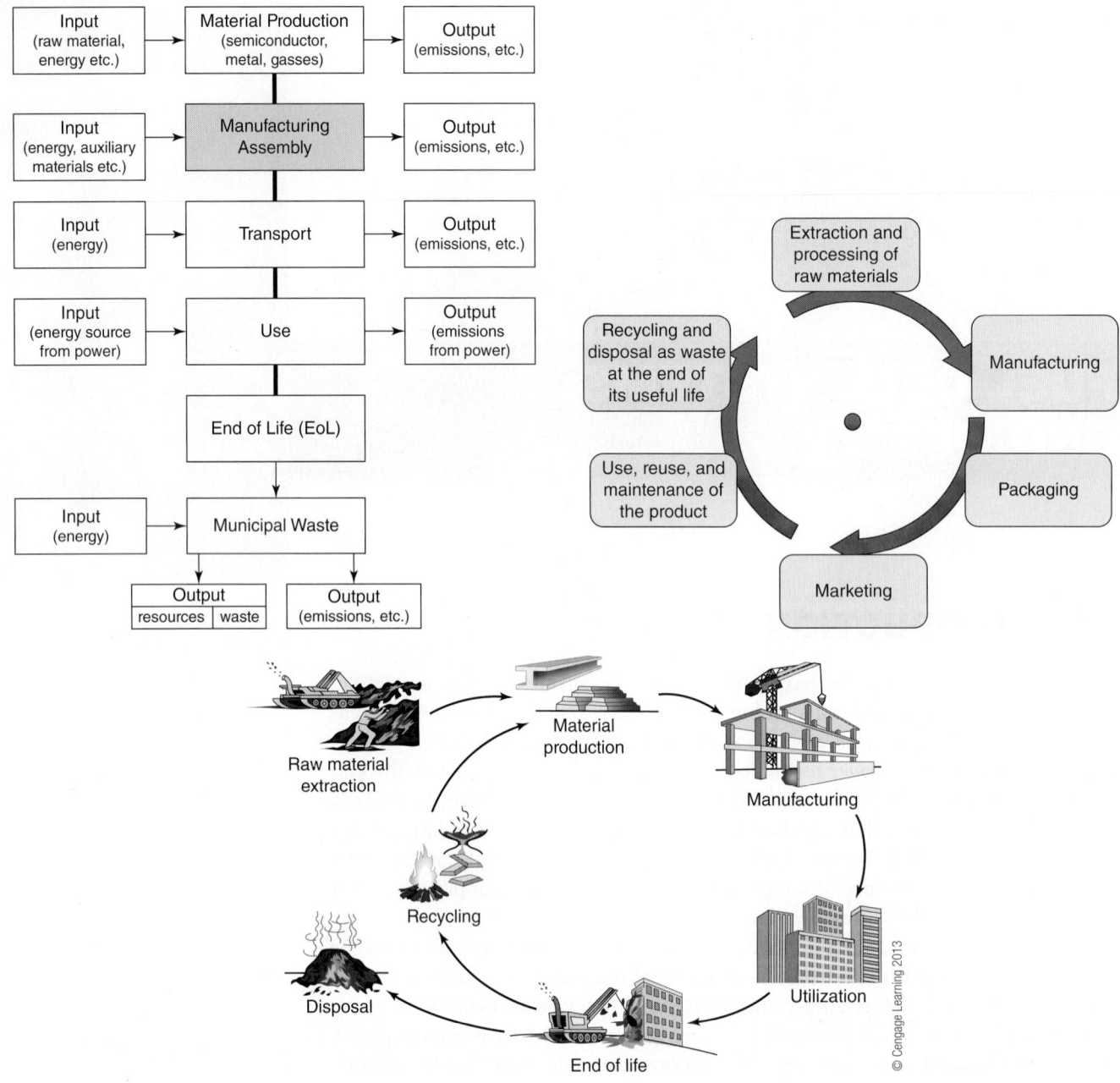

Figure 14-1 Life cycle assessment diagram.

in the less populated states. End of life usually means a product can no longer perform the duties for which it was purchased due to any number of reasons and thus must be removed from service, whether in a kitchen or a petroleum refinery. Some of the products can be directly recycled, some must go to a landfill (some landfills accept **toxic substances** and some do not) and others can be reprocessed into other products. All must be transported, and those costs are also factored in. Figure 14-2 lists common toxic substances found in electronic devices, for instance.

There are lists of what qualifies as a cost as well. Money is in fact a small part of the equation and in many cases is not even mentioned. Even a partial list is quite long. There are cancer-causing agents that in turn become medical costs for the greater society. There is what many call air pollution such as **smog** (respiratory organics) and **airborne particulates** (respiratory inorganics). There is the effect

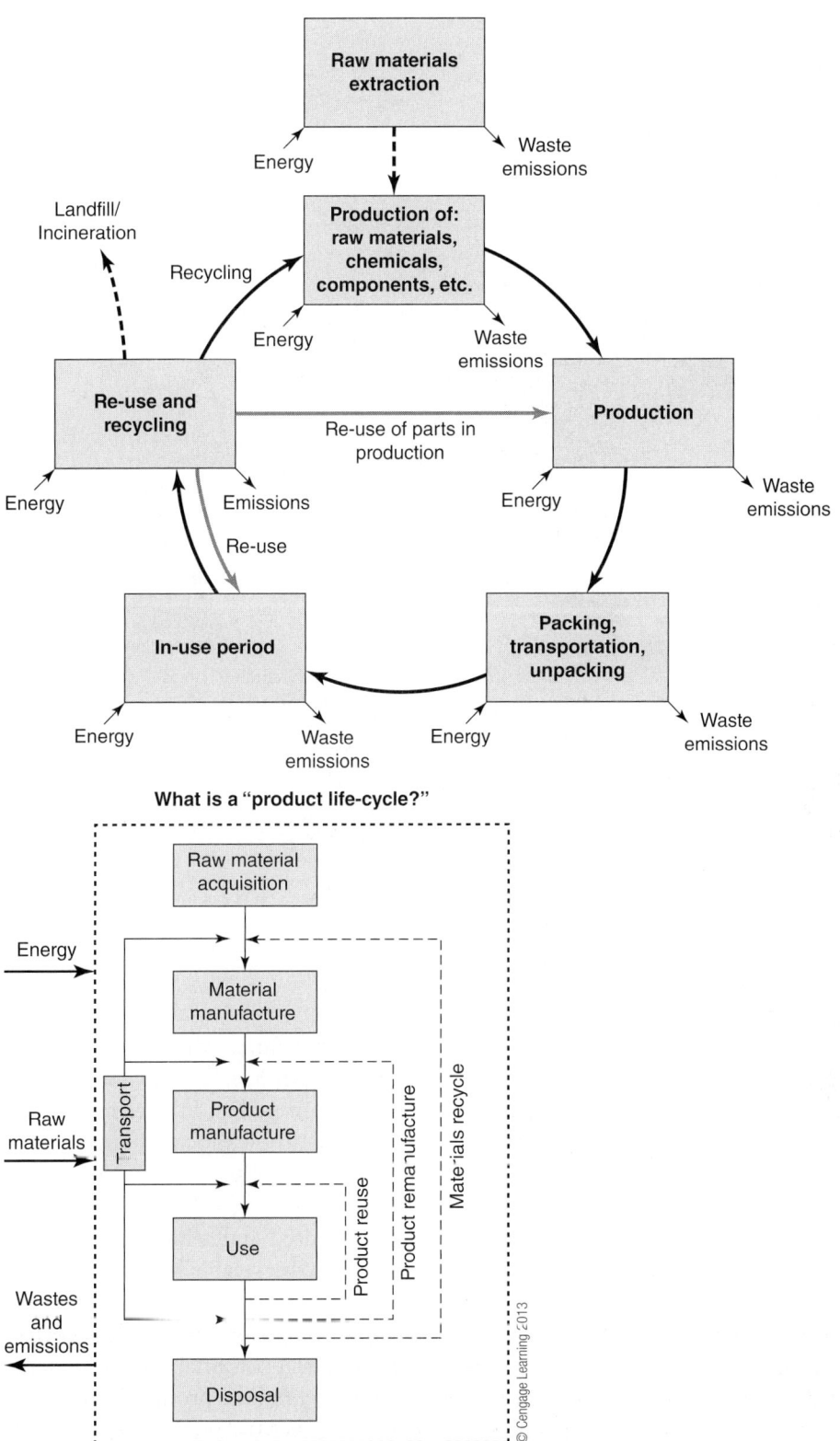

Figure 14-1 (Continued)

on the local and regional wildlife as the material enters the **food chain** in various ways. There is the effect on the local and regional plants as the material moves through the environment. There is the contribution to global climate change. There is the effect on local waterways and lakes such as **eutrophication** (excess nutrients promote plant growth in lakes, turning them eventually into swamps).

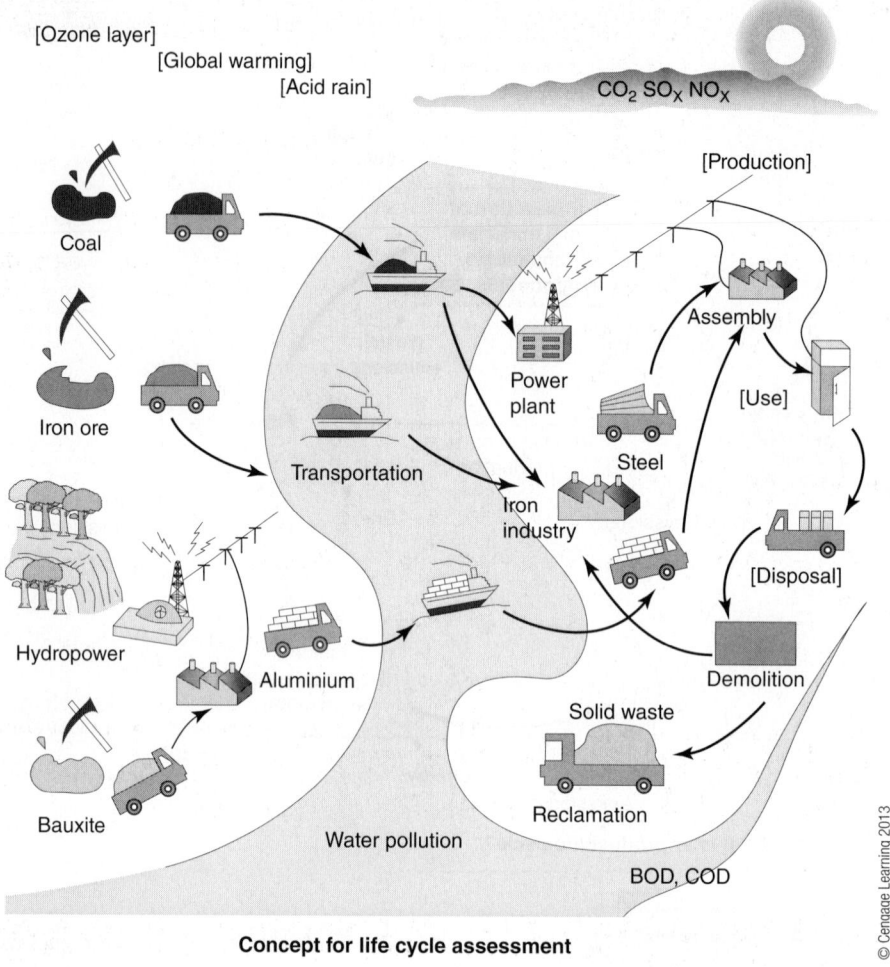

Figure 14-1 (Continued)

These can get quite detailed in some cases, for instance where permitting to construct new plants or mines is being proposed. In essence, each constituent (we will get what that means for fuel cells) has to be examined and the costs of making, using, and discarding that particular piece has to be allocated out. We can use the palladium catalyst in fuel cells as an example to illustrate this. Palladium is mined primarily in South Africa, with Russia producing substantial amounts, mainly from nickel mining operations, and the United States having one major mine in Montana. The palladium is considered fully recyclable but that is somewhat misleading. In the case of palladium, it only means that catalysts in fuel cells or automobile catalytic converters can be **reprocessed**, not that it can be **recycled**. In reprocessing, the product goes back through either its original process or some close version of it to produce a raw material. In recycling, the product is either reused or further processed into another product. Even in Europe, where recycling is mandated by law in many cases, only about half the catalytic converters are actually recycled with the rest presumably disposed of or lost when the older cars are shipped into eastern Europe for repair and resale.

Mining platinum or palladium (mining any metal) is rarely about producing one metal. Along with palladium comes platinum and rhodium and copper and nickel and sometimes gold and silver as well. The issue of allocating the environmental cost of platinum or palladium can be done based on the relative costs of the

N	Substance	Effects on Human Health
1	Arsenic	Lung and skin cancer
2	Brominated Flame Retardants	Hormonal problems
3	Cadmium	Long damage and eventual death
4	Chromium	Carcinogenic when it's inhaled
5	Lead	Anorexia, muscular pain, headache and other annoyances
6	Mercury	After Long-term exposure of many years, damages to the brain, kidney and fetus development
7	Polychlorate biphenyls	Harm to the endocrine system
8	PBDE	Harm to the endocrine system
9	Chlorine Fluorine Coal CFC	No effects
10	PVC	In low levels, no harm
11	Barium	Intake of this substance produce gastrointestinal disorders and muscular weakness
12	Beryllium	Respiratory inflammation known as the Beryllium Chronic disease
13	Cr+6	Cold, nose bleeding, ulcer and damage to sinuses
14	Lithium	Damages to the central nervous system
15	Nickel	Lung cancer and sinusitis
16	Antimony	No significant exposure with the use of electronic equipment
17	Selenium	Selenosis. hair loss, neurological problems
18	Zinc	Corrosive to skin and lungs

Chemical	Uses	Health effects (HE)/Environmental effects (EE)
Mercury	Fluorescent tubes, flat screen monitors	Accumulates in the food chain HE: highly toxic; sensory impairment, dermatitis, memory loss, muscle weakness; damage to central nervous system and kidneys EE: death, reduced fertility, slower growth and development in animals
Lead	Electrical solder, CRT monitor glass	HE: highly toxic; damage to nervous system, blood systems; kidneys, reductive organs EE: highly toxic; similar effects for animals and aquatic life
Cadmium	Contact switches; light sensitive resistors, nickel-cadmium batteries	HE: severe lung damage; kidney damage; bone toxicity EE: harmful to microorganisms and to ecosystem
Polyvinyl chloride (PVC)	Insulation on wires and cables	HE: respiratory problems/lung damage; poisonous when burnt
Beryllium	Heat insulation for CPUs and power transistors	HE: a known carcinogen to humans; lung damage—Chronic Beryllium Disease; beryllium related allergies and sensitivity
Bromated flame retardants (BFRs)	Flame retardant in most electronic devices	Build up in the environment. HE: impaired development of the nervous system, liver damage; damage to endocrine system
Hexavalent chromium	Corrosion protection	HE: highly toxic; causes cancer

■ Figure 14-2 Some toxic substances in electronic items.

Sr. no	Source of e-wastes	Constituent	Health effects
1	Solder in printed circuit boards, glass panels and gaskets in computer monitors	Lead (PB)	Damage to central and peripheral nervous systems, blood systems and kidney damage. Affects brain development of children.
2	Chip resistors and semiconductors	Cadmium (CD)	Toxic irreversible effects on human health. Accumulates in kidney and liver. Causes neural damage.
3	Relays and switches, printed circuit boards	Mercury (Hg)	Chronic damage to the brain. Respiratory and skin disorders due to bioaccumulation in fishes.
4	Corrosion protection of untreated and galvanized steel plates, decorator or hardner for steel housings	Hexavalent chromium (Cr) VI	Asthmatic bronchitis. DNA damage.
5	Cabling and computer housing	Plastics including PVC	Burning produces dioxin. It causes Reproductive and developmental problems; Immune system damage; Interfere with regulatory hormones
6	Plastic housing of electronic equipments and circuit boards	Brominated flame retardants (BFR)	Disrupts endocrine system functions
7	Front panel of CRTs		Short term exposure causes: Muscle weakness; Damage to heart, liver and spleen.
8	Motherboard	Beryllium (Be)	Carcinogenic (lung cancer) Inhalation of fumes and dust, causes chronic beryllium disease or beryllicosis. Skin diseases such as warts.

Figure 14-2 (Continued)

metals or the relative amount of the metals produced per ton of ore mined. That depends on the mining being used. Some mines are open pits, where all the rock is dug up and then processed to some extent. Some mines are underground, where veins of ore are followed and much of the rock left undisturbed. Open pits can economically recover even very small amounts of metal in the ore. Gold, for instance, can be mined in some cases (depending on the price of gold at any given time) down to 0.02 ounces of gold per ton of rock dug up (50 tons dug and processed to produce one ounce of gold, worth about $1700 at the time of this writing). All the metals produced are then allocated per ton of ore dug with that following down through the process of grinding, milling, smelting, and refining. Each step processes less and less material as the metals are concentrated along the way, but each process also tends to produce more concentrated waste. Not only the metal is an issue then, but all the material that had to be separated out as a nonuseful material and then dumped. Allocating the environmental cost of mining determines the environmental impact of the several hundred pounds of copper in a fuel cell as well as the precious metal catalysts used, the iron used in the steel, the aluminum in the motors, the alloying elements in the bolts, carbon in the bipolar plates, and any rare earth elements used.

This allocation can be used in a number of ways as well. It can go out as a per cell number, where the entire system manufacturing, transportation, installation, operations, maintenance, replacement of consumables (even the stacks themselves), final dismantling, and disposal is spread over however many cells are used over the entire life cycle. Each cell is then compared to other cells or competing technologies. This may not work well for direct comparison between other technologies like generators though, so the entire energy output (electricity and heat) over time might be calculated and then divided into the lifetime of the product. An average lifetime can also be used, so for instance an automobile may state a cost per mile figure that is based on the registrations of that line of cars across a country or several countries. Most of these assessments weigh more heavily on those items which have greater environmental impacts than others. Aluminum, for instance, takes less than 10 percent of the energy to recycle as it does to make from ore, so if significant recycled aluminum is used, then its impact on the product is less.

Many emerging technologies suffer from the lack of a recycling structure being in place so these assessments can be skewed for new materials or products. Fuel cell systems rely on a number of already existing technologies though, so much of their assessments are based on facilities already in place, such as steel, copper, or aluminum recycling, and to a large extent, only the fuel cells and parts of the reformers lack existing networks.

REGULATORY IMPACT AND PROCEDURES

There is a great deal of regulation involved in recycling, especially in Europe and Japan, which are probably the two primary fuel cell markets. While listing **regulations** is not the purpose of this chapter or textbook, the student no doubt has realized how important they are from how many times they have been mentioned. End of life management is one of the primary thrusts of environmental legislation with mandatory recovery and recycle targets set along with appropriate penalties. While fuel cells in particular are not yet included in this legislation in most countries, the stronger the market becomes, the more likely **end of life regulations** will be put in place. There is also an issue of **system regulations** (like a refinery or smelter) where there may be some situations where fuel cell installations are not allowed because they are not regulated, which may result in industry preferring regulations be put in place. An example of this is in the European automobile regulatory structure (85% of a car's weight has to be recycled) where SOFC systems may be kept out because they do not meet the overall automobile end of life requirements not because they cannot be recycled but because the required processes and regulatory paperwork do not yet exist to prove that recycling was done.

Life cycle assessments and end of life procedures rely on four primary routes: **reduce** the worst offenders, **reuse** whatever can be reused, **recycle** what cannot be directly reused, and finally **dispose** of what is left. Designers can reduce the volume of the worst offenders (these are the toxic materials or the one with the largest overall environmental impact) or replace them with materials that are less toxic or have less impact. Manufacturers can directly reuse components, going down to the level of nuts and bolts or at least directly return materials to primary reprocessing such as steel or aluminum. Recycling (segregation and purification) is usually split into three categories: **high value materials**, **low value materials**, and **hazardous materials**. High value materials are many times low in volume or weight but worth substantial sums like precious metals so they represent a lower risk due to small amounts but a higher return on recycling efforts in an economic sense. Many times, high value materials can help to cover the cost of other efforts in end of life issues. Low value materials are present in larger amounts but do not have the economic impact. These materials differ from reuse materials fed to primary processes in that they generally require treatment steps between disassembly and the actual recycling, such as the copper in wire that can be directly treated and the copper in circuit boards that needs to be separated first. Hazardous materials may or may not be recyclable but are on regulatory lists as representing known dangers for workers or to the environment if discharged directly. Any materials remaining are disposed of directly (or after some secondary processing in the initial recycling processes). These go to landfills for the most part.

FUEL CELLS AND SYSTEM EOL

When a fuel cell system is removed from service, it has to be dealt with in a formal end of life program. It is not complicated (well, the paperwork might be) but it is comprehensive and important to the company owning the systems. One very important fact needs to be addressed at this point. Unlike consumer items, when corporations buy or use something (equipment, land, buildings . . .), they own it forever. If they put it into a landfill, then it is logged in and tracked so that if a hazardous element is found in the groundwater 50 years later, all the companies who disposed of items containing that compound can be charged for the cleanup. If a mining company digs a big hole in the ground and leaves piles of waste rock, they have legal liability for those things forever. The only way to avoid this rather substantial liability is to either reuse or recycle the material. End of life issues are very important not only because of regulations and public opinion but also because **unlimited liability** stretching over centuries is probably not sound corporate policy. Much has been made of corporations walking away from environmental liabilities, but it should be noted that in fact, there are corporations that have been around for well over a thousand years, such as the Marinelli foundry in Italy and Ito Tekko metal works in Japan. Even bankruptcy and corporations being bought and sold do not eliminate liability, as governments will in fact pass liability on to **successor companies**. End of life planning and implementation has become a critical aspect of modern corporate strategy.

Fuel cells and systems are not really major issues when it comes to end of life. Given their balance-of-plant arrangements and internal chemical compositions, they rarely contain hazardous or toxic elements in any great amounts, and most can be dealt with using existing methods and processes. They do use expensive and sometimes scarce elements though, and that may prove to be an issue for some systems. SOFC in particular use several rare earth elements in the cell itself and have to compete for those resources while all the lower temperature systems use precious metal catalysts and the cost of those metals is very high and getting higher.

The simplest form of a fuel cell end of life sequence would be if the fuel cell portion was just going to be replaced. In that case, temporary utilities would be arranged, and a system shutdown planned but this would probably be a routine already established in plant procedures. The fuel cell stack would be replaced and any upgrades or improvements accounted for. The system would be brought back on line and then the **spent fuel cell** (the old one just removed) taken care of. Spent fuel cells are currently being returned to the manufacturer and probably refurbished.

There may be regulatory paperwork associated with such a changeover but other than that, the requirement would be minimal. The owner of the fuel cell system would have transferred legal ownership of the spent fuel cells to the manufacturer at that point. Subsequent refurbishing or recycling of the cell would then end the manufacturer's legal responsibilities.

The first step in a complete system end of life sequence would be the shut down itself. This may seem like an odd statement, but in fact this step can be extremely complicated in a fuel cell installation. The system will have been in place and operational for a number of years and this implies that it would have been successful and thus have supplied (and was still supplying) electricity and heat to the site. Well before it is turned off, the means to replace those critical **on-site utilities** has to be chosen, funds made available, a replacement site chosen and prepared, as well as installation and the changeover successfully completed. The changeover itself represents a substantial amount of time, effort, and money. Significant planning must be done to ensure any disruption to utilities is minimized or eliminated. If done during some type of shutdown, the incoming fuel must be stopped and rerouted to the new installation, plant utilities must still continue to contractors and testing be done on the new system. If the installation is at a campus setting such as a hospital or apartment complex, then it is even more complex, since utilities cannot be disrupted to these sites for more than several hours and the entire customer base served by the system would have to be informed. Any new technology (almost guaranteed as far as fuel cells are concerned) in the new installation would have to be taken into account, any new regulations would have to be met, updating to any revised codes and standards would have to take place, and any training needed for the replacement system done. In addition, a full end of life assessment on the old facility would have to be done, including an inventory of all subsystems, parts, materials, any spare consumables remaining, and possibly a full plan for dealing with all of these that is then submitted to a regulatory agency (maybe more than one).

Once the old system is removed from service, it must be disassembled. Since fuel cell systems consist of several subsystems, this could be an interesting exercise. The subsystems would first be evaluated for reuse, probably before the systems were even shut down. Blowers, compressors, external reformers, and other subsystems that might be of use (perhaps in the replacements system) would be marked and arrangements made. Keep in mind however that reusing equipment in an industrial setting is difficult, since for the most part shutting down a plant to remove, relocate, and replace something costs more in plant down time than is saved by using old equipment. In general, any still serviceable piece of equipment would either be sold to used equipment dealers or set aside as possible emergency equipment stores. Following that, the general scheme of recycle and disposal detailed in the above discussion of life cycle assessment would be followed.

It is not possible to catalog every possible piece of equipment that might be used in all the fuel cell systems and discuss their possible end of life destinations. A general listing of the many items from the associated subsystems can be given to provide a general guide to what may be encountered in an end of life situation for fuel cells in general. Some of the issues facing particular fuel cells can then be discussed as well.

There will be buildings or enclosures, support frames, and structures as well as wiring tracks and support structures to be dealt with. This would include concrete pads or pillars, foundations, walkways, and any type of walls. Some installations require leak pads to contain any liquids that might be lost from the installation during an emergency. The particular type of fuel cell installation would determine if a pad is needed, but given the water involved, it may be required. This has not been mentioned before, but there might also be an in-place **water treatment** plant that would be used for any final water discharge. The high temperatures and pressures usually combine with the materials and the deionized water to produce an effluent that is high in metals, particularly chrome, so that it would have to be handled as a part of operations prior to any **discharge**. Letting this water go into the general plant (hospital, apartment, or box store) water discharge system might not be allowed, so a small system could be required and it would require leak pad protection. That plant may well become part of the end of life dismantling as well. Figure 14-3 shows how water is treated industrially, many times for reuse within the operating plant.

There are several subsystems involved such as air, fuel, water, reformation, heat, steam, monitoring, and control. Most of the overall system material is going to be a combination of steel and stainless steel, with some systems reporting up to two-thirds of total weight being those two components and copper taking up another 10 to 15 percent.

Air systems generally will be composed of fans and blowers. These will be used primarily as compressing or air flow management agents but many systems will also have a vent-to-atmosphere capability for both the generated heat and any excess hydrogen, either from external reformers, supplied gas, or on-site storage.

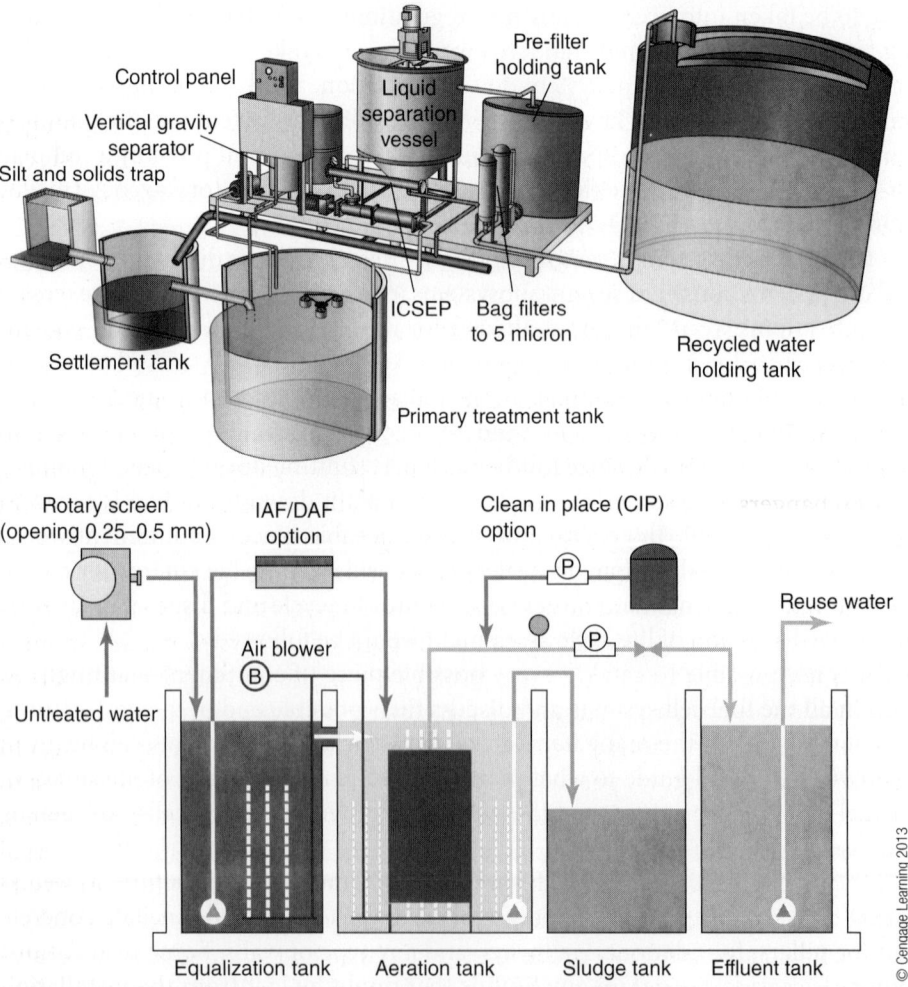

■ **Figure 14-3** A typical industrial water treatment scheme.

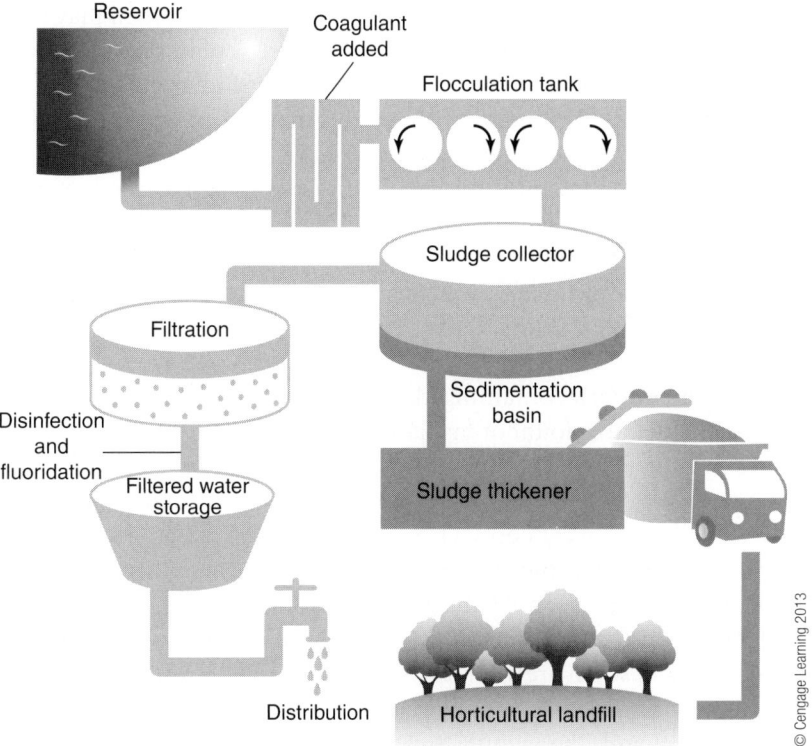

Figure 14-3 (Continued)

These fans will not be small, especially in larger systems where units weighing as much as a ton are reported. These would be multiple flow paths systems that can be used for cooling, air flow management, and venting and so represent major items. Each of these would have associated ducting that might be contaminated by sulfur or chloride compounds from any gas cleaning operations so testing would have to be arranged. It is important to realize that testing for hazardous or toxic chemicals by **certified laboratories** will be required for all parts and pieces. Regulators generally do not accept testing done by the responsible parties but require neutral third parties to test all material for contamination. Monitoring and control systems would also be in place for these fans. Other fans would vent any area into which fuel could leak as well as provide air flows to the reformer, control cabinets, heat exchangers, and the fuel cell stacks. Several different heat exchangers might be used as well. Each of these, along with ducting and electronics, would have to be evaluated individually with all testing and decommissioning documented.

The water system would have pumps to supply overall water, an entire treatment and purification system to provide the high quality water needed in most systems, and at least one major condensing system. Steam is an integral part of most systems so a separate steam handling system would be in place, consisting of piping, insulation, valving, monitoring, and control. The steam system would also have **steam ejectors**, possibly in several spots, used whenever steam has to be mixed with another gas stream, such as in desulphurizing systems, reformers, and humidification for the fuel input gas to the cells. There might be other internal loop systems such as a recirculation system for MCFC as well as the condensate loop that would operate in conjunction with the output gas system for the stacks.

The fuel system itself can prove to be costly in an end of life scheme. If a natural gas or industrial hydrogen source is used, it may have to be dismantled back to the main line. If it is a hybrid system using solar and wind to generate hydrogen which is then stored using one of the various methods, that entire system may have to be dismantled as well. An on-site reformation subsystem could possibly be reused in its entirety, but as mentioned that presents problems industrially as well as commercially or residentially. Technological advances may also render the subsystem obsolete, making reuse unacceptable. There are also sub-subsystems involved such as both high and low temperature shift converters and gas cleaning systems along with the consumables these use. Both of these will contain catalysts, with sulfur removal catalysts containing precious metals and shift converters probably containing copper, nickel, and zinc. This means the units have to be disassembled and the precious metals (or hazardous waste or toxic materials) separated. Metals, plastics (polymers), and ceramics have to be separated as well. The reformer itself follows, made of stainless steel, containing ceramic catalyst beds and nickel catalysts along with a substantial amount of insulation.

Cell stacks are next. There are the cells themselves with their electrodes and electrolytes, the catalysts, the seals, the outer housing units, insulators, tie rods, any end plates, and the electrical connections. Monitoring and control devices would also be in the stack. Just the stacks would represent the entire list of end of life issues. Commodity metals, precious metals, rare earth elements, toxic metals like hexavalent chrome, environmentally sensitive metals like copper, plastics, ceramics, electronics, and so on are all in the stack assembly, making it one of the most difficult to deal with. The ultimate solution to this type of assembly is to have it reused almost entirely, with the manufacturer taking it back, reusing everything but the cells themselves in a refurbished stack, and then sending the cell components to recycle where the metals (commodity, precious, and rare earth) can be recovered and the porous substrate material probably reused as a feed stock once it is cleaned.

The electrical output, control, and power conditioning systems are difficult during end of life closing because many of the parts such as wiring insulation and circuit boards contain both toxic and hazardous substances. The wiring, electronics, and other items can usually be disposed of using companies specializing is such issues, so as long as the parts are correctly disassembled, separated, packaged, and documented, this step may mean little more than making sure the paperwork to cover final end of life disposal in this manner is correctly filled out and filed in a timely manner.

Some parts of the thermal management system may well be left in place at end of life, since these systems usually tie into the general installation site. End of life for these may require piping and duct changes as a new system takes the place of the old system.

Finally, the many small parts have to be dealt with. It might seem ridiculous but inventory will be done and bolts, nuts, washers, fittings, valves, and a number of other parts may well be counted up and signed off on. These may go into a general warehouse inventory or sold as scrap, provided they were not in areas where contamination may have occurred.

Here is a short list of main materials that might be encountered in the end of life sequence:

- Various alloys of carbon steel, stainless steel, copper, and aluminum as well as nickel and precious metals attached to various substrates as catalysts.
- Carbon, carbon black or graphite, and possibly carbon nanotubes (particularly burdensome in disposal).
- Ceramics such as zeolite, zirconia, and alumina, many with alloying elements or catalysts attached.

- Plastics such as polytetrafluoroethylene, polypropylene, polyethylene, and polyethylene terephthalate, to name a few.
- Insulation materials.
- Wiring, computer, and electronic equipment.

There are individual items of concern as well. Plastics (polymers) are particularly difficult to deal with, especially any containing fluorine, such as the Nafion used in SPFC (PEM) cells. Carbon is difficult to deal with, mostly being incinerated but only if it has not been associated with other chemicals. The presence of fluorine in fuel cells makes burning carbon problematical. Certain metal oxides such as nickel (catalysts) or chrome (interconnects) require special treatment. Ceramics are difficult to do anything with as regards recycling or reuse due to their stability and the high temperatures (and thus cost) needed to process them. Many ceramics are recycled into concrete. Precious metal catalysts are worth so much money that recycling is almost always worthwhile but many companies tend to just sell the scrap to refineries and let them deal with the issue while the refineries do not appreciate raw scrap since they have to process it before putting into their stream. It may be that companies specializing as **middlemen** performing both stack disassembly and initial separation come into being, in which case technicians understanding the basics of fuel cells will be intimately involved in this type of work. Gas diffusion (bipolar/interconnect) plates may turn out to be problems to the industry since graphite and ceramic chrome alloys are difficult to reprocess, as are any resin impregnated items.

Finally it should be noted that end of life sequences represent a significant opportunity for engineers and technicians since they are the ones who will detail and perform the actual work. This is an ever expanding area and the student would be well advised to consider change occurring in what used to be called the scrap business.

KEY WORDS

Knowing the terminology used is critical when dealing with fuel cells. Following is a list of the important terms in this chapter, which are also in bold typeface within the chapter. It is recommended that students be required to submit definitions of some of these words as homework assignments in which they look the terms up in other books, articles, or on the Internet.

airborne particulates	hazardous materials	regulations
certified laboratories	high value materials	reprocessed
consumer items	life cycle	reuse
discharge	assessment	smog
dispose	low value materials	spent fuel cell
end of life	middleman	steam ejectors
end of life regulations	on-site utilities	successor companies
environmental	recycle	system regulation
impact	recycled	toxic substances
eutrophication	recycling programs	unlimited liability
food chain	reduce	water treatment

DISCUSSION QUESTIONS

1. At what point does an industrial products life cycle start?

2. What is the difference between recycling and reprocessing?

3. What are the four primary routes that can be followed when designing a product for end of life disposal?

4. What are the three recycling categories?

5. Is a fuel cell installation dismantled one the fuel cells themselves are worn out?

6. Can fuel cells or fuel cell systems be hot swapped at the end of life?

7. What are the three main components to be considered when arranging end of life for fuel cell installations?

DISCUSSION QUESTIONS AND ANSWERS

CHAPTER 1

1. What is the difference between a primary cell and a secondary cell?

Primary cells make use of irreversible chemical reactions while secondary cells make use of reversible chemical reactions. A primary cell produces electricity from a chemical reaction such that when the material used in the reaction is used up, the reaction stops and so does the production of electricity. Primary cells can only produce electricity; they cannot be recharged or renewed in any way. Secondary cells can be recharged by running electricity back through the cell (reversing the direction of the chemical reactions) and renewing the material used in the secondary cell.

2. What is the difference between the contact theory and the chemical theory?

In the chemical theory, electricity is produced during chemical reactions that free electrons from some part of the reactants and then are capable of supplying them to external wiring to allow electricity to flow. In the contact theory, electricity is produced when certain metals are moved within a magnetic field or the magnetic field is moved while the metal remains stationary.

3. Why are electrons negatively charged?

They must be or they are not electrons. The definition of electrons states that they must be negatively charged.

4. What is the difference between a proton and an electron?

A proton is a hydrogen atom that has lost an electron. An electron is smaller, lighter, and more mobile than a proton.

5. What is the difference between direct current and alternating current?

The difference between AC and DC depends on how the electrons behave. In DC, the electrons move in one direction in the wire, form the negative electrode to the positive electrode.

In AC, the electrons move forward and backward as the poles reverse, traveling only short distance between reversals.

6. What is economy of scale?

It is considerably cheaper to make many things at a time rather than to make one thing at a time. The larger the scale of your production, the cheaper it is to produce each individual thing in that production.

7. What is the difference between an anion and a cation?

An anion is an element or molecule that has gained one or more electrons and is negatively charged. A cation is an element or molecule that has lost one or more electrons and is positively charged.

CHAPTER 2

1. How many atoms of oxygen are in each of the following molecules?

H_2O	1
H_2SO_4	4
CH_4	0

2. In H_2SO_4 (sulfuric acid), are the oxygen atoms bonded to something or are they individual atoms in a group that are not bonded to any other atom?
All four are bonded to something else.

3. Express the following numbers in scientific notation.
910   9.1×10^2
730000 7.3×10^5
0.0024 2.4×10^{-3}
0.17 1.7×10^{-1}

4. The electrical charge on one electron is -1.602×10^{-19} coulombs and one amp requires 6.25×10^{18} electrons per second moving past your plug. Since your house has a 150 amp service, how many electrons would it take flowing through your main breaker in one second to trigger an overload and shut your power off?
6.25×10^{18} times 150 equals 938×10^{18} (9.38×10^{20}).

5. Why would the catalyst in the fuel cell shown in this chapter have to be conducting?
The electrons are freed at the catalyst, and if the catalyst was not able to conduct them through to the rest of the electrical circuit, there would be no electricity.

6. Which of the following is an atom and which a molecule?
H_2 molecule
Au   atom
H^+ atom
H_2O molecule
H_2SO_4 molecule
CH_4 molecule

CHAPTER 3

1. Under the HazCom standard, does the employer have the right to withhold information about hazards in the workplace until a person has accepted a job and discovers the hazard?
No.

2. Where would an employee look to find information about the hazards of a specific chemical used in the workplace?
Material Safety Data Sheets.

3. Combustible liquids can be either a solid or a liquid. Are explosive hazards a solid, liquid, or gas?
Can be any of the three mentioned.

4. Are irritants toxic?
Yes. Inflamed tissue can be a health hazard.

5. Does ingestion require penetration?
No. Ingestion cannot involve direct penetration of the skin.

6. If signs are required for a fuel cell installation detailing hazards, what codes would give guidance as to the types of signs?
The National Fire Protection Association code (NFPA 853) as well as the International Fire Code (27).

7. Name the parts to a fuel cell electrical system?
Disconnects, wiring methods and materials, grounding schemes, markings, connections to other circuits, outputs (usually differentiated between less than 600 volts and more than 600 volts), and ancillary equipment.

8. In a mixed system (AC and DC combined), are grounding systems required for both forms of current?
No, the DC systems are required to be bonded to the AC systems with only a single grounding bar allowed.

9. What is the purpose of the NFPA?
To "provide fire prevention and fire protection requirements for safeguarding life and physical property associated with buildings or facilities that employ stationary fuel cell power systems."

10. Are codes the same as laws?
The codes are meant to be advisory but can be granted the force of law by governmental agencies such as zoning boards or by specific contracts between individuals.

11. Why are definitions critical in codes and standards?
Many codes are international in nature and the exact meaning of words used in contractual obligations may determine liability.

12. Are exceptions allowed under most codes?
All codes contain nonstandard clauses that allow for exceptions.

13. How far away from an air intake can exhaust fans be located?
4.6 meters (15 feet).

14. Look up several of the chemicals mentioned in the preceding chapters and find their MSDS sheets. Do this as new chemicals are introduced in future chapters as well.

CHAPTER 4

1. Why would a low temperature fuel cell better handle on-off cycling than a high temperature fuel cell?
Cooling down and reheating a high temperarture cell may take longer than the on-off cyle itself.

2. What are the five major types of fuel cells?
Alkaline (AFC), Solid Polymer (SPFC), Phosphoric Acid (PAFC), Molten Carbonate (MCFC), and Solid Oxide (SOFC).

3. Why would a PAFC system usually operate at lower temperatures?
So the liquid acid does not boil off.

4. What is a hydroxide ion?
One atom of oxygen that is chemically bound to one atom of hydrogen but that also has one extra electron so it has a negative charge.

5. What is the name of a chemical reaction that absorbs heat while it is occurring?
An endothermic reaction.

6. What is 17 parts per million (ppm) expressed as a percentage?
17/1,000,000 or 0.000017 or 0.0017%.

7. In the reaction 2H$_2$ + O$_2$ ↔ 2H$_2$O, how many molecules are reacting and how many molecules are being produced?

Three molecules (2 hydrogen gas molecules of H$_2$ and 1 oxygen gas molecules of O$_2$) are reacting to form two molecules (2 water molecules of H$_2$O).

8. What is Direct Internal Reforming?

Where the input fuel gas is reformed (the input hydrocarbon fuel is chemically broken apart to create carbon and pure hydrogen) as part of the cell and not in a separate system.

9. Why would a PAFC need a reservoir plate?

Because the liquid will evaporate over time and since the unit is sealed, there is no other way to replace the lost phosphoric acid.

10. What is ohmic loss?

When electrons move through a material that does not conduct electricity very well, they lose energy, usually because the material heats up, and that energy is not available as current to be used.

11. What is an inorganic substance?

Something that does not contain carbon.

12. What is the difference between a bipolar plate and an interconnect plate?

They are the same thing. Interconnect plates are often mentioned when SOFC systems are being discussed, while bipolar plates are often mentioned with SPFC (PEM) systems.

13. How many electrons are transferred between the reactants and the products in the reaction H$_2$ + O^{-2} ↔ H$_2$O?

Two, as indicated by the −2 above and to the right of the oxygen symbol (O).

CHAPTER 5

1. Where is the reaction chamber of an internal combustion engine?
The cylinder.

2. Rank following cells according to how long it takes the system to come to full power: SOFC, SPFC(PEM), and AFC (shortest to longest).
SPFC(PEM), AFC, and SOFC.

3. What is the difference between the electrolyte in AFC systems and that in PAFC systems?
Both are liquid but PAFC are acids while AFC are bases.

4. Look up the coefficient of thermal expansion for zirconia and for 316L stainless steel.

5. Will an electron move toward a cathode or toward an anode?
Electrons are negatively charged, so they will move toward the positively charged electrode, which is the cathode. Anodes are negatively charged.

6. Will water move toward the Polytetrafluoroethylene (PTFE) in Nafion or toward it?
PTFE is hydrophobic, so it repels water.

7. In an AFC system, will adding more of the KOH electrolyte make the cell more acidic or more basic?

More basic, since the OH^- molecule will go into solution and a base has more OH^- than H^+.

8. Name the parts in an SOFC membrane electrode assembly.

Interconnect, wire mesh to gather electrons, anode, electrolyte, and cathode.

9. If you put a drop of liquid potassium hydroxide (KOH) on a countertop, what would it do?

Since KOH has a low surface tension, the liquid drop would spread over the top and not bead up.

10. If you had a powder made of a material that melted at 1000°C and wanted to sinter it to form a solid, would you use a temperature over the melting temperature or under it?

Sintering is done at temperatures under the melting point.

11. If you used pure oxygen instead of air in a fuel cell, how much less gas would you be able to use?

Since air is about 20% oxygen, using pure oxygen would allow you to use about one-fifth less gas.

12. Name three required design elements that might be used when designing an electrode.

Gas must be able to diffuse through the electrode to reach the catalyst sites. Catalysts must be widely dispersed to maximize the three-phase requirements of fuel cell reactions. The electrolyte (liquid potassium hydroxide in the case of AFC systems) must be in contact with the catalysts. Some way to transfer both electrons (electricity) and ions (OH^+ in the case of AFC) to the reaction sites on the catalyst must be in place. The combination of materials used in an electrode must withstand gas, electrolyte, temperature and pressure.

13. What is the difference between indirect internal reformation and direct internal reforming?

In DIR, the input fuel gas is reformed using the same catalysts sites as those used by the main fuel cell reactions, whereas in IIR the reformation is done first using outer catalysts sites, and then the fuel gas moves further into the electrode to use other catalyst sites for the main fuel cell reactions.

14. How many atoms are in one molecule of $LaMnO_3$?

Five. One lanthanum, one manganese, and three oxygen.

CHAPTER 6

1. What would happen if a SPFC(PEM) cell became dehydrated?

The cell would stop producing electricity, since the electrolyte must have water to pass the ions between the cathode and anode.

2. What type of cooling system is used to maintain the right operating temperature in a low temperature, recirculating AFC system stack?

The alkaline electrolyte itself would be used to cool the stack.

3. Why is the cooling system of an SOFC system more complicated than the cooling system of a SPFC(PEM) system?

Because SOFC systems operate at much higher temperatures than SPFC (PEM) systems.

4. What gas pressure is used in the electrolyte area of a SOFC?
Since SOFC electrolytes are solid, there is no gas pressure.

5. If I put a lid on a pot of water, does the water have to reach a higher temperature to boil or a lower temperature?
If you increase the pressure above a liquid, you will also increase the temperature needed to boil (move a liquid molecule off the liquid surface and into the gas above the liquid), so the boiling temperature goes up.

6. Which is more expensive to produce, internal manifolding or external manifolding?
Internal manifolding.

CHAPTER 7

1. How is power conditioning handled differently in fuel cell systems than in grid supplied systems?
Fuel cell systems generally do power conditioning at the source of the current (the fuel cell itself), where grid systems usually condition the power where it is to be used rather than where it is produced.

2. Name three different methods to produce hydrogen.
Reforming fuel gasses, electrolysis, gasification, liquid reforming, or biological production.

3. Name two important issues that currently limit the widespread adoption of fuel cells.
The availability of fuel and the cost of produced electricity.

4. Which of the following energy sources are considered distributed sources: fuel cells, batteries, solar power, and wind power?
All of them.

5. Why are fuels like diesel or gasoline not considered good candidates for reforming in fuel cells?
They contain too many contaminants.

6. Why is there more than one step to reforming hydrocarbons?
One step has to break the carbon and hydrogen apart, another has to get rid of the carbon monoxide that is left over, and others have to deal with any contaminants in the gas stream.

7. Which of these hydrocarbons would be considered heavy: CH_4, C_2H_6, or $C_{10}H_{22}$?
$C_{10}H_{22}$ since it has a large number of carbon atoms.

8. Can fuel cells be shut down in an emergency?
Not really. Low temperature systems will suffer but not be destroyed because thermal cycling is bad for all the parts; high temperature systems stand a good chance of being destroyed if the cooldown is too fast.

9. Name some of the advantages internal reforming has over external reforming.
IR requires less equipment, makes direct use of the heat generated by the fuel cell, effectively eliminates several subsystems, and tends to equalize the supply of hydrogen going to the cells so that manifolding and gas distribution schemes are not needed.

10. **The chapter says that hydrogen must be cooled to 22°K to be in the liquid form. What is 22°K in degrees Celsius and degrees Fahrenheit?**

22 K is −251.15 Celsius and −420.07 Fahrenheit.

CHAPTER 8

1. **What are two of the most important features that distinguish one type of fuel cells from another?**

The electrolytes used within the cell and the temperatures the cells operate at.

2. **If the operating pressures and temperatures of a fuel cell are increased, what happens?**

The voltage and current produced by the cell increase but efficiency and life expectancy tend to decrease.

3. **Why can't fuel cells used in automobiles add Combined Heat and Power cycles to make the cells more efficient?**

The heat cannot be used for anything, since automobiles do not typically use heat for anything other than warming the cabin up.

4. **Name two reasons the ceramic matrix holding a MCFC electrolyte might crack in service.**

If the temperature rises or falls too fast within the cell, expansion and contraction might crack the brittle ceramic or if the pressure at one electrode is much greater than at the other electrode, the pressure gradient might crack it as well.

5. **Why are MCFC systems not considered mass produced items?**

Because the seals cannot be set until the operating temperature is first reached, so the system will not work until it is fully activated.

6. **In lower temperature systems like PAFC, how much of the total system cost pays just for the platinum catalyst?**

Up to 20% of the total system cost.

7. **Why do fuel cells produce variable voltages?**

Because it is not possible to produce exactly the same number of chemical reactions in any given time frame, so that the number of electrons flowing in the system has to fluctuate somewhat.

8. **Discuss some of the problems that might arise as fuel cell stacks get larger.**

The larger the cell (or stack) the more problems high temperatures will cause. The larger the cell, the more pressure changes there will be internally, causing fatigue cycling that may severely shorten the life span of a cell. The larger the cell, the more material there is inside it, increasing weight and thus stress within the cell. The larger the cell, the more likely it was produced as a custom installation, and thus the more likely it will have non-standard features that are not documented or spare parts that are not easily obtained in a timely manner. The larger the stack, the less mechanically stable the stack will be and the worse the seals will be due to small deformations in the cells. The larger the cell or stack, the less reliable the gas distribution will be. The larger the cell or stack, the more important the cooling system will be. The larger the stack, the higher the internal gas pressure that will be needed to guarantee gas flow to anode and cathode. The larger the stack, the more difficult it will be to access the internal workings of each individual cell.

CHAPTER 9

1. What is the primary feed fuel currently used in stationary fuel cell installations?

Natural gas.

2. Coal-fired power plants in England generate approximately 14 gigawatt-hours of electricity each year and the loss on transmission for the English national grid is 2.3%. How many watt-hours of electricity are lost each year form the coal fires' generating capacity?

14 gigawatt-hours is 14,000,000,000 watt-hours. 2.3% of 14,000,000,000 is 322,000,000 watt-hours lost on transmission.

3. What associated units might be included in a smaller stationary system used for consumer applications and set up for plug-and-play use?

At the least it will have air and water filters, control units, a reformer unless higher temperature units are set up to do internal reformation, some type of fuel cleaning unit (to take sulfur out in particular), heat exchangers, tie ins to the residential heating system, a cooling water system, power conditioning that includes AC to DC converters, and a burner to combust any fuel gas not used in the fuel.

4. In a small, stationary consumer system meant to service a single house, would it be better to use a CHP subsystem or a cogeneration subsystem with the unit?

A CHP would be more appropriate due to cost and size constraints in consumer applications of this sort.

5. For a unit to be considered as a primary utility source, what needs must it meet?

They must be extremely reliable, operating both for extended times between scheduled maintenance and for what is essentially the life of the building overall.

6. What are two primary requirements of data centers?

These centers need two things: very high quality electrical supplies and reliable cooling.

7. The computers in data centers generate large amounts of excess heat. How might this heat be used if fuel cells are installed in a data center?

It can preheat fuel and air supplies, be combined with the higher temperature waste heat streams of the fuel cell to provide both heat and cooling, as well as be used in turbines to increase efficiency.

8. Why is maintaining a synchronized spinning reserve a very expensive proposition for a utility?

Since spinning reserves are essentially generators turning but not producing electricity unless needed, the initial capital cost, the energy needed to turn the generator, and all the associated costs such as maintenance and repair must be paid for by some other generator actually producing electricity that someone is paying for.

CHAPTER 10

1. What are some of the difficulties mobile systems need to overcome that stationary systems do not generally have to deal with?

They must run for short periods of time, perhaps hours, but rarely for more than that. They must be able to start up and shut down from cold (sometimes very cold) conditions thousands of times. They must supply power very rapidly. They must respond to a wide range of load requirements. They have to carry their own fuel.

2. Why are low temperature fuel cells more suited to mobile applications than high temperature cells?

The main reasons for this are the time it takes for the systems to come up and then shut down as well as the efficiency of the system.

3. What are some of the common design elements that might be seen in mobile systems?

There must be some form of energy storage. There must be the engine (fuel cell). There must be a way to transmit the energy created by the engine to whatever will use that energy. There must be temperature control. There must be access to the internal workings.

4. What are two major problems in transporting liquid methanol?

It is very corrosive and has an affinity for water that makes transporting it difficult.

5. What are the two fuels actually used at the fuel cell?

Low temperature cells use hydrogen; high temperature cells use hydrogen and can also use carbon monoxide (CO).

6. Starting up a cold fuel cell presents many problems. One of the most serious is in dealing with any trace contaminants in reformed fuels. What is one way to deal with this problem?

One of the best ways to ensure clean initiation of the reactions is to use pure hydrogen from a stored, high quality source rather than reformed input fuel to begin the sequences.

7. Why is trucking hydrogen considerably more expensive than trucking gasoline or diesel?

Even liquid hydrogen is much lighter than gasoline and thus there is less energy value in each truckload delivered, not to mention the high cost of turning hydrogen into a liquid for shipping.

8. What are the three types of reformation methods?

Steam, partial oxidation, and autothermal.

9. What are some of the problems associated with ramping temperatures up at high rates?

If there is an inner material that expands more than the material surrounding it, then the outer material can split. If one material connected in line with another expands more, then the two can be bent or even buckle. If a single, large block of material expands, the outer part will get hotter faster, expand more, and crack away from the inner, cooler part. As a gas heats, it expands (think hot air balloons) and that expansion has to be taken into account when pipes and valves are used to move the gas. Any manifolding such as that done in interconnects (bipolar plates) complicates the matter since the material is also expanding at the same time.

10. Why is there no such thing as true zero emission technologies?

Because all such systems must use materials and energy in the fabrication or fueling process that produce pollutants and in some cases toxic pollutants.

CHAPTER 11

1. In general, what types of things would be shown in drawings?
Buildings, utilities, equipment, process flows, and even material balances.

2. Which are the simplest of all the diagrams used to present process data in a graphical format?
Block flow diagrams.

3. What are Process Flow Diagrams used for?
PFDs are used to show the relationships and flow paths between the major equipment in a system and to provide information on both standard operating values for that equipment as well as upset conditions (minimum and maximum allowed).

4. Process Flow Diagrams use symbols to represent a number of different things within a process; name four standards that detail which symbols are used in these diagrams.
ISO 14617-6, IEC 61346, ANSI/ISA-S5.1, and DIN 30600.

5. What are Process Control Diagrams used for?
They provide information on the equipment being used to control the process as well as the values of the control variables that can be expected in the process.

6. What do Piping and Instrumentation Diagrams usually show?
They show piping and process flow along with major and minor equipment as well as all installed instrumentation.

7. P&ID drawing sets are almost always used along with PFD drawing sets. How should the two tie together?
PID layouts should mimic the process flow as much as possible, since one of their primary purposes is to provide a graphical representation of what is going on in an operating facility or equipment cluster (such as a fuel cell system).

CHAPTER 12

1. Liquids and solids generally do not mix together. Why is not a good idea then to put a piece of steel in water?
Even though they might not mix, chemical reactions often occur where the liquid and solid meet. The oxygen in water will react with the iron in steel to form iron oxide, or rust.

2. Why is balancing the liquid pressure and the gas pressure important in a fuel cell like an MSCF system?
In MCFC systems, the molten electrolyte has to be in contact with the electrodes or ions and electrons will not flow to produce electricity. If the gas pressure is greater than the liquid pressure, the hydrogen or the air (oxygen) will push the electrolyte away from the electrode stopping the reaction. If the liquid pressure is too great, then it will push the electrolyte too far into the electrode, flooding the part of the electrode with the catalyst on it and keeping the fuel gas from coming in contact with the catalyst; the reaction will stop.

3. Of the three types of pressure, solid, liquid, and gas, which is more important in fuel cells?
Gas pressure.

4. Search for MCFC system piping and instrumentation diagrams on the Internet and look for pressure control loops or chamber sizes as well as pressure drops. Pick one such system and detail it as a homework assignment.

5. What is one of the main issues in low temperature systems where a pressurized cell is going to be used?

Increasing the pressure of the gas means that the stream will change temperature and that has to readjusted to match the operating temperature of the cell.

6. What are 3 atmospheres of standard pressure expressed as pounds per square inch (psi)?

44.1 psi.

7. Why do electrodes have to be porous?

So gas can enter them, pass through to contact the catalyst sites, react, and then move into the electrolyte.

8. Why does transferring a gas flow from a large volume to a smaller volume increase the gas pressure?

The gas molecules are forced closer together so more of them hit the channel walls and since pressure is the result of gas molecules hitting the walls of a container, that increases the pressure.

9. Why are the edges of the electrodes important in fuel cell design?

Since they are porous, they must be sealed so gas will not bleed around the edges of the cells.

10. Why is thermal management usually more important in liquid electrolyte cells than in solid electrolyte cells, no matter which operating temperature is used?

If the temperature gets too high in liquid electrolyte cells, the liquid will boil, eliminating its usefulness as an electrolyte and probably damaging the cell from the resulting gas expansion.

11. How would a fuel cell use excess heat in a bottoming cycle?

A bottoming cycle uses excess heat outside the process that generates it, so preheating incoming gas or heating water to make steam or running a turbine or transferring the heat to heat exchanger to be used to warm a building would all be good candidates for a bottoming cycle.

12. Can only one bottoming cycle be used with a fuel cell?

No, multiple bottoming cycles can be used and generally are. Any heat vented is money wasted, so it is recovered as is possible.

13. What functions do seals perform in a fuel cell?

The electrodes have to be sealed from the electrolyte to prevent flooding (or drying out) which require seals keeping solids, liquids, and gases in place for some types of fuel cells and gases from gases in others. Seals must be set in place but also allow "sliding" to accommodate expansion, contraction, and operating movement within the stack. The anode gas has to be sealed from the cathode gas. The input fuel gas has to be sealed from the atmosphere as well as from the fuel used at the anode. The cooler gasses have to be kept from the hotter gasses, in the fuel cell as well as in reformers and heat exchangers.

14. If high quality electricity is needed, what are the three general requirements for the delivery system?

Deliver it over the shortest possible path using the highest quality materials made by the best possible workers.

15. How does the conditioning of grid-generated electricity differ from distribute generation electricity?
Grid generated is conditioned at the user site while distribute generation is typically conditioned at the generating site.

16. Why are grid systems operated at high voltages?
The more electrons there are to move, the more energy is required and since the grids are so large, the number of electrons they need to keep moving is enormous so the voltage has to be high.

17. Do all fuel cells use the same power conditioning systems?
No, the conditioning configuration will depend on the application being served.

CHAPTER 13

1. What would be done if the incoming fuel for a fuel cell contained contaminates beyond the level called for in the system specifications?
Install a separate gas cleaning system before the fuel reaches the fuel cell.

2. Why would the codes for pressure piping and boilers, fire and insurance requirements, and local, regional, and federal regulations change the time initial discussions are begun and final installation takes place?
Neither the fuel cells nor the codes governing their installation, use and decommissioning are mature so change is to be expected as the systems do mature.

3. Why should government incentives to encourage the use of fuel cell systems not be included in long term cost estimates?
The incentives may change or disappear.

4. Which fuel cells are the most mature of the large scale systems?
PAFC.

5. When considering start-up times for the group of systems making up a fuel cell installation, which would be the most critical?
The one taking the longest time to come to full operations.

6. What is the most complicated aspect of systems operations in fuel cell systems?
Shutting down.

7. Why is it best to detail all time requirements of shutdown sequences in fuel cells?
Each system will take a different amount of time to cool and opening one part of the fuel cell may cause problems in the BOP or cool the system too fast. Safety issues are also important in this respect, since some systems run at very high temperatures and can cause severe damage if not completely cooled.

CHAPTER 14

1. At what point does an industrial products life cycle start?
When the raw materials it is made from are produced.

2. What is the difference between recycling and reprocessing?

In reprocessing, the product goes back through either its original process or some close version of it to produce a raw material. In recycling, the product is either reused or further processed into another product.

3. What are the four primary routes that can be followed when designing a product for end of life disposal?

Reduce the compounds having the worst environmental impact as much as possible, reuse whatever can be reused, recycle what cannot be directly reused, and finally dispose of what is left.

4. What are the three recycling categories?

High value materials, low value materials, and hazardous materials.

5. Is a fuel cell installation dismantled one the fuel cells themselves are worn out?

No, most fuel cells and stacks can be replaced so the system itself can continue to be used.

6. Can fuel cells or fuel cell systems be hot swapped at the end of life?

Probably not, since they generate electricity and run at elevated temperatures, the risks to infrastructure and workers is too great.

7. What are the three main components to be considered when arranging end of life for fuel cell installations?

Steel, stainless steel, and copper.

INDEX

A

Absorption chillers, 195–196
Absorption route, 29
AC. *See* Alternating current (AC)
Accessing fuel cell, 30
Access, unauthorized, 39
Acid, nitric, 1
Activation energy, 82, 218
Active thermal management, 21
Acute dose levels, 43
Acute exposure, 43
Administrative controls, 29
AFC. *See* Alkaline fuel cells (AFC)
Airborne particulates, 286
Air pollution. *See* Smog, Airborne particulates
Alcohol, 150
Alkaline, 56
Alkaline electrolyte. *See* Alkaline based electrolyte
Alkaline based electrolyte, 56, 97
Alkaline fuel cells (AFC), 11, 53–58, 84–85
   applications of, 171–172
   characteristics of, 171–172
   components of, 84–85
   electrodes, 110–113
   electrolytes, 97–99
   fuel cell stack in, 261
   gas systems and pressures in, 139–140
   heat generation in, 254
   reactions in, 55–58
   thermal management in, 131–132
   water management in, 124–127
Alloying element, 114
Alternating current (AC), 8
Alternative energy sources, 199
Ambient humidity, 84
American National Standards Institute (ANSI), 38
Ammonia (NH_3), 77, 150, 168
Amp, defined, 1, 14
Ampacity of circuit, 34
Anaerobic digesters, 156, 274
Analysis equipment in vehicles, 215
Anions, 10
Anode, defined, 15, 21
Anode electrode, 53
ANSI. *See* American National Standards Institute (ANSI)
ANSI diagramming symbols, 232
Architectural drawings, 229
Asphyxiation hazards, 44
Atmospheres, 136
Atmospheric pressure, 51, 116
Atom, 4, 13, 55
Automatic control, 234
Automobile applications of fuel cells, 207, 209–215
Autothermal reformation method, 150, 161, 219
Auxiliary power system, 206

B

Backflooding, 124
Back-pressures, 251
Back-up power generator, 192
Balanced equation, 55, 160
Balance-of-plant (BOP), 98, 134, 199, 269–270
Base. *See* Alkaline
Base electrodes, 10
Basic solution, 84
Batteries, 2, 51
BFD. *See* Block flow diagrams (BFD)
Biological group, 29
Biological production of hydrogen, 149
Bipolar plates. *See* Interconnect (bipolar plates)
BLEV. *See* Boiling Liquid Expanding Vapor (BLEV)
Block flow diagrams (BFD), 230–231
Block heating stations, industrial, 275
Boiling Liquid Expanding Vapor (BLEV) explosions, 45
Boiling point, 136
Boiling water method, 255
Borohydrides, 168
Bottoming cycle, 256
Breach and release in SOFC system, 135
Brittle, 69
Building codes, 197
Butane, 150, 163, 167
By-product, 53, 72, 129, 156

C

CAD. *See* Computer Aided Drafting (CAD)
Cadmium, 86
Calibrations, 238
Callaud gravity cells, 3
Capital cost, 179
Carbon, 112
Carbonate ion, 60, 62–63
Carbonates, 58, 97, 99
Carbon dioxide, 17, 88, 115, 133, 142, 143, 150, 159, 163, 166, 198, 221, 247
Carbon monoxide, 60, 93, 116, 127, 149, 157, 159, 165, 206, 217, 270
Carburization, 264
Carcinogens, 29
Catalyst
   AFC, 126
   conducting, 16
   defined, 6, 16, 20
   reformation reaction, 160, 163, 164, 194, 217
   MCFC, 128

Catalyst (*Continued*)
  nickel as, 128, 159, 165, 296
  palladium as, 288
  partial oxidation, 150
  platinum as, 17, 20, 22, 122, 166, 173
  precious metals as, 52
  SOFC, 72, 117
  SPFC (PEM), 122
Catalytic reactions, 16, 20, 83, 160
Catastrophic failure, 172
Cathode, defined, 15, 21
Cathode electrode, 53
Cathode oxidation reaction, 84
Cathode support, 89
Cations, 10
CCHP. *See* Combined Cooling, Heating, and Power (CCHP)
Cell potential, defined, 15
Cell stacks, recycling of, 296
Cell voltages, 60
Centralized generation model, 266
Ceramic electrolyte, 88
Ceramics, 89, 176, 296, 297
Cermet, 117
Certified laboratories, 295
Chemical attack, 95
Chemical group, 29
Chemical hazards, 28
Chemical path, 82
Chemical reactions, 1, 5, 81. *See also* Reformation in reactions
Chemical reduction reaction, 110
Chemical symbol, defined, 13
Chemical theory, 4
Chloride compounds, 86
Chronic dose levels of hydrogen, 43
Chronic effects of methanol, 43
Circuit, electrical, 83
Circuit overcurrent protection, 31
Closed-cup flash point, 39
Code and standards, 27, 30–48. *See also* NFPA codes and standards
  definitions, 38–39
  electrical systems, 31–37
  inspection, 45–47
  installation, 37
  safety and fire markings, 31–32
  siting, construction and installation requirements, 39–45
  using multiple, 37–38
Coefficient of thermal expansion (CTE), 91, 182
Cogeneration systems, 78, 83. *See also* Solid oxide fuel cell (SOFC)
Cold start, AFC system, 84
Combined Cooling, Heating, and Power (CCHP), 194
Combined heat and power designs, 153
Combustible hazards, 28
Command and control centers, 238
Compartmentalized stacks, 224

Competitive price, 206
Components, fuel cell, 84–93, 121–145. *See also* Electrodes; Electrolytes
  AFC, 84–85, 124–127, 131–132, 139–140
  end of life of, 296–297
  MCFC, 85–88, 127–128, 132–133, 140–143
  PAFC, 88, 128–129, 133–134, 143–144
  SOFC, 88–93, 129, 134–135
  SPFC, 121–124, 129–131, 136–139
Compressed gas, 136, 246
Compressed gas hazards, 28
Compressor, 247
Computer Aided Drafting (CAD), 231
Computer modeling, 250
Concentration, 21, 126
Condensing water, 113
Conditioning, power system, 265, 266–269
Conducting catalyst, 16. *See also* Hydrogen
Conducting ions, 116
Conduction, 56
Conductors, 251
Consumer items, 285
Contact site, 252
Contact theory, 4
Continuous *versus* on-call operations, 176
Contraction, 61
Control equipment, 215, 232
Control loops, 233
Control variables, 232
Control, interlock, 237
Convection, heat, 131
Convection systems design, 132
Conventional fuel cell liquid loop, 128
Coolers. *See* Absorption chillers
Corrosion, 64
Corrosives, 10, 29
Cost of power economics, 134
Costs of fabrication and disposal, mobile application issue, 206
Counter-ion, 94
Covalent bonding, 16
Cracking, 161
Crossing lines, 238
Cryogenic burns, 44. *See also* Health hazards
Cryogenic liquids, 167
CTE. *See* Coefficient of thermal expansion (CTE)
Current, 6, 56, 265–266
Current collector, 112
Current densities, 109, 173
Cycles per second, 200
Cycles, unloading, 55

D

Daniel cells, 3
Data centers, 197

DC. *See* Direct current (DC)
DC current generation, 6–8
DC-DC converter, 269
DC grid systems, 8
DC to AC conversion. *See* Inverter
Decision points, 234
De-energizing protocols, 32
Dehumidification circuit, 98
Dehydration, 122
Detection equipment, 43
Diamond structure, 144
Diatomic gases, 4, 17. *See also* Hydrogen, Nitrogen
Diffusers, 256
Diffusion, 21, 251
Dipole, 94
DIR. *See* Direct internal reformation (DIR)
Direct current (DC), 6
Direct internal reformation (DIR), 60, 114, 165
Direction of stress flow, 243
Direct Methanol Fuel Cells (DMFC), 160
Discharge, water, 294
Disconnects, electrical, 35
Dismantling and decommissioning of fuel cells, 292–297
Disposal of fuel cell, 291
Distributed generation model, 153, 266
DMFC. *See* Direct Methanol Fuel Cells (DMFC)
Documents, PID operation, 237
Doping, 69, 101
Drags, water, 126
Dry reforming, 159
Dump choppers, 32
Dynamic systems, 42

E

Economies of scale, 9, 147
Efficient conversion of hydrocarbon, 114
Efficiency, 51, 84, 116, 152
  of AFC, 171–172
  of MCFC, 176–177
  of PAFC, 177
  of SOFC, 179
  of SPFC, 172–173
Electrical disconnects, 35
Electrical systems of fuel cells, 31–37
Electrically conductive pathway, 104
Electrically neutral species, 55
Electricity. *See* Power generation
Electrocatalyst layer, 111
Electrode assembly, 104
Electrode stability, 85
Electrode weeping, 112
Electrodes, 1, 81, 103–117
  defined, 14
  MCFC, 113–115
  PAFC, 115–116

poisoned by contaminants, 114
  SOFC, 116–117
  SPFC (PEM), 104–110
Electrolysis, 149
  water management in AFC and, 125
  generating hydrogen from, 91, 109, 146, 224
  maintenance issues in fuel cells and, 281
Electrolytes, 3, 15, 53, 93–103
  AFC, 97–99
  MCFC, 99–100
  PAFC, 100–101
  SOFC, 102–103
  SPFC (PEM), 93–96
Electromagnetic field theory, 6
Electrons, 4, 14, 55, 81
Electron channels, 251
Electro-osmotic drag, 95, 123
Elevation control, 238
Emergency locations, 238
Emergency services management, 31
End of life, 285–297. *See also* Dismantling and decommissioning of fuel cells
  fuel cells and subsystem, 292–297
  of main materials, 296–297
  regulatory impact and recycling procedures, 291–292
Endothermic reactions, 21, 56, 150
Energy density of methanol, 220
Engineering controls, 29
Engineering drawings, 229
Engines, 81
Environmental impact, 285, 289–290
Equipment cluster, 237
Equivalency in NFPA 853, 38
Ergonomic group, 29
Ethane, 150
Eutrophication, 287
Evaporation, 116
Examinations and testing, NFPA, 46
Exchanged, heat, 260
Exothermic reactions, 21, 56, 113
Expansion, 61
Explosive hazards, 28
Explosive limits of hydrogen, 43
External manifolding, 140, 141
External reformation, 113, 217, 219–221
  of hydrogen, 162–164
Extrude, 89

F

Fabrication drawings, 229
Faraday's coil, 6–7
Faraday's disc generator, 7, 8
Faraday's rod, 6–7
Filter beds, 147
Fire suppression, 30
Fire walls, 31

Fixed rotor, 258
Flammable hazards, 28
Flash point, 39
Flooding, 124
Flow channels, SOFC, 92
Flow diagrams, 229
Flow field plates, 252
Flow patterns, 92
Flow, defined, 244
Flushing lines, 237
Flywheels, 200
Following process, 149
Food chain, 287
Fouling, 117
4-electron reactions, 59
Fractured markets, 216
Free energy. *See* Gibbs free energy
Freezing, 85
Frequency regulation, 199, 200
Friction, 244
Fuel cell applications, 187–203. *See also Specific applications*
Fuel cell assembly, 261
Fuel cell components, 10
   management of, 121–145
Fuel cell engineering, 273–277
Fuel cell heat balance, 254–256. *See also* Thermal management
Fuel cell operations, 277–278
Fuel cell stack, 106, 296
Fuel cell stacking, 260–265
   in AFC, 261–262
   in MCFC, 262–263
   in PAFC, 263–264
   in SOFC, 264–265
   in SPFC, 264–265
Fuel cell systems, 147–170, 171–185
   dismantling and decommissioning of, 285–291
   engineering issues in, 273–277
   characteristics and applications of, 171–185
   installation of, 37, 273–277
   issues in, 174–175
   maintenance issues in, 277–282
   operational issues in, 277–278
   power and control of, 241–270
   process and instrumentation of, 229–238
   size of, 207–216
   and subsystems, 147–168
Fuel cells
   advantages/disadvantages of, 52
   batteries *vs.*, 1–4
   codes and standards, 30–48
   cost of, 151–152, 173–174, 270
   design of, 22–24
   efficiency of, 152
   electrical installation of, 37
   electrical systems, 31–37
   electricity production by, 190
   engineering of, 273–277
   fuel distribution in, 27, 250–253
   history of, 1–11
   maintenance of, 278–282
   operations, 277–278
   pressure in, 242–250
   safety and markings of electrical system of, 30–37
   siting and construction requirements of, 39–42
   temperature in, 254–260
   types of, 53–78
   usefulness of, 269–270
Fuel cells and cars, 210–212
Fuel cells codes and standards, 30–31
Fuel cells in space vehicles, 54
Fuel delivery system, 43, 218
Fuel distribution in fuel cells, 250–253
Fuel gas, 57
Fuel generation system, 155
Fuel hazards of fuel cells, 42–45
Fuel production to use in fuel cells, 148–168
   cleaning of fuel stream, 165–166
   external reforming of hydrogen, 162–164
   hydrogen storage and, 167–168
   internal reforming of hydrogen, 164–165
   methods used in, 148–156
   reforming reactions, 156–162
Fuel reformation subsystems, 149
Fuel shut-off systems, 30
Fuel stream, cleaning of, 165–166
Fueling of fuel cells, 216–222
Fuels in fuel cells, 27, 148–168
Full operating power, 210

G

Gas, 83
Gas crossover, 141
Gas detection, 31
Gas diffusion membrane, 104
Gas diffusion plates, 148, 251
Gas flow, 246
Gasification, 149
Gas manifolding for PAFC, 143
Gasoline, 156, 221
Gasoline blends, 221
Gas phase, 135
Gas pressure, 243
Gas systems and pressures, 135–144
   AFC, 139–140
   MCFC, 140–143
   PAFC, 143–144
   SPFC (PEM), 136–139
Gas-tight, 253
Gas turbines, 51
Gemini fuel cell, the, 107, 108

General Duty Clause, 28
General installation codes of fuel cells, 30
Generators, 6, 51
Gibbs free energy, 19, 23
Graphite structure, 144
Greenhouse gases, 198, 225
Grid, 87
Grid feeds, 276
Grid mains, 31
Grid systems, 155
Grounding methods, 31
Grove battery, 2
Guaranteed components, 238

H

H_3PO_4. See Phosphoric acid (H_3PO_4)
H_2S. See Hydrogen sulfide (H_2S)
Hazard operability studies, 232
Hazardous materials, 292
Hazards, 28. See also specific hazards
HazCom standard, 28
HCl. See Hydrogen chloride (HCl)
Health and safety risks of methanol, 220
Health hazards, 28, 288–290
Health effects, toxic substances and, 289–290
Heat, 21
Heat-aided systems, 247
Heat balance in fuel cells, 254–256
Heat engines, 254
Heat recovery, 256–260
Heat transfer, 260
"Heavier" hydrocarbons, 160
"Heavy" hydrocarbons, 222
Hertz (Hz), 200
High pressure tanks, 44
High temperature creep, 91, 114
High temperature fuel cells, 60
   advantages/disadvantages of, 52
   medium applications of, 194–196
High value materials, 291
Hole, 102
Hot spots, 86, 93
Hot zone design, 254
Human health hazards. See Health hazards
Humidification, 125, 137, 221, 279, 280, 295
Humidity, 123
Hybrid fuel cell/turbine systems, 180–181
Hybrid system, 148
Hybrid systems storage, 216
Hydrate, 129
Hydrocarbon based fuels, 149
Hydrocarbons, 52, 88
Hydrocarbon to hydrogen conversion. See Reforming reactions
Hydrogen, 1, 156–168
   cleaning of, 165
   external reforming of, 162–164
   dry reforming of, 159
   internal reforming of, 164–165
   production of, 143–156
   safety concerns of, 43
   steam reforming of, 160
   storage, 167–168
Hydrogen chloride (HCl), 77
Hydrogen damage, 42–44
Hydrogen embrittlement, 43
Hydrogen fuel cell, basic principles of, 16–18
Hydrogen and oxygen reaction, 5
Hydrogen sulfide (H_2S), 77
Hydrogen's explosive limits, 43
Hydronium ion, 94
Hydrophilic area, 95
Hydrophobic, 95
Hydrostatic leak test, 46
Hygroscopic materials, 124
Hz. See Hertz (Hz)

I

ICE systems in load cycling, 224
IGBT. See Insulated gate bipolar transistors (IGBT)
IIR. See Indirect internal reformation (IIR)
Impregnation of carbon paper, 115
Indirect internal reformation (IIR), 114, 165
Industrial diagrams, common, 229–239
   block flow diagram, 230–231
   piping and instrumentation diagram, 234–238
   process control diagram, 232–234
   process flow diagram, 231–232
Industrial hygiene requirements, 43
Ingestion route, 29
Inhalation route, 29
Initiation of reactions, 218
In-line measurement, 238
Inorganic substances, 67
Input fuels, 217
Input streams, 42
Inspection, fuel cell installation, 45–48
   hydrostatic leak test, 46
   NFPA and, 45–46
   random radiography, 46
   ultrasonic, 46–48
Inspector, 46
Instrumentation diagrams, 229
Insulated gate bipolar transistors (IGBT), 269
Insulators, defined, 16
Installation issues in fuel cell systems, 273–277
Interconnect (bipolar plates), 72, 91, 92, 93, 104, 117, 251
Interlock systems, 31
Internal combustion engines, 51
Internal manifolding, 140, 141
Internal pressure, 86
Internal reformation, 87, 113, 217, 218–219
   of hydrogen, 162–164

Internal (ohmic) resistance, 86
Internal *versus* external manifolding, 253
Interstitial alloying, 116
Inverter, 34, 148, 265, 269
Ion exchange membrane, 53
Ionic bonding, 16
Irritants, 29
Issues, in fuel cell system
   disposal, 292–293, 296
   electrical installation, 37
   installation, 273–277
   instrumentation, 280
   maintenance, 278–282
   operational, 277–278

K

Kilowatt hours, 188
Kinetics, 20, 56, 134. *See also* Thermal management

L

Laminar flow, 127, 252
Lanthanum, 69
Lanthanum alloys, 179
Lanthanum chromite (LaCrO$_3$), 72
Large fuel cell systems, 199–202
Lead, 86
Lead acid storage battery, 3–4
Leclanché cells, 3
Leveling control, 238
Life cycle assessment (LSA), 285, 286
Linear systems, 179
Line of sight, required, 192, 238
Liquid electrolytes, 88, 99
Liquid reforming, 149
Load, 87
   cycling, 278
   in mobile systems, 222–224
Lockout-tagout locations, 234
Loop, 58
Loss on transmission, 188
Low temperature fuel cells
   advantages/disadvantages of, 52
   mobile applications of, 206
Low value materials, 291
LSA. *See* Life cycle assessment (LSA)

M

Main shutoff valves, 236
Maintenance call-out abilities, 208
Maintenance issues in fuel cells, 278–282
Management of fuel cell components
   gas systems and pressure, 135–144
   thermal, 129–135
   water, 121–129
Manganese, 69
Manganese alloys, 179

Manual control, 234
Manual shut-off valves, 45
Markings, 30
Mass, defined, 14
Mass production, 173, 274
Mass transfer resistances, 109
Material balances, 230
Material Safety Data Sheets, 28
Maturity of fuel cells, 54
MCFC. *See* Molten Carbonate fuel cells (MCFC)
MEA. *See* Membrane Electrode Assembly (MEA)
Mean-time-between-failures, 278
Mechanical pressures, 242
Medium scale systems, 193–199
Megawatts, 64, 86
Melting of electrolytes, 61
Membrane electrode assembly, 98, 148
Membrane Electrode Assembly (MEA), 104, 105, 252
Mercury, 86
Metallic bonding, 16
Metal oxide semiconductor field effect transistors (MOSFET), 269
Methanation, 166
Methane, 128, 149, 159
Methanol, 22, 42, 43, 58, 156, 160, 165, 216, 220, 241
Micro fuel cells, 208
Microelectronics fuel cells, 208
Microns, 70
Microturbines, 257
Middleman, 295
Military applications of fuel cells, 187
Milliamps, 60
Mitigation of health hazards, 28
Mixed systems, grounding of, 36
Mobile fuel cell, applications of, 191, 205–227
   cooling parameters of, 206
   fueling for, 216–222
   high temperature systems, 206
   low-temperature systems, 206
   pollution in, 224–225
   size and, 207–216
   technologies in, 215–216
   variable operating conditions and load cycling of, 222–224
Mobile system, 205
Modular subsystems, 147
Molecular chains, 149
Molecules, 13, 17
Molten carbonate fuel cells (MCFC), 53, 59–64, 85–88
   applications of, 175–177
   characteristics of, 175–177
   components of, 85–88
   electrodes, 113–115
   electrolytes, 99–100
   fuel cell stack in, 262–263
   gas systems and pressures, 140–143
   heat generation in, 254–255

issues in, 99–100
pressure in, 242, 243, 247–248
reactions in, 62–64
thermal management, 132–133
water management, 127–128
Moving platform, 207

N

$NaBH_4$. *See* Sodium borohydride ($NaBH_4$)
Nafion®, 84, 93, 150, 178, 297
Naptha, 156, 163, 216
National Fire Protection Association (NFPA)
 codes and standards of, 30–48
Negatively charged electrons, 94, 115
Neutrons, 14
NFPA. *See* National Fire Protection Association (NFPA)
NH_3. *See* Ammonia (NH_3)
Nickel alloys, 113
Nickel felt, 117
Nickel oxides, 179
Nitrogen, 17
Nitrogen complexes (NO_x), 192
Nonmetallic substances, 67
Non-noble metals, 60
Nonspinning reserves, 199
NO_x. *See* Nitrogen complexes (NO_x)
Number generation, 234

O

Occupational Safety and Health Administration (OSHA), 27
Off-the-grid alternatives, 191
Ohm, 1
Ohmic loss, 65, 83, 116
Ohmic resistance, 110, 134
Onboard fuel storage, 219
On-site utilities, 293
Operating life of SOFC, 70
Operating temperature, fuel cells, 51
Operational issues in fuel cell system, 277–278
Operations plan for medium systems, 194
Optimization testing, 278
OSHA. *See* Occupational Safety and Health Administration (OSHA)
Output compatibility, 31
Output streams, 42
Overcurrent protection, 34
Overperform reaction, 268
Overpotential, 82, 109
Overvoltage, 139
Owner's responsibility, 46
Oxidation, chemical, 15
Oxidation/reduction reaction. *See* Redox reaction
Oxide cells, defining factors for, 103
Oxidizing agents, 10
Oxidizing atmosphere, 114, 143

Oxygen, 1, 4–5, 9–10, 17–18, 20, 43, 122–123, 149, 159, 160, 172, 198, 243, 247
Oxygen sensors, 181

P

Packed bed filters, 159
PAFC. *See* Phosphoric acid fuel cells (PAFC)
Palladium membranes, 65, 147
Parallel, 99
Parasitic losses, 249
Partial oxidation reforming reactions, 150, 160, 219
Partial pressures, 137
Particulate emissions, 198
Particulates, 63, 86
Parts per million (ppm), 58, 86, 97, 166
Pascals, 171
Passive thermal management, 21
Path of least resistance, 251
Peak load cycling, 55, 155
PEM. *See* Proton Exchange Membrane (PEM)
Pendant molecule, 94
Penetration route, 29
Perovskites structure, 111
Personal communication, size for, 209
Personal protective equipment, 29
Phosphoric acid (H_3PO_4), 64
Phosphoric acid fuel cells (PAFC), 53, 64–66, 88
 applications of, 177–179
 characteristics of, 177–179
 components of, 88
 electrodes, 115–116
 electrolytes, 100–101
 fuel cell stack in, 263–264
 gas systems and pressures, 143–144
 heat generation in, 255
 mature designs of, 178
 reactions in, 66–67
 thermal management in, 133–134
 water management in, 128–129
Physical group, 29
PID. *See* Piping and instrumentation diagram (PID)
Piping and instrumentation diagram (PID), 229, 234–238
 components of, 237–238
Piping diagrams, 229
Planar design, 70, 89
Plasma deposition process, 70
Platinum black, 115
Platinum loading target, 106
Platinum, 106. *See also* Conducting catalyst
Plug-in systems, 43
Pneumatic symbol, 233
Point-of-use site, 188
Poisoned by contamination, 60, 109
Pole, 265
Polishing step in external reformation, 218

Pollutants, 81. *See Specific pollution*
Pollution in fuel cells, 224–225
Polyethylene, 93
Polymer electrolyte, 55
Polymers, 178
Polytetrafluoroethylene, 93
Pores, 58
Porosity, 250
Porous ceramic matrix, 176
Porous electrodes, 58
Porous stainless steel, 262
Positively charged electrons, 94, 115
Potassium hydroxide, 97
Potassium ion, 56
Potential measurement, 268
Pound per square inch pressure, 242
Pounds force, 245
Pounds weight, 245
Power and control of fuel cells, 241–270
Power and heat system, AFC, 86
Power conditioning, 148
Power delivery, 265
Power demand, 217
Power densities, 84
Power generation, 4–8, 187–202, 241–269
    cost for, 174–175
    chemical reaction for, 55–58, 59, 62–64, 66–67, 72–78
    conditioning of, 148, 266–269
    current, 265–266
    environmental concerns to, 198
    equipment, 51
    stationary applications and, 187–191
Power up cycling in vehicles, 215
Pressure in fuel cells, 242–250
Pressure piping, 30
Pressure Swing Absorption, 166
Pressure vessels, 44, 282
Pressure, differential gas, 10
Pressurized water method, 255
Primary batteries, defined, 2, 15
Primary power system, 206
Process control diagrams, 232–234
Process flow diagrams (PFD), 229–230
Process piping, 30
Process vessels, 231
Programmable device, 238
Programmable logic controllers, 280
Propane, 150
Proton exchange membrane (PEM), 11, 17
Protonation of water, 94
Protons, 5, 14, 94
Psi, 136
Purging, 86
Purified fuels, 63
Pyrolysis gasifiers, 156, 158
Pyrophoric hazards, 28

Q
Quality control subsystem, 147, 178

R
Ramping, 277
Random radiography, 46
Raney metals, 111
Range of operations, 222
Rare earth elements, 282
Raw materials, 136
Reactants, 83
Reaction chamber, 51
Reaction kinetics, 267
Reaction sequences of fueling system, 217
Reaction zones, 279
Reactive, chemically, 111
Reactive power reserves, 199
Recirculating electrolyte systems, 55, 84, 124
Recycle (primary route of end of life), 291
Recycle gas loops, 247
Recycling programs of fuel cells, 285, 288
Redox reaction, 15
Reduce (primary route of end of life), 291
Reducing agents, 10
Reducing atmosphere, 114, 143
Reduction, defined, 15
Reference numbers, 232
Reform gases, 60
Reformate systems, 58
Reformation in reaction, 62, 156–162
Reformed fuels, 109, 217
Reforming fuel gases, 149. *See also* Reformation in reaction
Regulations, OSHA, 28
Regulations and recycling of fuel cells, 291–292
Remote generating sites, 188, 192
Replacement schedule for medium systems, 194
Reprocessed, automobile catalytic converters, 288
Reproductive toxins, 29
Reservoir plate, 65
Resistance to electron flow, 261
Reuse (primary route of end of life), 291
Reverse cycling, 223
Revolutions per minute (RPM), 257
Risk assessments, 175
Rotating rotors, 258
RPM. *See* Revolutions per minute (RPM)

S
Safety
    and fire markings, 31–32
    hydrogen's concern, 43
    of maintenance personnel, 282
    NFPA and, 38
    occupational, 27–28
    PCD and, 232
    PID and, 235

release, 238
   in steam reforming reaction, 159
     training, 29, 30, 230
Sampling locations, 234
Saturation, 122
Scalability, 52
Scalable system, 81, 155
Scheduled power shutdowns, 194
Scientific method, 1
Scientific notation, 13, 268
Sealing surface, 261
Secondary (storage) batteries, 15
Self-repair (reprogramming), 208
Self-reporting, 209
Self-sustaining diesel combustion reaction, 217
Semiautomatic control, 234
Series, 99
Service access, 31
Shear stresses, 244
Shutdown procedures for hydrogen gas, 44
SiC. See Silicon carbide (SiC)
Silicon carbide (SiC), 65, 101
Singular prepackaged self-contained power systems, 37
Sintering, 70
Siting, interconnection and installation requirements of fuel cells, 39–42
Size limitations, 182
Size, of fuel cells, 207–216
Slurry casting, 89, 90
Small stationary systems, 191–192
Smog, 286
Sodium borohydride ($NaBh_4$), 57
SOFC. See Solid oxide fuel cell (SOFC)
Software diagnostics, 208
Solid oxide fuel cell (SOFC), 53, 67–78, 88–93
   applications of, 179–185
   characteristics of, 179–185
   components of, 88–93
   electrodes, 116–117
   electrolytes, 102–103
   fabrication of, 69–70
   fuel cell stack in, 264–265
   heat generation in, 255–256
   hybrid systems, 75–77
   issues in, 184
   planar design of, 70, 73–74, 91
   reactions in, 72–78
   thermal management in, 134–135
   tubular design of, 70, 71, 91
   water management in, 129
Solid polymer fuel cell (SPFC), 53, 58–59, 93–96
   applications of, 172–175
   characteristics of, 172–175
   electrodes of, 104–110
   electrolytes, 93–96
   fuel cell stacks in, 264–265
   gas systems and pressures, 136–139
   heat generation in, 254
   humidity in, 123
   reactions in, 59
   thermal management in, 129–131
   water management in, 121–124
SO_x (sulfur complexes), 192
Space vehicles, fuel cells in, 54
Spark initiators, 81
Spent fuel cell, 292
SPFC. See Solid Polymer fuel cells (SPFC)
Spinels structure of AFC, 111
Spinning reserves, 199
Spray casting, 89, 90
Stabilized zirconia structures, 101
Stable, 150
Stage sintering methods, 89
Standard operating conditions of AFC, 84, 280
Standard Operating Procedures of PID, 235
Standard operating values of PFD, 231
Start-up issues in fueling system, 217
Static systems, 42
Stationary fuel cell systems, 37, 187–203
   applications of, 187
   large, 199–202
   list of potential sites for, 193
   medium, 193–199
   military applications of, 187
   small, 191–193
Stationary installations of fuel cells, 88
Steady state, 123
Steam ejectors, 295
Steam method, 219
Steam reformation, 88, 128, 150, 160
Steam turbines, 257–260
Stop-go control process, 234
Storage batteries, 3
Storage cells, 3
Storage, hydrogen, 167–168
Strain gauges, 234
Stress distribution in metals, 242
Stresses, types of, 244–245
Stress riser, 242
Strontium, 69
Strontium alloys, 179
Subcycle response schemes, 275
Substitutional alloying, 116
Substrate, 69
Subsystem, end of life of, 293–295
Successor companies, 292
Sulfur compounds, 86
Supply chain risks in methanol system, 215
Surface area, 10, 104
Surface tension, 98, 137, 138
Surveillance devices, size for, 209
Symbols for process flow diagrams, 231

Synchronized, 200
System grounding, fuel cell, 33
System regulations, fuel cell, 291
System start-up, fuel cell, 277
Systems integration, fuel cell, 280

T

Tape casting, 89, 90, 70, 100
Technology, early adoption of, 216
Teflon®, 84, 93, 111, 178, 273
Temperature of fuel cells, 254–260
Temperature ramping, failures in, 223–224
Temperature sensitive, 51
Thermal efficiency, 154
Thermal expansion, 69, 182
Thermal management in fuel cells, 129–135, 254
    in AFC, 131–132
    end of life of, 296
    in MCFC, 132–133
    in PAFC, 133–134
    in SOFC, 134–135
    in SPFC (PEM), 129–131
Thermal reactions, 160
Thermodynamics, 19
Thin films *vs.* thick films, 103
Three-phase area for fuel cells, 98
Thyristors, 269
Tie-ins, installation, 277
Tin, 86
Toxic. *See* Toxic substances
Toxic substances, 29, 286
    carcinogens, 29
    in electronic items, 289–290
    reproductive, 29
Trace amounts, 97
Transient temperature differentials, 133
Transportable engines, 207
Transportation fuels, 216
Transport medium, 67
Troubleshooting procedures, 238
Tubular designs, 70, 89
Tubular systems, 179
Turbine based generators, 88
Turbine systems, 256
Turbines, 51, 248
Turbocharging, 248
Turbulent flow, 126, 127, 252
Types of fuel cells, 53–78

U

UL. *See* United Laboratories (UL)
UL. *See* Underwriters Laboratories (UL)
Ultrasonic examination, 46
Ultrasonic inspection, 47
Underwriters Laboratories (UL), 38
Underperform reaction, 268
Unidirectional flow, 266
United Laboratories (UL) markings, 36
Unlimited liability, 292
Unstable hazards, 28
Upset condition
    in PFD, 231
    in PID, 237
Upset temperatures, 134

V

Vacuum, 224
Valves, 45
Vapor pressure, 112
Variable loading in vehicles, 215
Variable operating conditions in mobile systems, 222–224
Variation in line voltage, 268
Viscosity, 43
Viscous liquid, 91
Volt, 1
Voltage, 6, 14, 21
Voltaic pile (cell), 1, 2, 15
Volt/amp fluctuation, 267
Voltilization, 64
Volume, 140

W

Warning symbols, 29
Waste water treatment facilities (WWTF), 274
Water gas shift reaction, 128, 159
Water gas shift reactor, 221
Water management in fuel cells, 53, 121–129
    in AFC, 124–127
    in MCFC, 127–128
    in PAFC, 128–129
    in SOFC, 129
    in SPFC (PEM), 121–124
Water reactive hazards, 28
Water system, end of life of, 295
Water treatment plant, 294, 295
Watt, 1, 15
Weight, defined, 14
Work, 83
Work-in-progress, 232
Workplace hazards, 28, 43
Work practices, 29
WWTF. *See* Waste water treatment facilities (WWTF)

Y

Y_2O_3. *See* Ytria (Y_2O_3)
Ytria (Y_2O_3), 69
Yttrium stabilized zirconia, 101–102

Z

Zero emission vehicles, 221
Zirconia (ZrO_2), 69
Zirconium molecules, 69
ZrO_2. *See* Zirconia (ZrO_2)